John **Widdowson**
Rebecca **Blackshaw**
Meryl **King**
Simon **Oakes**
Sarah **Wheeler**
Michael **Witherick**

AQA
GCSE
(9–1)

GEOGRAPHY

Approval message from AQA

This textbook has been approved by AQA for use with our qualification. This means that we have checked that it broadly covers the specification and we are satisfied with the overall quality. Full details of our approval process can be found on our website.

We approve textbooks because we know how important it is for teachers and students to have the right resources to support their teaching and learning. However, the publisher is ultimately responsible for the editorial control and quality of this book.

Please note that when teaching the *AQA GCSE Geography* course, you must refer to AQA's specification as your definitive source of information. While this book has been written to match the specification, it cannot provide complete coverage of every aspect of the course.

A wide range of other useful resources can be found on the relevant subject pages of our website: www.aqa.org.uk.

AN HACHETTE UK COMPANY

The Publishers would like to thank the following for permission to reproduce copyright material.

Every effort has been made to trace all copyright holders, but if any have been inadvertently overlooked, the Publishers will be pleased to make the necessary arrangements at the first opportunity.

Although every effort has been made to ensure that website addresses are correct at time of going to press, Hodder Education cannot be held responsible for the content of any website mentioned in this book. It is sometimes possible to find a relocated webpage by typing in the address of the home page for a website in the URL window of your browser.

Hachette UK's policy is to use papers that are natural, renewable and recyclable products and made from wood grown in sustainable forests. The logging and manufacturing processes are expected to conform to the environmental regulations of the country of origin.

Orders: please contact Bookpoint Ltd, 130 Milton Park, Abingdon, Oxon OX14 4SB. Telephone: (44) 01235 827720. Fax: (44) 01235 400454. Email education@bookpoint.co.uk Lines are open from 9 a.m. to 5 p.m., Monday to Saturday, with a 24-hour message answering service. You can also order through our website: www.hoddereducation.co.uk

ISBN: 978 1 471 859 922

© 2016, John Widdowson, Rebecca Blackshaw, Meryl King, Simon Oakes, Sarah Wheeler, Michael Witherick

First published in 2016 by
Hodder Education
An Hachette UK Company
Carmelite House
50 Victoria Embankment
London EC4Y 0DZ
www.hoddereducation.co.uk

Impression number   10 9 8 7
Year   2020 2019

Cover photo © Frederic Bos–Fotolia.com

Illustrations by DC Graphic Design Limited and Barking Dog Art

Typeset in Gotham Light 9pt

Printed in Italy

A catalogue record for this title is available from the British Library.

Contents

Paper 1: Living with the physical environment

Section A: The challenge of natural hazards

Section B: The living world

Section C: Physical landscapes in the UK

Paper 2: Challenges in the human environment

Paper 3: Geographical applications

Text credits

p. 5 Data from *Scientific American*, 19 August 2011; **p. 28** Graphs republished with permission of American Association for the Advancement of Science, from Changes in Tropical Cyclone Number, Duration, and Intensity in a Warming Environment, P. J. Webster, G. J. Holland, J. A. Curry and H.-R. Chang, *Science*, 16 September 2005, Vol. 309, permission conveyed through Copyright Clearance Center, Inc.; **p. 32** *bl* Image provided by weather underground; **p. 38** *tr, bl* Data from Met Office, http://www.metoffice.gov.uk/climate/uk/interesting/nov2009; **p. 42** *t, b* Data from Met Office, http://www.metoffice.gov.uk/climate/uk/summaries/actualmonthly; **p. 45** Graph reproduced by permission of The Royal Society of Chemistry; **p. 46** Graph from The Science Museum, http://www.sciencemuseum.org.uk/ClimateChanging/ClimateScienceInfoZone/Exploringwhatmighthappen/2point2/-/media/ClimateChanging/FindOutMore/Images/twotwoonethree_SolarOutputSince1900.ashx; **p. 47** *bl* Map from NASA, http://earthobservatory.nasa.gov/IOTD/view.php?id=1510; **p. 49** Graph from Climate Audit, http://climateaudit.org/2009/09/25/spot-the-hockey-stick-n-2/; **pp. 50–51** Map from Met Office, http://www.metoffice.gov.uk/climate-guide/climate-change/impacts/human-dynamics; **p. 52** Text from United Nations Environment Programme, http://www.unep.org/climatechange/mitigation/Energy/tabid/104339/Default.aspx; **p. 54** Text from Food and Agriculture Organization of the United Nations, Module 6: Conservation and Sustainable Use of Genetic Resources for Food and Agriculture, p. 184. Reproduced with permission; **p. 55** Table from Securing London's Water Future: The Mayor's Water Strategy, October 2011. Used with permission from Greater London Authority (GLA); **p. 70** *t* Graph from Rhett A. Butler/mongabay.com, *b* Graph from Forest for Climate, http://www.forestforclimate.org/News/Page-11.html; **p. 71** Map from World Wide Fund for Nature. Retrieved from http://panda.maps.arcgis.com/apps/webappviewer/index.html?id=67478a42d7554ae598e4169b70e9f093; **p. 73** Graph from Federation of American Scientists, 2011; **p. 95** Data from tearfund.org; **p. 97** Map from Global Environment Facility; **p. 105** Map from Alaska Humanities Forum; **p. 107** Map from The Financial Times Ltd; **p.109** Graph from Rob Gamesby, http://wwwcoolgeography.co.uk; **p. 113** Map from World Protected Areas Database, UNEP-WCMC (2005). Russian Data Protected Areas Digitized by WWF in 2005 from Official Sources. Wilderness from Analysis by UNEP-WCMC (2001); **p. 175** Icons from Flood Codes and Messages User Guide. Flood warning icons. Used with permission from Environment Agency; **p. 199** Graph from Rob Gamesby, http://wwwcoolgeography.co.uk; **p. 204** Map from Lindsay Sawyer, ETH Zurich; **p. 206** *t, m, b* Data from United Nations Development Programme, Human Development Report 2015: Work for Human Development – Nigeria; **p. 227** *t* Graph from London's Sectors: More Detailed Jobs Data. Used with permission from Greater London Authority (GLA), *b* Graph from London's Sectors: More Detailed Jobs Data Working. Used with permission from Greater London Authority (GLA); **p. 228** *tr* Graph from Transport for London, Building Our Capital: Five Years of Delivery by London Underground,

March 2015, *b* Map from Crossrail Ltd, http://www.crossrail.co.uk/route/maps/; **p. 234** Graph from Homes for London: The London Housing Strategy, 2013. Used with permission from Greater London Authority (GLA); **p. 236** Map from GLA – The London Plan Consolidated with Alterations since 2004 (2008). Used with permission from Greater London Authority (GLA); **p. 247** Map from Draft: Bristol Cycle Strategy, p. 11. Used with permission from Bristol City Centre; **p. 269** *l, r* Data from IMF; **p. 270** *b* Map from University of Texas; **p. 277** Map from www.stratfor.com; **p. 282** Graph from The National Archives, 170 Years of Industrial Change across England and Wales, http://www.ons.gov.uk/ons/rel/census/2011-census-analysis/170-years-of-industry/170-years-of-industrial-changeponent.html; **p. 286** Map from Grant Thornton International Ltd (GTIL), http://www.grant-thornton.co.uk/en/Where-growth-happens/Dynamic-growth-corridors/; p. 293 Map from Department for Business Innovation and Skills, 2014–2020 Assisted Areas Maps: The Government's Response to the Stage 2 Consultation, April 2014; **p. 294** Map from https://en.wikipedia.org/wiki/High_Speed_2#/media/File:HS2_classic_network.png; **p. 301** Graph from Office for National Statistics, Statistcal Bulletin: Migration Statistics Quarterly Report, May 2015; **p. 303** Graph from ITU World Telecommunication/ICT Indicators Database; **p. 308** Map from wfp.org; **p. 309** Map from Sound Vision Productions, http://burnanenergyjournal.com/wp-content/uploads/2013/03/WorldMap_EnergyConsumptionPerCapita2010_v4_BargraphKey.jpg; **p. 324** Map from ChartsBin.com; **p. 334** Flow chart from Geographyfieldwork.com; **p. 338** Graph from Reducing Food Loss and Waste: Creating a Sustainable Food Future, Installment Two. Retrieved from http://www.wri.org/publication/reducing-food-loss-and-waste. Used with permission from World Resources Institute; **p. 341** Text from Practicalaction.org; **p. 348** Graph from Igor A. Shiklomanov, State Hydrological Institute (SHI, St Petersburg) and United Nations Educational, Scientific and Cultural Organisation (UNESCO, Paris), 1999; **p. 350** *t* Map from China is Moving More than a River Thames of Water Across the Country to Deal with Water Scarcity. Retrieved from http://www.unep.org/dewa/vitalwater/article46.html, *b* Map from Asia Pacific Memo; **p. 354** *l* World Map Showing Access to Improved Drinking Water. Used with Permission by United Nations Environment Programme (UNEP), *r* Map Showing Access to Sanitation. Retrieved from http://www.unep.org/dewa/vitalwater/jpg/0215-0-sanitation-tot-EN.jpg. Used with Permission by United Nations Environment Programme (UNEP); **p. 356** *l* Map from Seeking Alpha, http://seekingalpha.com/article/187166-china-both-pollutes-and-tackles-pollution-in-quest-for-clean-coal?page=2, *r* Map from BP Statistical Review & World Energy 2015; **p. 357** Map from Maplecroft, 2011; **p. 358** Graph from BP Global, http://euanmearns.com/global-energy-trends-bp-statistical-review-2014/.

Maps on **pages 155**, **161**, **187, 188** and **288** reproduced from Ordinance Survey mapping with permission. © Crown copyright 2016 Ordinance Survey. License number 100036470.

Introduction to this book

The AQA GCSE Geography course has three main units, with two or three sections within each part. Most of your study is compulsory, although there are some optional topics (see the table below). Your teacher will tell you which options you will be studying.

Unit	Covered in book	Optional content
3.1 **Living with the physical environment**	Section A CH 1–4	No options
	Section B CH 5–8	You will study **one** of either: 7 Hot deserts **or** 8 Cold environments
	Section C CH 9–12	You will study either: CH 10 Coastal landscapes and CH 11 River landscapes **or** CH 10 Coastal landscapes and CH 12 Glacial landscapes **or** CH 11 River landscapes and CH 12 Glacial landscapes
3.2 **Challenges in the human environment**	Section A CH 13–16	No options
	Section B CH 17–20	No options
	Section C CH 21–25	You will study one of: CH 23 Food **or** CH 24 Water **or** CH 25 Energy
3.3 **Geographical applications**	Section A CH 26 Issue evaluation	No options
	Section B CH 27 Fieldwork and geographical enquiry	No options
3.4 **Geographical skills**	Throughout	N/A

Features of this book

The following features of this book have been designed to help you make the most of your course.

- **Key learning** boxes at the top of each spread provide a useful overview of the learning objectives.
- **Activities**, linked to the text and resources on each spread, to develop your understanding.
- **Extension questions** (numbered in orange) in most activity boxes to stretch your thinking.
- **Going Further** activity boxes to challenge you further.
- **Geographical Skills** boxes to give you opportunities to practise and develop your skills.

- **Get out there!** Fieldwork suggestions to show you how the knowledge you learn can be applied to your fieldwork.
- **AQA-specific key terms** in purple and other key terms in black throughout the chapters, all defined in the glossary to help you learn and demonstrate your geographical vocabulary.
- **Case studies and examples** with up-to-date information to illustrate geographical concepts.
- **Practice questions** in the Question Practice pages, which include a range of topical questions and practical advice from skilled teachers to help you prepare for your exam.

How you will be assessed

At the end of two years, you will be assessed on your knowledge through these final exams:

	Unit 3.1	Unit 3.2	Unit 3.3
Length of exam	1 hour and 30 minutes	1 hour and 30 minutes	1 hour and 15 minutes
Total marks	88 (incl. 3 for SPaG)	88 (incl. 3 for SPaG)	76 (incl. 6 for SPaG)
Proportion of course	35%	35%	30%

Paper 1 (Unit 3.1)

Questions will be on ...	Which questions to answer
3.1.1 The challenge of natural hazards 3.1.2 The living world 3.1.3 Physical landscapes in the UK 3.4 Geographical skills	Section A: all (33 marks) Section B: all (25 marks) Section C: any two from Coastal landscapes, River landscapes and Glacial landscapes) (30 marks)

Paper 2 (Unit 3.2)

Questions will be on ...	Which questions to answer
3.2.1 Urban issues and challenges 3.2.2 The changing economic world 3.2.3 The challenge of resource management 3.4 Geographical skills	Section A: all (33 marks) Section B: all (30 marks) Section C: one on Resources and one from Food, Water and Energy (25 marks)

Paper 3 (Unit 3.3)

Questions will be on ...	Which questions to answer
3.3.1 Issue evaluation 3.3.2 Fieldwork 3.4 Geographical skills	Section A: all (37 marks) Section B: all (39 marks)

For Paper 3, the questions are synoptic. This means that questions will cover a number of different aspects of the issue so there will be some links to other parts of the core specification.

In all three exam papers, you will receive extra marks for correct spelling, punctuation, grammar and specialist terminology in your exam answers (SPaG). The marks for each question or part of a question will be shown in brackets: the number of marks should guide the allocation of your time.

You will be faced with answering different question types: multiple-choice, short answer, levels of response and extended prose. There will be some fairly demanding command terms in the exam questions that will require you to apply your knowledge and understanding.

You can practise your exam skills by using the Question Practice pages at the end of each section in this book, as well as the activities in each chapter.

Case studies and examples

The following maps show which pages of this book cover case studies and examples. Throughout your course you will cover case studies and examples in the United Kingdom (UK), higher income countries (HICs), newly emerging economies (NEEs) and lower income countries (LICs).

Case studies and examples in the UK

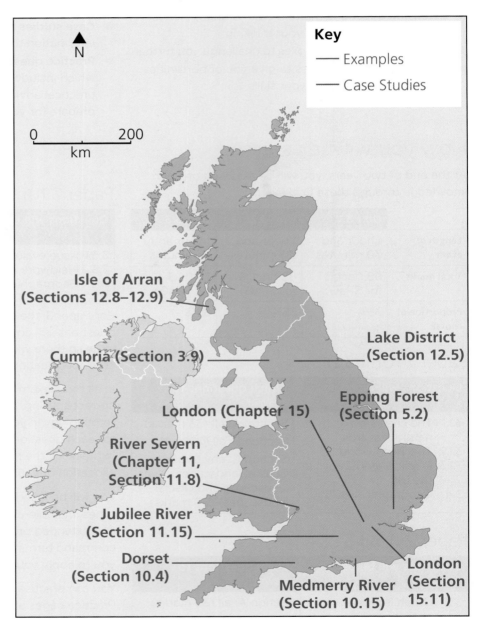

Case studies and examples across the world

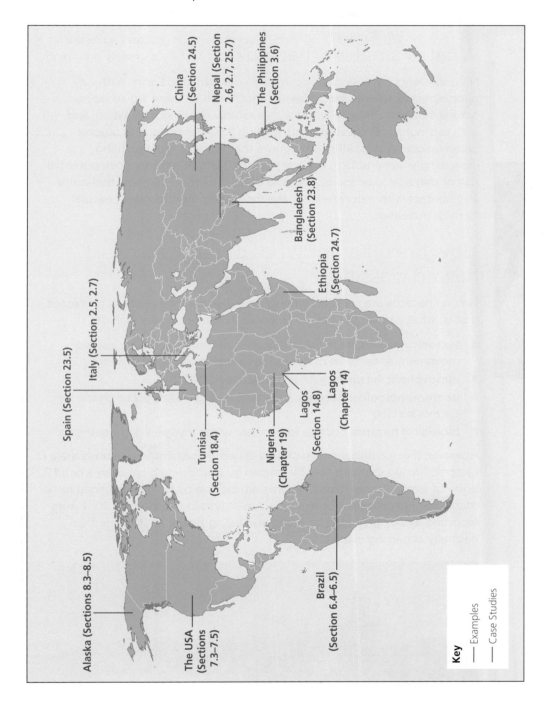

China (Section 24.5)

Nepal (Section 2.6, 2.7, 25.7)

The Philippines (Section 3.6)

Bangladesh (Section 23.8)

Ethiopia (Section 24.7)

Spain (Section 23.5)

Italy (Section 2.5, 2.7)

Lagos (Section 14.8)

Lagos (Chapter 14)

Tunisia (Section 18.4)

Nigeria (Chapter 19)

Alaska (Sections 8.3–8.5)

The USA (Sections 7.3–7.5)

Brazil (Section 6.4–6.5)

Key

— Examples

— Case Studies

1 Natural hazards

★ KEY LEARNING

➤ What natural hazards are
➤ Different types of natural hazards
➤ Factors that affect hazard risk

▲ Figure 1.1 Helping people after an avalanche, Afghanistan, 2014

Defining natural hazards

What is a natural hazard?

Natural events have always occurred on our dynamic Earth. Without people, natural events would be just that, events – there would be no natural 'hazards'. Yet in a world with a rapidly growing population, and with technological developments leading to faster travel and quicker communication, it is difficult to ignore that humans are becoming increasingly vulnerable to **natural hazards**. Natural hazards pose potential **risk** of damage to property, and loss of life. The more humans that come into contact with natural events, the more the potential risk of natural hazards increases.

How are different types of natural hazard classified?

Natural hazards are most commonly classified by their physical processes, that is, what caused the hazard to occur. These processes include:

- **tectonic hazards**, such as earthquakes or tsunamis, which involve movement of tectonic plates in the Earth's crust
- **atmospheric hazards**, such as hurricanes
- **geomorphological hazards**, such as flooding, which occur on the Earth's surface
- **biological hazards**, such as forest fires, which involve living organisms.

However, these categories are closely linked. For example, tsunamis are a tectonic hazard, but can also be caused by a landslide displacing a large body of water. Some natural hazards are caused by human influence rather than a natural process. For example, forest fires in California in 2014 were recorded as being caused by arson and falling power lines rather than naturally occurring events.

➤ Figure 1.2 Winter storms, Cornwall, 2014

Where do natural hazards occur?

Some regions around the world are more vulnerable to natural hazards than others. The different colours of the countries in Figure 1.3 show the likelihood of a natural disaster occurring, based on historical data.

The year 2010 was a bad one for natural hazards. There were earthquakes in Haiti and Chile, volcanic eruptions in Iceland and Indonesia, flooding in China and avalanches in Pakistan. The natural hazards that occurred around the world in 2010 are located on the map in Figure 1.3.

What factors affect hazard risk?

Globally, the incidence of natural hazards is increasing. As a result of human influences, such as global warming, **deforestation** and **urbanisation**, the frequency and magnitude (strength) of natural hazards are increasing. With eight billion people expected to populate the planet by 2024, it is inevitable that the risk will increase as there are even more people to interact with natural events. However, the risk of natural hazards is also made worse by the locations in which people live, whether out of choice or necessity.

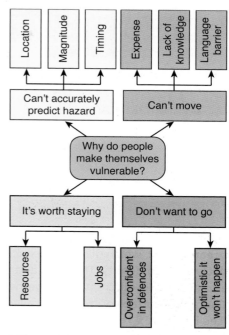

▲ Figure 1.4 Why people live near hazardous areas

▲ Figure 1.3 World natural hazard risk and biggest natural disasters in 2010

→ Activities

1 Distinguish between a natural hazard and a natural event.

2 Use the key on Figure 1.3 to list which hazards the UK could experience.

3 Categorise the different hazards shown in Figure 1.3 into tectonic, atmospheric, geomorphological and biological.

4 Explain why some hazards are more difficult to categorise than others.

5 Describe the pattern of natural hazard risk in Figure 1.3.

6 Use Figure 1.4 to explain factors affecting hazard risk.

7 Suggest why humans could have caused the hazards in Figure 1.1 and Figure 1.2.

2 Tectonic hazards

⭐ KEY LEARNING

➤ The Earth's structure
➤ Why tectonic plates move
➤ The location of earthquakes and volcanoes
➤ The relationship between earthquakes, volcanoes and plate margins

The distribution of earthquakes and volcanoes

What is the Earth's structure?

The Earth's internal structure is divided into layers. At the centre is the **core**, which is extremely hot and under a lot of pressure. The inner core is solid; the outer core is liquid. The **mantle** that surrounds the core is made of solid material that can flow very slowly. The upper portion of the mantle is a weak layer called the **asthenosphere**, which can deform like plastic. The outermost layer of the Earth is the **crust**, which is very thin compared to the thickness of the mantle and the core.

There are two types of crust: **continental crust** and **oceanic crust**, which differ in chemical composition, thickness and density. The crust and upper mantle are also chemically different, but together they form a rigid shell at the surface of the Earth called the **lithosphere** (see Figure 2.1). The lithosphere is broken into several major fragments, called **tectonic plates**, which move very slowly over the upper mantle. The movement of plates can be tracked from space using GPS.

Where two plates meet it is known as a **plate margin**. There are three types of plate margins, which describe the different ways that the plates are moving. They are called **constructive**, **destructive** and **conservative plate margins** (see Sections 2.2 to 2.4). The interaction between the different tectonic plates and the mantle beneath leads to the triggering of **earthquakes** and volcanic activity.

How do tectonic plates move?

What exactly makes tectonic plates move is still being explored by scientists. There is a lot of evidence to show that the plates do move, including evidence from GPS measurements over time.

One theory is called **convection**. The core's temperature is around 6,000 °C. This causes magma (molten rock) to rise in the mantle and sink towards the core when it cools. The currents flow beneath the lithosphere, building up lateral pressure and carrying the plates with them. However, only limited evidence of convection currents has so far been found.

The more accepted explanation is called **ridge push** and **slab pull**. At constructive margins **ocean ridges** form high above the ocean floor. Beneath ocean ridges the mantle melts; the molten magma rises as the plates move apart and cools down to form new plate material. As the lithosphere cools, it becomes denser and starts to slide down, away from the ridge, which causes plates to move away from each other. This is called ridge push. Additionally, at destructive margins (see Section 2.3) the denser plate sinks back into the mantle under the influence of gravity, which pulls the rest of the plate along behind it. This is called slab pull.

Where are earthquakes and volcanoes located?

Earthquakes and volcanoes are not randomly distributed. Figure 2.3 shows the distribution of earthquakes and **volcanoes**. The pattern matches where plate margins are located (see Figure 2.2). Earthquakes are found at all three types of plate margins, whereas volcanoes are found at two, constructive and destructive (see Sections 2.2 to 2.4.)

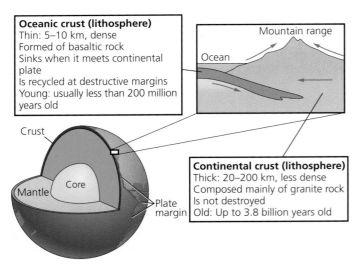

Oceanic crust (lithosphere)
Thin: 5–10 km, dense
Formed of basaltic rock
Sinks when it meets continental plate
Is recycled at destructive margins
Young: usually less than 200 million years old

Mountain range

Ocean

Crust

Mantle · Core

Plate margin

Continental crust (lithosphere)
Thick: 20–200 km, less dense
Composed mainly of granite rock
Is not destroyed
Old: Up to 3.8 billion years old

▲ Figure 2.1 A cross-section through the Earth

Earthquakes and volcanoes are found both on land and in the sea. For example, there is a chain of volcanoes and earthquakes which runs along the west coast of North and South America. The large band of volcanoes and earthquakes which circles the Pacific Ocean is known as the Ring of Fire.

Not every earthquake and volcano lies along plate margins. Some also occur in the middle of plates. These are known at 'hot spots', where the Earth's crust is thought to be particularly thin. For example, Hawaii has formed due to volcanic eruptions at a hot spot that is located far from the edge of the Pacific plate.

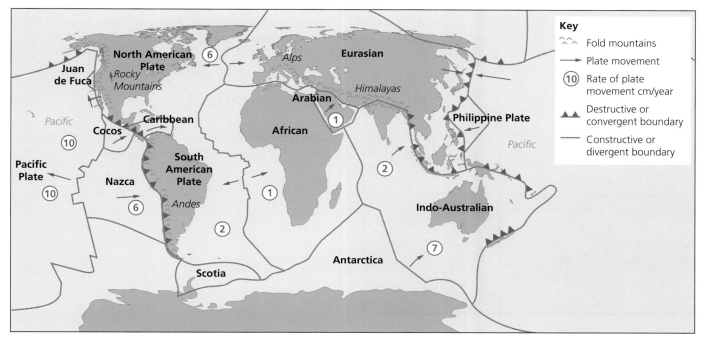

▲ Figure 2.2 World map of plate margins

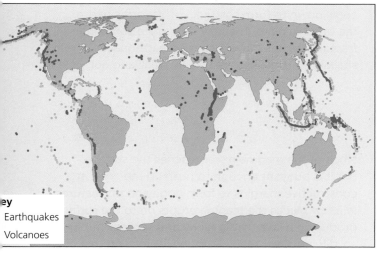

▲ Figure 2.3 Tectonic hazards around the world

→ Activities

1   Define what a tectonic plate is.

2   Contrast the characteristics of continental and oceanic crust.

3   Describe the distribution of earthquakes.

4   Describe the distribution of volcanoes.

5   Do you agree with these statements? Explain your answer.

   a) 'Earthquakes and volcanoes occur together.'

   b) 'Earthquakes and volcanoes occur away from each other.'

   c) 'Earthquakes and volcanoes occur along plate margins.'

   d) 'Earthquakes and volcanoes occur away from plate margins.'

6   Use the diagram in Figure 2.1 to explain what causes the tectonic plates to move.

7   Explain what hotspots are.

Constructive plate margins

How do plates move at constructive margins?

Constructive plate margins occur when tectonic plates move apart from each other. Most tectonic plates move a few centimetres a year. This may not sound much, but over time this has meant that whole continents have moved position. As shown in Figure 2.2 in Section 2.1, the Eurasian plate and North American plate form a constructive plate margin: they are moving away from each other at a rate of five centimetres per year.

Why are earthquakes and volcanoes found at constructive plate margins?

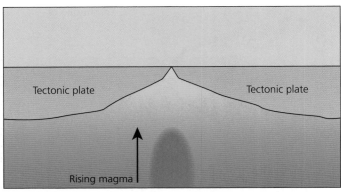

1   At constructive margins, the upper part of the mantle melts and the hot molten magma rises.

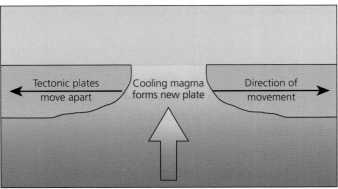

2   As the tectonic plates are moved away from each other by slab pull, ridge push or a combination of these, the molten magma rises in between and cools down to form solid rock. This forms part of the oceanic plate. The new solid plate sometimes fractures as it is moved, causing earthquakes. These shallow earthquakes are usually small and not violent.

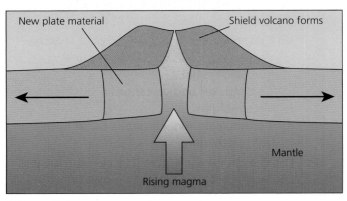

3   Much of the magma never reaches the surface but it is buoyant enough to push up the crust at constructive margins to form ridge and rift features. In a few places the magma erupts on to the surface, producing a lava that is runny and spreads out before solidifying. Over many eruptions a volcano that typically has a wide base and gentle slopes, known as a **shield volcano**, is formed.

▲ Figure 2.4 Constructive plate margins

On land, constructive plate margins form a steep-sided valley, known as a **rift valley**, where the land drops as plates move apart. One example is the Mid-Atlantic Ridge. As the Eurasian and North American plates moved apart, magma rose to the surface to form an island, Iceland. Thingvellir (Figure 2.5) has a visitor centre with a path leading to the Almannagjá fault, where a stretch of the plate margin can be viewed. The strain put on the land as the tectonic plates move away from each other is splitting Iceland in two, and causes cracks or faults to form on either side. Hundreds of small earthquakes occur in Iceland on a weekly basis.

▲ Figure 2.5 Mid-Atlantic Ridge, Thingvellir, Iceland

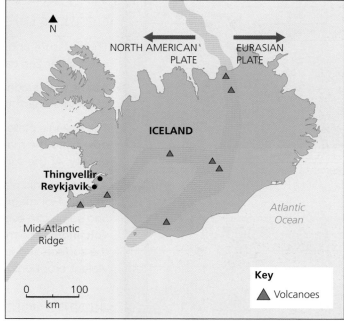

▲ Figure 2.6 Map of Iceland

➜ Activities

1 Describe the direction in which plates move at constructive plate margins.

2 Use Figure 2.2 in Section 2.1 to identify the plate names at two different constructive plate margins.

3 Are earthquakes experienced at constructive plate margins? Explain your answer.

4 What is the evidence in Figures 2.5 and 2.6 that this is a constructive plate margin?

5 Explain why there are so many volcanoes located in Iceland.

Geographical skills

Draw a sketch of Figure 2.5 and label human and physical features.

Destructive plate margins

How do plates move at destructive margins?

Destructive plate margins occur when tectonic plates move towards each other and collide. The effect this has depends on what kinds of plates are colliding:

- If two continental plates collide, they are both buoyant and so cannot sink into the mantle. As a result, compression forces the plates to collide and form mountains.
- If an oceanic and a continental plate move towards each other, the denser oceanic plate is **subducted** and sinks under the continental plate and into the Earth's mantle, where it is recycled. Earthquakes, fold mountains and volcanoes occur.

KEY LEARNING

➤ How plates at destructive margins move
➤ Why earthquakes and volcanoes are found at destructive plate margins

Why are earthquakes and volcanoes found at destructive plate margins?

The pressure and strain of an oceanic and continental plate moving towards each other can cause the Earth's crust to crumple and form fold mountains.

As the plates converge, pressure builds up. The rocks eventually fracture, causing an earthquake, which can be very destructive.

The oceanic plate, as the denser of the two plates, is subducted or pulled down into the mantle beneath the lighter (and thicker) continental plate under gravity. At the surface this creates a deep ocean trench (see Figure 2.7b).

As the oceanic plate sinks deeper into the mantle, it causes part of the mantle to melt. Hot magma rises up through the overlying mantle and lithosphere, and some can eventually erupt out at the surface producing a **linear belt of volcanoes**. The magma becomes increasingly viscous (sticky) as it rises to the surface, producing composite volcanoes which are steep sided and have violent eruptions.

Japan's volcanoes

Japan is prone to earthquakes and volcanic eruptions. It has 118 **active volcanoes** (ten per cent of the global total), which is more than almost anywhere else in the world. The band of volcanoes in Japan form part of the Ring of Fire (see page 5) which surrounds the Pacific Ocean. There are so many because Japan lies on the margin of four plates: the Eurasian, North American, Pacific and Philippine (see Figure 2.2 in Section 2.1). The Pacific plate subducts beneath the North American plate and the Philippine plate, and the Philippine plate subducts beneath the Eurasian plate. Many parts of Japan have experienced earthquakes due to the pressure built up in the plates as they move at this destructive plate margin.

Also formed at the destructive plate margin of the Pacific and Philippine plate is an ocean trench, known as the Mariana Trench. It is 10,994 metres deep, which is deeper than the tallest mountain, Mount Everest, at 8,848 metres. The Mariana Trench is the deepest known part of the Earth's oceans.

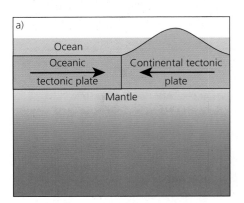

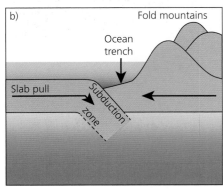

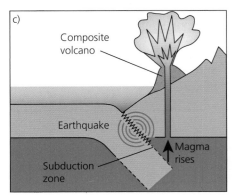

▲ Figure 2.7 Destructive plate margins

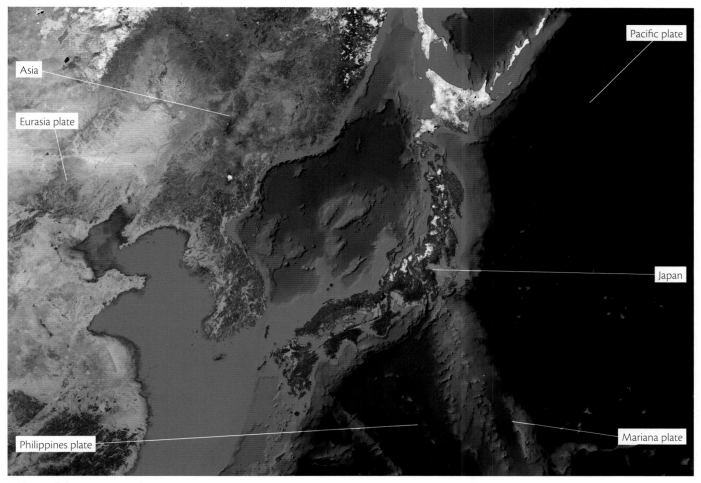

Asia

Eurasia plate

Pacific plate

Japan

Philippines plate

Mariana plate

▲ Figure 2.8 Satellite image of Japan. The darker the blue, the greater the depth

➔ Activities

1   Describe the direction in which plates move at destructive plate margins.

2   Use Figure 2.2 in Section 2.1 to determine which of the following are destructive plate margins:

   ■ Eurasian and Philippine plates

   ■ Nazca and Pacific plates

   ■ Nazca and South American plates

   ■ North American and Eurasian plates.

3   What landforms are found at destructive plate margins?

4   Explain how earthquakes and volcanoes are formed at destructive plate margins.

Geographical skills

Draw an annotated sketch map of Figure 2.8 (above). Annotate the image to show where you would expect earthquakes, volcanoes and mountains to be formed – explain why.

⭐ KEY LEARNING

➤ How plates at conservative margins move

➤ Why earthquakes are found at conservative plate margins

Conservative plate margins

How do plates move at conservative margins?

A conservative plate margin occurs when tectonic plates move parallel to each other. The two plates can move side by side, either in the same direction but at different speeds, or simply in the opposite direction to one another.

Why are earthquakes found at conservative plate margins?

One theory is that pressure might build up at the margin of the tectonic plates as they are pulled along behind a plate being subducted elsewhere. As the plates move past each other, friction causes them to become stuck. Pressure builds up and up until eventually the rock fractures in an earthquake. However, volcanoes are not formed at conservative plate margins. Magma cannot rise to fill a gap as there is no gap created between the tectonic plates, and therefore there is no new land formed. Neither is there any land destroyed, because there is no tectonic plate subducted into the mantle.

The San Andreas Fault stretches 800 kilometres through the state of California in the USA. It is found along the margin between the North American plate and the Pacific plate. These tectonic plates are sliding past each other in roughly the same northwest direction, but at different speeds (see Figure 2.10). The North American plate moves at approximately six centimetre per year, whereas the Pacific plate moves at approximately ten centimetres per year. Fifteen to twenty million years ago, Los Angeles would have been south of where San Diego is now. If the plates continue to move at the same speed, in 20 million years' time, Hollywood in Los Angeles will be adjacent to the Golden Gate Bridge in San Francisco.

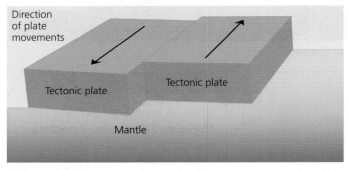
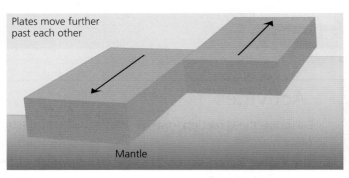

▲ Figure 2.9 Movement of tectonic plates at a conservative plate margin

California experiences thousands of small earthquakes every year. One of the biggest earthquakes to hit California was in San Francisco in 1906. It measured 7.8 on the **Richter scale**. (The largest earthquake ever recorded was in Chile, in 1960, which measured 9.5.) Approximately 700 people died and the damage caused was estimated to cost over US$500 million.

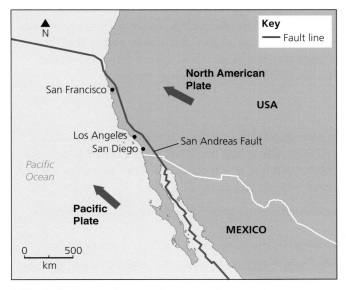

▲ Figure 2.10 Map showing the San Andreas Fault, California, USA

▲ Figure 2.11 The San Andreas Fault

→ Activities

1 Use Figure 2.2 in section 2.1 to identify a different conservative plate margin to the San Andreas Fault.

2 a) Are the following statements about conservative margins true or false?
   i) Two tectonic plates move away from each other.
   ii) Two tectonic plates move parallel past each other.
   iii) Two tectonic plates move towards each other.
   iv) Earthquakes are found at conservative margins.
   v) Volcanoes are found at conservative margins
   b) Explain your answer for each statement.

3 Describe and explain the similarities between the photographs in Figure 2.5 in Section 2.2 and Figure 2.11 on this page.

4 Complete the following summary table about the three types of plate margins.

Margin type	Type of plates (oceanic or continental or either)	Direction of plate movement	Earthquakes (tick or cross)	Volcanoes (tick or cross)
Constructive				
Destructive				
Conservative				

Geographical skills

The Richter scale is logarithmic. This means each increase of one on the scale means the power is increased by 10, not 1. So a magnitude 6 earthquake is 10 times more powerful than a magnitude 5 earthquake.

Calculate how much more powerful a magnitude 8 earthquake is than a magnitude 4 earthquake.

✪ KEY LEARNING

➤ Primary and secondary effects of an earthquake
➤ How people responded to the L'Aquila earthquake

Earthquake in L'Aquila, Italy (2009)

What were the effects of the earthquake?

On 6 April 2009, an earthquake measuring 6.3 on the Richter scale struck L'Aquila in the Abruzzo region of Italy (Figure 2.12). The earthquake's epicentre was seven kilometres northwest of L'Aquila. L'Aquila experienced a range of impacts which affected the wealth of the area and of the community (**economic impacts**), the lives of members of the community (**social impacts**) and the landscape (**environmental impacts**).

Primary effects

As a direct result of the earthquake, an estimated 308 people were killed, 1,500 were injured and 67,500 were made homeless. It struck at 3.32 a.m., so most people were asleep in buildings which collapsed. Approximately 10,000–15,000 buildings collapsed, including:

- many churches, medieval buildings and monuments with considerable cultural value
- the Basilica of St Bernardino, the National Museum and Porta Napoli
- San Salvatore Hospital, which was so severely damaged that patients had to be evacuated as it could not cope with injured victims
- several buildings in L'Aquila University, with some fatalities in its student accommodation.

Overall, the EU reported US$11,434 million of damage to L'Aquila.

▼ Figure 2.12 L'Aquila in the region of Abruzzo, Italy

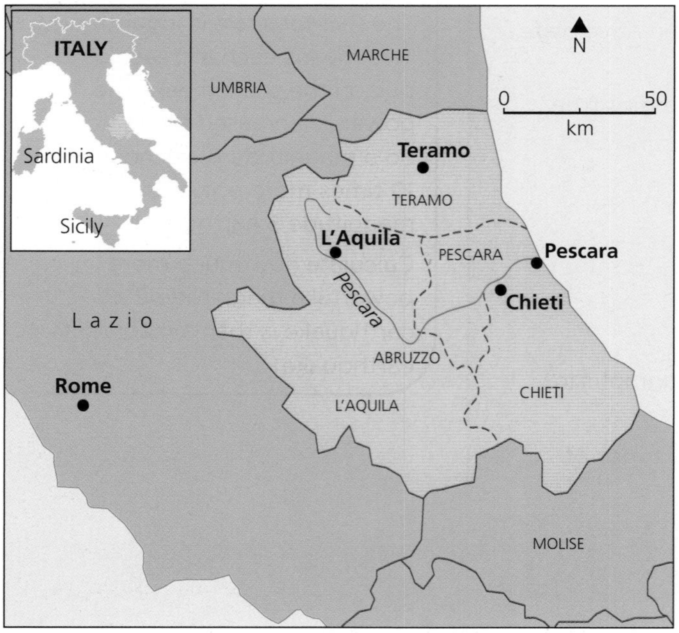

Secondary effects

Some effects of the earthquake occurred later and indirectly as a result of the initial earthquake itself. These are some of the **secondary effects**:

- Aftershocks triggered landslides and rockfalls, causing damage to housing and transport.
- A landslide and mudflow was caused by a burst main water supply pipeline near the town of Paganio.
- The numbers of students at L'Aquila University has decreased.
- The lack of housing for all residents meant house prices and rents increased.
- Much of the city's central business district was cordoned off due to unsafe buildings. Some 'red zones' still exist, which has reduced the amount of business, tourism and income.

What were the responses to the earthquake?

Immediate responses

There were a range of **immediate responses**. For those made homeless, hotels provided shelter for 10,000 people and 40,000 tents were given out. Some train carriages were used as shelters. The then prime minister, Silvio Berlusconi, reportedly offered some of his homes as temporary shelters.

- Within an hour, the Italian Red Cross was searching for survivors. They were helped by seven dog units, 36 ambulances and a temporary hospital. Water, hot meals, tents and blankets were distributed. The British Red Cross raised £171,000 in support.
- Mortgages and bills for Sky TV, gas and electric were suspended. The Italian Post Office offered free mobile calls, raised donations and gave free delivery for products sold by small businesses.
- L'Aquila was declared a state of emergency, which sped up international aid to the area from the EU and the USA. The EU granted US$552.9 million from its Solidarity Fund for major disasters to begin rebuilding L'Aquila. The Disasters Emergency Committee (DEC), a UK group, did not provide aid because it considered Italy a more developed country which had the resources to provide help, and had the help of the EU.

▲ Figure 2.13 Aftermath of the earthquake in L'Aquila, Italy, 2009

Long-term responses

Long-term responses included a torch-lit procession, which took place with a Catholic mass on the anniversary of the earthquake, as remembrance. Residents did not have to pay taxes during 2010. Students were given free public transport, discounts on educational equipment and were exempt from university fees for three years. Homes took several years to rebuild and historic centres are expected to take approximately 15 years to rebuild.

Additionally, in October 2012, six scientists and one government official were found guilty of manslaughter as they had not predicted the earthquake. They were accused of giving residents a false sense of confidence and seriously underestimating the risks. They each received six years in prison and were ordered to pay several million euros in damages. However, in November 2014, the verdict was overturned for the six scientists.

→ Activities

1. Use Figure 2.12 to describe the location of L'Aquila.
2. Define the primary and secondary effects of an earthquake.
3. a) Give three examples of primary effects likely in any earthquake around the world.
   b) Give three examples of secondary effects likely in any earthquake around the world.
4. Complete the table of effects for this earthquake.

Economic	Environmental	Social

5. Which do you think was the worst (a) economic, (b) social and (c) environmental effect, and why?
6. Explain how four of the responses described would help to manage the effects.
7. Justify why both immediate and long-term responses were needed.

→ Going further

Research why the guilty verdict was overturned for the six scientists found guilty of manslaughter.

⭐ KEY LEARNING

➤ Primary and secondary effects of the earthquake

➤ How people responded to the Gorkha earthquake

Earthquake in Gorkha, Nepal (2015)

What were the effects of the earthquake?

On 25 April 2015, a 7.8 magnitude earthquake struck the Gorkha district in Nepal (Figure 2.14). The earthquake's epicentre was in Barpak, 80 kilometres northwest of the capital, Kathmandu.

Primary effects

The immediate **primary effects** included:

- A total of 8,841 dead, over 16,800 injured and 1 million made homeless.
- Historic buildings and temples in Kathmandu, including the iconic Dharahara Tower (Figure 2.15), a UNESCO World Heritage Site, in which 200 people were estimated to be trapped, were destroyed; there were no compulsory building standards in Nepal, so many modern buildings collapsed.
- The destruction of 26 hospitals and 50 per cent of schools. (Save the Children estimated 29,000 more people would have been killed if the earthquake had struck during school hours.)
- A reduced supply of water, food and electricity.
- 352 aftershocks, including a second earthquake on 12 May 2015 measuring 7.3 magnitude.

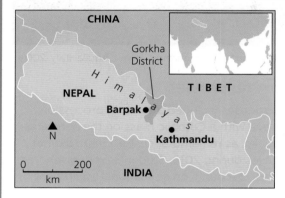

▲ Figure 2.14 Nepal, Asia

Secondary effects

The earthquake triggered an avalanche on Mount Everest. It swept through Everest Base Camp, which is used by international climbing expeditions. Out of the nineteen who died, several were tourists and the rest were members of an ethnic group in Nepal called Sherpas. Sherpas work as porters, guides and cooks. They are aware of the dangers of Mount Everest, but tourism can provide them with an income to help lift them out of poverty.

In 2014, the World Travel and Tourism Council reported that tourism was 8.9 per cent of Nepal's GDP and provided 1.1 million jobs. It was expected to increase by 5.8 per cent in 2015, but until Nepal has recovered from the earthquake, tourism, employment and income will shrink.

The earthquake happened just before the monsoon season, when rice is planted. Rice is Nepal's staple diet, and two-thirds of the population depend on farming. Rice seed stored in homes was ruined in the rubble, causing food shortages and income loss.

What were the responses to the earthquake?

Immediate responses

Nepal requested international help. The UK's DEC (see page 13) raised US$126 million by September 2015 to provide emergency aid and start rebuilding the worst-hit areas.

Temporary shelters were set up. The Red Cross provided tents for 225,000 people. The United Nations (UN) health agency and the World Health Organization (WHO) distributed medical supplies to the worst-affected districts. This was important as the monsoon season had arrived early, increasing the risk of waterborne diseases.

Nepal's mountainous terrain and inadequate roads made it difficult for aid to reach remote villages. 315,000 people were cut off by road and 75,000 were additionally unreachable by air. Sherpas were used to hike relief supplies to remote areas. Facebook launched a safety feature so people could indicate they were 'safe'. Several companies did not charge for telephone calls.

▲ Figure 2.15 Dharahara (Bhimsen) Tower, Kathmandu before and after the earthquake

▲ Figure 2.16 Tents set up near Kathmandu airport, Nepal

Long-term responses

Nepal's government (along with the UN, EU, World Bank, Japan International Cooperation Agency and Asian Development Bank) carried out a Post-Disaster Needs Assessment. It reported that 23 areas required rebuilding, such as housing, schools, roads, monuments and agriculture. Eight months after the earthquake, the Office for the Coordination of Humanitarian Affairs (OCHA) reported that US$274 million of aid had been committed to the recovery efforts.

The Durbar Square heritage sites were reopened in June 2015 in time to encourage tourists back for the tourism season. Mount Everest was reopened for tourists by August 2015 after some stretches of trail were re-routed. By February 2016 the Tourism Ministry extended the climbing permits that had been purchased in 2015 to be valid until 2017, so that climbers would return and attempt Everest again.

A recovery phase started six months later by the Food and Agriculture Organization of the United Nations (FAO). To expand crop production and growing seasons individuals were trained how to maintain and repair irrigation channels damaged by landslides in the earthquake.

Nepal's recovery needs are US$6.7 billion, roughly a third of the economy. Early estimates suggest that an additional three per cent of the population has been pushed into poverty as a direct result of the earthquakes. This translates into as many as one million more poor people.

▲ Figure 2.17 Post-Disaster Needs Assessment (*Source: Worldbank.org press-release 23/06/2015*)

➜ Activities

1 Use Figure 2.14 to describe the location of Nepal.

2 Draw a sketch of the photo in Figure 2.15 or Figure 2.16 that shows the effects of the earthquake. Annotate the effects.

3 Identify the economic, environmental and social effects for the Gorkha earthquake.

4 Which do you think was the worst (a) economic, (b) social and (c) environmental effect and why?

5 Why do you think the tents in Figure 2.16 were set up near Kathmandu airport?

6 Describe three immediate and three long-term responses to an earthquake.

7 Research other 'before and after' images of the earthquake. Describe the differences.

⭐ KEY LEARNING

➤ How the effects and responses compare

➤ Why effects and responses are different

Comparing the Italy and Nepal earthquakes

Do effects and responses to earthquakes differ?

Earthquakes occur whether the country is a high-income country (HIC), newly emerging (NEE) or a low-income country (LIC). Tectonic events do not discriminate by wealth. However, the effects of an earthquake differ due to the ability to predict, protect against and prepare for the hazard. Equally, a country's ability to manage the effects and devastation can be affected by its level of wealth and resources.

How did Italy's and Nepal's earthquakes compare?

	Italy	Nepal
Income level	High income	Low income
GDP	US$2.144 trillion (2014)	US$19.64 billion (2014)
GNI per capita	US$34,280 (2014)	US$730 (2014)
Magnitude	6.3 magnitude	7.8 magnitude
Time of day	3.32 am (local time)	11.56 am (local time)
Deaths	308	8,841 (19 on Mount Everest)
Homeless	67,500	1 million
Hospitals damaged	San Salvatore Hospital	26 hospitals
Sites damaged	Basilica of St Bernardino, National Museum, Porta Napoli and L'Aquila University	World Heritage sites, e.g. Dharahara Tower, the Patan and Bhaktapur Durbar Square.
Cost of damage	US$1.1 billion	US$5.15 billion
Amount of aid	US$552.9 million from EU	US$274 million from EU

There may be more damage in a poorer country but it doesn't cost as much to repair.

Economically developed countries can respond faster to earthquakes.

Poorer countries always suffer more in an earthquake.

Poorer countries rely on international help.

It doesn't matter where the country is – the impacts are always devastating.

▲ Figure 2.18 Attitudes to earthquakes in HICs and LICs

Why do effects and responses differ?

Earthquakes never occur in exactly the same circumstances. Figure 2.19 shows some factors that are responsible for the different effects and responses in different countries. Many are influenced by the wealth of the country concerned.

Depth of focus
The shallower the focus the more energy to cause damage.

Population density
The more people the more potential risk of injuries and fatalities.

Building density
The more buildings the greater the likelihood some will collapse.

Plate margin
Some plate margins are more destructive than others.

Distance from epicentre
The closer to the epicentre the greater the magnitude will be.

Magnitude
The stronger the earthquake the greater the impact.

Medical facilities
The more medical resources the easier it is for victims to get treatment.

Transport infrastructure
The better the transport network the quicker it is to evacuate and for aid to arrive.

Monitoring and predicting
More resources available to monitor and predict earthquakes should reduce the risks.

Time and day of the week
Buildings and roads that collapse when they are empty reduce casualties.

Time of year
Climate in different seasons can make impacts and responses harder (e.g. monsoon).

Construction standards
Buildings with strict building regulations have less chance of collapsing.

Resources and finance
The more resources and money available the quicker it is to rebuild homes and businesses.

Earthquake-prone area
Planning, preparation and protection is more likely in areas that expect earthquakes.

Emergency services
More skilled emergency services, armies and volunteers reduce casualties.

Type of event
Secondary effects, e.g. tsunamis, avalanches and landslides, cause further devastation.

Corruption
Corrupt governments and organisations divert aid and supplies away from areas that need it most.

Training
More training so the public and emergency services know what to do reduces casualties.

▲ Figure 2.19 Factors affecting the impact and responses to an earthquake

→ Activities

1. Draw a large Venn diagram, as shown, to categorise which effects and responses are similar and which are different in the earthquakes in Italy and Nepal. You will need to use the information in Sections 2.5 and 2.6. Colour-code the effects and responses.

2. Sort the reasons for varying impacts in Figure 2.19 into two categories: those influenced by wealth and those not influenced by wealth.

3. Suggest other ways that the impacts in Figure 2.19 could be categorised.

4. Use Figure 2.19 to explain reasons for the differences in:
   a) effects in Italy and Nepal
   b) responses in Italy and Nepal.

5. Decide if you agree or disagree with the statements in Figure 2.18 and give reasons for your answers.

6. Are earthquakes more devastating in high-income countries (HICs) or low-income countries (LICs)? Justify your answer.

Risking it

Do people live in areas prone to volcanic eruptions and earthquakes?

Despite all the dangers, millions of people still live in hazard-prone areas. As the world's population rises, more people will live in volcanic and earthquake-prone areas. Figure 2.2 on page 5 (Section 2.1) shows that cities are found in the same locations (and within close proximity) to earthquake and volcano zones. Approximately eight per cent of the 7 billion people who live in the world live near volcanoes, and 50 per cent of the 320 million people in the USA are living at risk of earthquakes.

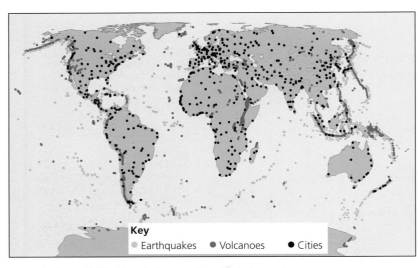

Key
● Earthquakes   ● Volcanoes   ● Cities

▲ Figure 2.20 Global cities and tectonic hazards

Why do people continue to live in hazardous areas?

Geothermal energy

In volcanically active areas, geothermal energy is a major source of electrical power: steam is heated by hot magma in permeable rock, then boreholes are drilled into the rock to harness the super-heated steam to turn turbines at power stations. It is renewable energy – it will not run out, and it will reduce greenhouse gases and the likely effects of climate change.

Hellisheidarvirkjun (or Hellisheidi) power plant is the largest geothermal power station in Iceland and the second largest in the world. It provides electricity and hot water for the capital, Reykjavik. Geothermal energy produces approximately 30 per cent of Iceland's total electricity.

Farming

Lava and ash erupting from volcanoes kill livestock and destroy crops and vegetation. After thousands of years, weathering of this lava releases minerals and leaves behind extremely fertile soil, rich in nutrients. Land can be farmed productively in these areas to provide a source of food and income. Volcanic soils are found on less than one per cent of the Earth's surface, but support 10 per cent of the world's population.

▲ Figure 2.21 Hellisheidarvirkjun geothermal power plant, Iceland

Mining

Settlements develop where valuable minerals are found, as jobs are created in the mining industry. It is not just dormant and extinct volcanoes that are mined, but also active volcanoes.

Kawah Ijen is an active volcano in East Java, Indonesia. Its crater is one of the biggest sulphuric lakes in the world. Sulphur is sold, for example, to bleach sugar, make matches, medicines and fertiliser. However, mining in active volcanoes is dangerous:

- Miners can afford little protective clothing.
- Hydrogen sulphide and sulphur dioxide gases burn their eyes and throat and cause respiratory diseases.
- In the last 40 years, 74 miners have died from the fumes.
- Loads of sulphur weighing 100 kilograms are carried up and down the rocky and slippery mountain paths.

Nevertheless, miners can earn an average of six dollars per day (more than on a coffee plantation), so miners continue to live and work in these dangerous areas.

▲ Figure 2.22 Hey, great news! I've finally decided to sell you my house on the island!

Tourism

Tourists visit volcanoes for the spectacular and unique views, relaxing hot springs, adventure, and, for thrill seekers, the sense of danger. More than 100 million people visit volcanic sites every year. The revenue they generate benefits the locals and the countries they are in.

Family, friends and feelings

People do not wish to leave because their friends and family are there. It is often cheaper and easier to stay, especially when the risks may not be perceived as dangerous enough, or residents are in denial that a disaster may occur.

▲ Figure 2.23 Mining sulphur at Kawah Ijen crater, Indonesia

→ Activities

1 How does the map in Figure 2.20 demonstrate that people are at risk from tectonic hazards?

2 Why would a farmer consider the benefits of living near a volcano different in the short term rather than in the long term?

3 Draw a sketch of the photograph in Figure 2.23. Annotate with the following labels:

crater, lake, sulphur, hydrogen sulphide and sulphur dioxide gases, little protective clothing, hand-carried loads of sulphur, rocky paths.

4 Describe three benefits of living near a volcano.

5 Explain why it may have taken time for the character in Figure 2.22 to make that decision.

Risk management

Can the risks of earthquakes be reduced?

Prediction, **protection**, **planning** and **monitoring** all aim to reduce the damage that earthquakes and volcanic eruptions cause to people and property.

Monitoring and prediction

It is possible to predict the general locations where earthquakes are most likely to happen, as they occur along plate margins. However, it is extremely difficult to predict their time, date and exact location. The following show some ways that technology is used to try to monitor and predict tectonic hazards:

- **Seismologists** use radon detection devices to measure radon gas in the soil and **groundwater**, which escapes from cracks in the Earth's surface.
- Sensitive seismometers are used to measure tremors or **foreshocks** before the main earthquakes.
- Earthquake locations and their times are mapped to spot patterns and predict when the next earthquake will occur.
- Smart phones have GPS (Global Positioning System) receivers and accelerators built in. They can detect movements in the ground, which are analysed to potentially warn others further away.
- Animals are believed to act strangely when an earthquake is impending.

Protection

Buildings made of brick or buildings with no reinforcement collapse more easily during an earthquake. Designing buildings and strengthening roads and bridges to withstand earthquakes provides protection. This is also called **mitigation**. Figure 2.24 shows features of earthquake-resistant designs.

Unfortunately, earthquake-resistant buildings and infrastructure are extremely expensive, so it is usually not possible to adapt existing buildings. The aim of earthquake-resistant buildings is to ensure that people are not injured or killed – so although this might be achieved, the building may still need to be repaired or even rebuilt.

Planning

Planning and preparing what to do during and after an earthquake helps the authorities, emergency services and individuals to act quickly and calmly, so there is less chaos and fewer injuries and deaths:

- Furniture and objects can be fastened down so they are secure from toppling over.
- Residents can learn how to turn off the main gas, electricity and water supplies to their properties.
- Preparing emergency aid supplies, how they would be distributed and where evacuation centres will be saves lives, as food, water, medicine and shelter are accessed faster.
- On 1 September each year, the Japanese practise earthquake drills on a national training day. This marks the anniversary of the Tokyo earthquake in 1923, which killed 156,000 people.
- The American Red Cross provides an earthquake safety checklist to help people plan and prepare for earthquakes in their homes, at work and in schools.

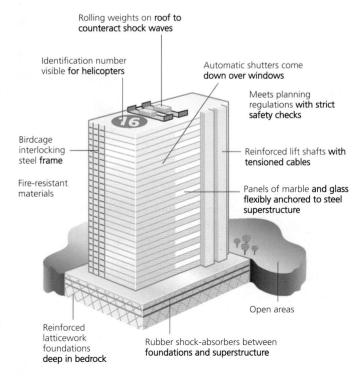

Rolling weights on **roof** to counteract shock waves

Identification number visible **for helicopters**

Automatic shutters come **down over windows**

Meets planning regulations **with strict safety checks**

Birdcage interlocking steel **frame**

Fire-resistant materials

Reinforced lift shafts **with tensioned cables**

Panels of marble **and glass flexibly anchored to steel superstructure**

Open areas

Reinforced latticework foundations **deep in bedrock**

Rubber shock-absorbers between **foundations and superstructure**

➤ Figure 2.24 Earthquake-resistant building design

Can the risks of volcanic eruptions be reduced?

Monitoring and prediction

It is easier to predict volcanic eruptions than earthquakes. Volcanoes usually give advance warning signals that they are going to erupt. However, the exact time and day of the eruption is still difficult to predict.

- Satellites (GPS) and tiltmeters monitor ground deformation (changes in of the volcano's surface).
- Seismometers measure small earthquakes and tremors.
- Thermal heat sensors detect changes in the temperature of the volcano's surface.
- Gas-trapping bottles and satellites measure radon and sulphur gases released.
- Scientists measure the temperature of water in streams and rivers to see if it has increased.

Protection

Protecting against a volcanic eruption is extremely difficult. Buildings cannot be designed to withstand the lava flows, **lahars** or weight of debris and ash falling on roofs, especially if this mixes with water. Therefore people need to **evacuate** their homes to a safe location under the instruction of the authorities.

Planning

An evacuation plan is one of the most effective methods of protection against an eruption. Authorities and emergency services need to prepare emergency shelter, food supplies and form evacuation strategies. Exclusion zones can be designated so that no one is allowed to enter where people are considered vulnerable and in danger. Additionally, residents can be educated about preventing unnecessary injury and loss of life. They can practise advice to cover their eyes, nose and mouth to prevent being irritated by gas fumes. If residents are not evacuated, they are taught to seek shelter or go indoors to avoid the dangers of falling ash and rock.

▲ Figure 2.25 Monitoring volcanic activity

→ Going further

1. Is prediction, protection or planning the most useful in reducing the risks of a tectonic hazard
2. Why may monitoring, prediction, protection and planning differ in different parts of the world?

→ Activities

1. Define the terms prediction, protection, planning and monitoring.
2. True or false: (a) Earthquakes can be predicted. (b) Volcanoes can be predicted. Explain your answers.
3. Use Figure 2.24 to describe how each feature in the earthquake-resistant building reduces injury and loss of life.
4. Describe how the risks of living in an earthquake-prone area may be lessened by (a) individuals and (b) governments?
5. Devise your own earthquake safety checklist and explain your reason for each item.
6. Complete the information in the following table.

Monitoring and predicting a volcanic eruption		
Changes a volcanologist would observe	Equipment used to monitor volcano	Explain why the changes mean a volcanic eruption is imminent

7. Suggest what the man in Figure 2.25 is doing.
8. 'Predicting tectonic hazards is a waste of time'. To what extent do you agree with this statement?

3 Weather hazards

Global atmospheric circulation

What are the features of global atmospheric circulation?

Global atmospheric circulation helps to explain the location of world climate zones (see Figure 3.1) and the distribution of weather hazards. In Chapter 5 you will also learn how the Earth's climate zones govern the pattern of global **ecosystems**.

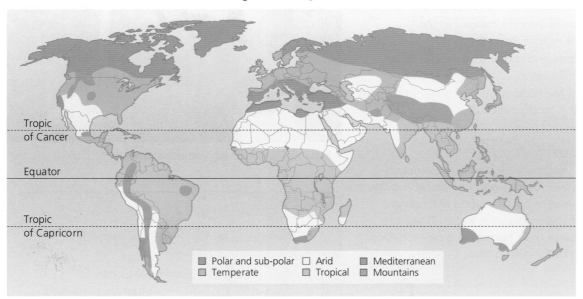

◼ Polar and sub-polar	☐ Arid	◼ Mediterranean
◼ Temperate	◼ Tropical	◼ Mountains

⌃ Figure 3.1 World climate zones

The most important influence on worldwide variations in climate is **latitude**. Because of the curved surface of the Earth, the Equator receives much higher **insolation** than the polar latitudes. The parallel rays of the Sun are spread thinly when they strike the Earth's surface at high latitudes, whereas at low latitudes sunlight is more highly concentrated (Figure 3.2).

As a result, air at the Equator is heated strongly. It becomes less dense and rises to a high altitude. This creates a global climate zone of low pressure, the equatorial zone. After rising, the air spreads out and begins to flow towards the North and South Poles.

Meanwhile, the low insolation received at polar latitudes results in colder, dense air and high pressure. As the air sinks towards ground level, it spreads out and flows towards the Equator.

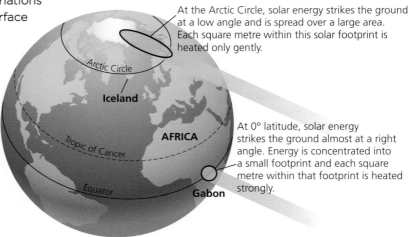

At the Arctic Circle, solar energy strikes the ground at a low angle and is spread over a large area. Each square metre within this solar footprint is heated only gently.

At 0° latitude, solar energy strikes the ground almost at a right angle. Energy is concentrated into a small footprint and each square metre within that footprint is heated strongly.

⌃ Figure 3.2 Solar heating of the Earth varies with latitude

Taken together, the low pressure belt at the Equator and the high pressure belt at the Poles provide the basis for a simple **convection cell** to operate. Global atmospheric circulation, however, is a little more complex than this. Three convection cells operate, not just one, as Figure 3.3 illustrates. As well as pressure belts at the Poles, there are areas of high pressure at the Tropics of Cancer and Capricorn. Air sinks towards the ground there. As it descends, the air warms. The result is high pressure and hot, dry desert conditions. This circulation of air between the tropics and the Equator is called the Hadley cell.

Global circulation involves three cells because the Earth rotates on its axis. The movement generates strong, high-altitude winds which wrap around the planet like belts. These winds flow towards the east, as the Earth spins, and interact with the convection cells. Figure 3.3 shows two particularly strong high-altitude currents of air called **jet streams**.

The exact position of the jet streams and convection cells varies seasonally. Seasons arise because the Earth is tilted on its axis. Each year, as the planet journeys around the Sun, insolation rises and falls at each latitude. In high polar latitudes the Sun does not even rise during the winter. In southern Europe, temperatures rise steeply in summer before falling in winter.

▼ Figure 3.3 The three global convection cells and the position of the high-altitude jet streams

How do global pressure and surface winds influence precipitation?

Global pressure and surface wind patterns influence precipitation in several important ways:

- Rainfall is high and constant throughout the year near the Equator. As hot air rises, it cools slightly. Water vapour is converted into droplets of convectional rain.
- The low-pressure zone around the Equator is called the intertropical convergence zone (ITCZ). Air rises and triggers bursts of torrential rain. Sometimes, the ITCZ grows a 'wave' of low pressure which extends further than usual. Tropical storms develop along these waves. Once they gain energy, they can travel even further away from the Equator (pages 24–5).
- Rainfall is often higher in coastal areas in Western Europe due to the movement of the jet stream over the Atlantic. Rain-bearing weather systems called depressions (also known as cyclones) follow the jet stream, often bringing stormy conditions to the UK's west coast (page 38).
- Rainfall is often low around the Tropics of Capricorn and Cancer. Dry air descends there as part of the Hadley cell, resulting in **arid** conditions.
- Precipitation is also very low in polar regions and falls mostly as snow. The cold air has a limited ability to hold water vapour.

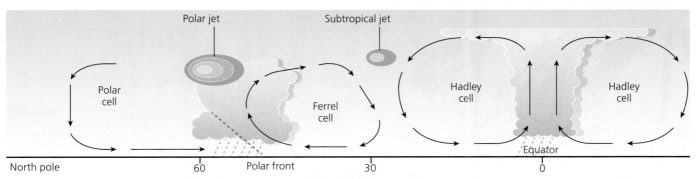

Polar jet   Subtropical jet

Polar cell

Ferrel cell

Hadley cell

Hadley cell

Equator

North pole   60   Polar front   30   0

→ Activities

1 Look at Figure 3.1. Which climate zone is the UK in? Identify all the different climate zones in the USA and Russia (you may need an atlas).

2 Copy and fill out the table.

Cell name	Characteristics	Influence on climate

3 Look at Figure 3.2. (a) Describe what insolation means and (b) explain why it varies with latitude.

4 a) Describe what is meant by an arid climate.
   b) Explain why arid environments can be found in many different places, including the tropics and the poles.

5 Look at Figure 3.3. Examine how the position of the jets streams corresponds with the location of the global convection cells.

⊗ KEY LEARNING

➤ Tropical storms and why they occur

➤ Why tropical storms are distributed where they are

➤ How tropical storms relate to global atmospheric circulation

The global distribution of tropical storms

What are tropical storms?

Tropical storms are a natural hazard. A tropical storm occurs when tropical warm air rises to create an area of intense low pressure, much lower than the depressions experienced in the UK. As the warm, moist air reaches high altitudes, powerful winds spiral around the calm central point, creating the 'eye of the storm', and the warm air cools and condenses into heavy rainfall and thunderstorms.

A UN report on global climate stated that from 2001 to 2010, more than 500 tropical storm disasters killed nearly 170,000 people, affected over 250 million people, and caused damage estimated at US$380 billion. The devastation caused by tropical storms is clear to see in Figure 3.4.

What are tropical storms called?

Tropical storms have different names depending on the location in the world (see Figure 3.6). Their location is their only difference. They are known as:

- hurricanes in the Atlantic and Eastern Pacific Oceans (such as Hurricane Katrina, 2005)
- **typhoons** in the west of the North Pacific Ocean (such as Typhoon Haiyan, 2013)
- **cyclones** in the Indian and South Pacific Oceans (such as Cyclone Pam, 2015).

Each tropical storm has its own unique international name. These names are predetermined by the World Meteorological Organization. They are alphabetical and alternate in gender. For example, the 2020 name list starts Arthur, Bertha, Cristobal, Dolly, Edouard, Fay and so on. Tropical storms are more recognisable and engaging for the public when given names rather than co-ordinates. Names repeat every six years unless a large loss of life or cost in damage would make it insensitive to repeat them.

▲ Figure 3.4 Port Vila, Vanuatu after Cyclone Pam in 2015

Why are tropical storms distributed where they are?

The 80–100 tropical storms that take place every year are caused by particular conditions. Tropical storms occur in the tropics (mainly where the intertropical convergence zone (ITCZ) lies) – broadly south of the Tropic of Cancer and north of the Tropic of Capricorn.

Tropical storms are usually found in areas of low latitude, between 5° and 30° north and south of the Equator. Here, a higher insolation means temperatures are higher than at the poles (see Section 3.1). The sea temperature must be above 27°C, to a depth of approximately 60–70 metres. This provides the heat and moisture that causes the warm air to rise rapidly in this low-pressure region. Latent heat is then released which powers the tropical storm. The warmest seasons are between summer and autumn when it is most typical for tropical storms to develop. The seasons vary depending on their location (see Figure 3.3). Additionally, there is usually low wind shear (wind which remains constant and does not vary with height), so that the tropical storm clouds

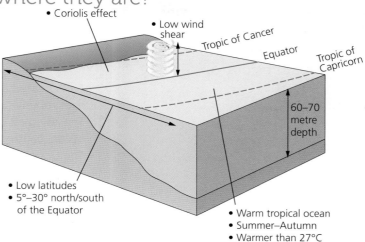

▲ Figure 3.5 Conditions for a tropical storm

can rise to high levels without being torn apart. Finally, tropical storms do not develop along the Equator, because the Coriolis effect (see Section 3.3) is not strong enough here for tropical storms to spin.

▼ Figure 3.6 The global distribution of tropical storms

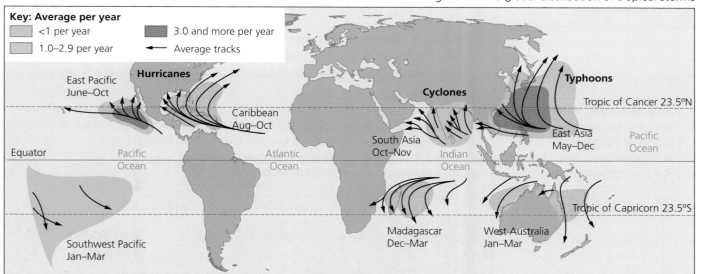

→ Activities

1 Define the term tropical storm.

2 Why is the tropical storm in Figure 3.4 called a cyclone? (Hint: find Vanuatu in an atlas.)

3 Describe the global pattern of tropical storms in Figure 3.6.

4 Explain why the following conditions are needed for a tropical storm to form:
   a) temperature above 27°C
   b) not along the Equator
   c) low wind shear
   d) over the ocean.

5 Use Figure 3.6 to locate the regions where you would expect people to be most vulnerable to tropical storms.

6 Research a recent tropical storm. Identify evidence which demonstrates that tropical storms are natural hazards that pose risk of damage to property and life.

In a spin

How does a tropical storm form?

Tropical storm formation follows a particular sequence:

1   Air is heated above the surface of warm tropical oceans. The warm air rises rapidly under the low-pressure conditions.
2   The rising air draws up more air and large volumes of moisture from the ocean, causing strong winds.
3   The Coriolis effect causes the air to spin upwards around a calm central eye of the storm.
4   As the air rises, it cools and condenses to form large, towering cumulonimbus clouds, which generate torrential rainfall. The heat given off when the air cools powers the tropical storm.
5   Cold air sinks in the eye, therefore there is no cloud, so it is drier and much calmer.
6   The tropical storm travels across the ocean in the prevailing wind.
7   When the tropical storm meets land it is no longer fuelled by the source of moisture and heat from the ocean so it loses power and weakens.

What are the structure and features of a tropical storm?

Figure 3.7 shows a cross-section of the structure of a tropical storm. The satellite image in Figure 3.11 (see page 32) also show the swirling wind and cloud around the central circular eye of the storm where there is no cloud.

A At the start of a tropical storm, the temperature and air pressure fall. Air rises and clouds begin to form. It becomes windy.
B As the tropical storm continues, the air pressure falls more rapidly, wind increases, cumulonimbus cloud forms and there is heavy rainfall.
C There is a period of calm with no wind or rain at the eye of the storm. The Sun appears, so it gets warmer. Air pressure is very low.
D Wind and heavy rainfall increase dramatically again, the temperature drops and air pressure begins to rise.
E As the tropical storm ends, the air pressure and temperature rise. Wind and rainfall subside.

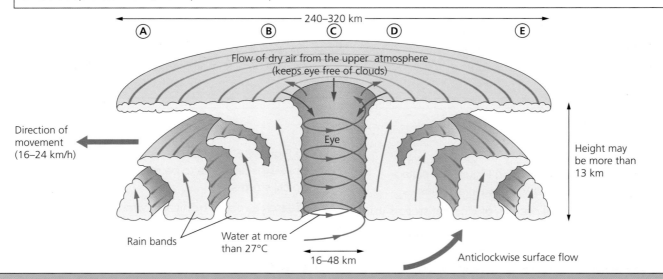

▲ Figure 3.7 What would it be like to be in a tropical storm?

Why does a tropical storm spin?

The **Coriolis effect** bends and spins the warm rising air. The spinning can be seen in satellite images (such as in Figure 3.11 on page 32). Hurricanes in the northern hemisphere bend to the right, which causes the clouds to swirl anticlockwise, whereas cyclones in the southern hemisphere swirl in a clockwise direction.

▼ Figure 3.8 The Coriolis effect

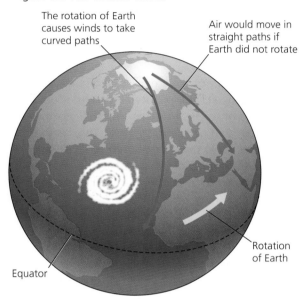

The rotation of Earth causes winds to take curved paths

Air would move in straight paths if Earth did not rotate

Rotation of Earth

Equator

What direction do tropical storms travel?

Tropical storms travel from east to west due to the direction in which the Earth spins. When they hit land, they lose their energy source from the sea that powered them. As they pass over land, friction also slows them down. As they lose energy they change direction. This exact direction and speed is unknown. However, tropical storms in the northern hemisphere track north and tropical storms in the southern hemisphere track south (Figure 3.6 on page 25). An average tropical storm has a lifespan of approximately one to two weeks.

What is the Coriolis effect?

Winds blow from areas of high pressure to areas of low pressure. They do not blow in straight lines across the Earth but are affected by the Coriolis effect. As the Earth rotates it causes the wind to bend. This is because the Earth has a curvature, with the Equator far wider than the poles. Therefore the Earth has to spin faster at the Equator. This difference in speed means that wind bends as it blows across the Earth. This is known as the Coriolis effect.

→ Activities

1 Draw a sequence of at least three diagrams, with captions, to show the formation of a tropical storm.

2 What is the eye of the storm?

3 Using the satellite image on page 32, state which hemisphere the tropical storm is in and how you reached your answer.

4 Sketch a larger version of the cross-section of a tropical storm in Figure 3.7. Annotate it with the sequence of weather conditions that would be experienced. (Include information about wind, rain, clouds, temperature and air pressure.)

5 Write a short paragraph to explain what causes tropical storms to spin.

6 What happens to a tropical storm when it reaches land? Explain why.

→ Going further

Look up the definition of troposphere. What does this have to do with tropical storms?

⊕ KEY LEARNING

➤ How climate change might affect tropical storms

Climate change and tropical storms

How might climate change affect tropical storms?

Climate change (see Chapter 4) will alter the conditions that cause tropical storms to form:

- As the temperature increases, sea levels will rise due to **thermal expansion**. The impact of rising sea levels will mean **storm surges** are expected to become higher.
- A warmer atmosphere will mean the air can hold more moisture. Heavy rainfall is expected to increase. Therefore flooding during a tropical storm is expected to be more destructive.

Intensity

People do not know exactly what the impact of climate change on tropical storms will be. However, there is evidence of a link between warmer oceans and the intensity (destructive power) of tropical storms. Tropical storms are expected to become more intense, by 2–11 per cent, by 2100. The number of the most severe category 4 or 5 tropical storms (see Section 3.5) has increased since the 1970s (Figure 3.9). Predictions suggest that every one degree Celsius increase in tropical sea surface temperatures will mean a 3–5 per cent increase in wind speed.

Key a)
(Left) The total number of category 1 storms (blue), the sum of categories 2 and 3 (green), and the sum of categories 4 and 5 (yellow) in five-year periods. The black line is the maximum hurricane wind speed observed globally (measured in metres per second). The horizontal dashed lines show the 1970–2004 average numbers in each category.

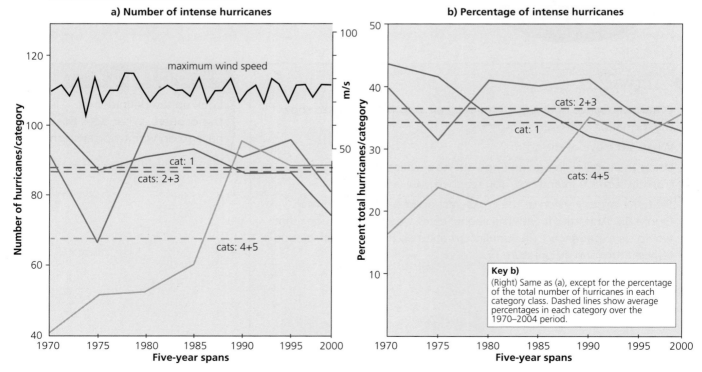

Key b)
(Right) Same as (a), except for the percentage of the total number of hurricanes in each category class. Dashed lines show average percentages in each category over the 1970–2004 period.

▲ Figure 3.9 Climate change affecting intensity. Intensity of hurricanes according to the Saffir-Simpson scale (categories 1 to 5).

Frequency and distribution

The overall frequency of tropical storms occurring is expected to either remain the same, or decrease, as a result of climate change. However, the number of more severe tropical storms (categories 4 and 5) is expected to increase, while category 1–3 storms will decrease (see Figure 3.9).

The regions where tropical storms are experienced are not expected to change significantly as a result of climate change.

Uncertainty

Wind speed monitoring has only become more accurate in recent decades, so the use of previous data – which is less accurate – to decide how tropical storms are affected by climate change is questionable. Predicting the impact of climate change is unreliable, as the rate of and impact of climate change in the future is uncertain. Indeed, potential risk to life and property has already increased due to population growth and building in coastal locations, even without factoring in climate change.

→ Activities

1   What conditions will climate change affect that cause tropical storms to form?

2   Is climate change expected to affect:
    a)   the distribution of tropical storms?
    b)   the frequency of tropical storms?
    c)   the intensity of tropical storms?
    In each case explain why.

3   Explain how climate change may make the impact of tropical storms worse.

4   What makes the link between climate change and tropical storms uncertain?

5   What benefit do tropical storms have on the planet?

→ Going further

Study Figure 3.9. Why is it wrong to say the number of category 4 and 5 tropical storms is increasing in Graph B, but correct to say this in Graph A?

The effects of and responses to tropical storms

What are the effects of a tropical storm?

Tropical storms have significant effects on people and environments. The impacts of a tropical storm are strong winds, torrential rainfall and storm surges (when the sea level rises rapidly and particularly high due to the storm). Landslides and tornadoes can also be caused by tropical storms.

Wind speeds are at least 119 kilometres per hour. They can demolish houses across whole towns and villages, destroy infrastructure such as electricity power lines, and wipe out crops. These are primary effects. The amount of destruction will depend on the storm's strength and how well people and property are protected.

Flooding is caused by the heavy rain and storm surges. Storm surges can be up to five metres and are driven by the wind pushing seawater onto the coastline. Heavy rainfall can be up to 500 millimetres falling within 24 hours. The cause of death for the majority of victims is due to flooding.

Aid is hampered as roads are flooded. Torrential rain can also trigger landslides, causing further devastation. Furthermore, water supplies can be contaminated with seawater, sewage and industrial waste, which increases the risk of waterborne diseases such as cholera. These are secondary effects.

How are tropical storms measured?

Tropical storms are measured using the Saffir–Simpson hurricane wind scale (Figure 3.10). This scale was updated in 2012 and is now based on wind speed, which is considered easier to understand. (Previously, storm surge, flooding impact and central pressure were included, but surges and flooding are influenced too much by local conditions.) The higher the category scale, the higher the intensity of the tropical storm.

What are the responses to a tropical storm?

As tropical storms can generally be predicted, warning systems provide crucial information regarding strong winds, heavy rain and storm surges, which are broadcast to the public. This allows vital time to prepare and protect property (see Section 3.7). A common immediate response for predicted disasters is evacuation, to higher ground (away from the potential impact of storm surges) or even to emergency storm shelters. Shelter provided by public buildings, or tents provided by international aid, are also necessary where homes are extensively damaged or even destroyed.

Distributing emergency food and water is essential for survivors in the aftermath of a tropical storm. High-income countries (HICs) are more likely to have the resources available to do this, though during large-scale disasters international help is often necessary and welcomed. Aid may be hindered if roads have become blocked by debris or fallen trees, or flooded. Equally, if there is large-scale devastation, it can take longer for aid to reach where it is needed, especially in more remote locations.

Wind category 1
Winds: 119–153 km/h
Very dangerous winds will produce some damage

Well-constructed frame homes could have some damage to the roof, shingles, vinyl siding and gutters. Large branches of trees will snap and shallowly rooted trees may be toppled. Extensive damage to power lines and poles will likely result in power outages that could last a few to several days.

Wind category 2
Winds: 154–177 km/h
Extremely dangerous winds will cause extensive damage

Well-constructed frame homes could sustain major roof and siding damage. Many shallowly rooted trees will be snapped or uprooted and block numerous roads. Near-total power loss is expected with outages that could last from several days to weeks.

Wind category 3
Winds: 178–208 km/h
Devastating damage will occur

Well-built frame homes will incur major damage or removal of roof decking and gable ends. Many trees will be snapped or uprooted, blocking numerous roads. Electricity and water will be unavailable for several days to weeks after the storm passes.

Wind category 4
Winds: 209–251 km/h
Catastrophic damage will occur

Well-built frame homes can sustain severe damage, with loss of most of the roof structure and/or some exterior walls. Most trees will be snapped or uprooted and power poles downed. Fallen trees and power poles will isolate residential areas. Most of the area will be uninhabitable for weeks or months.

Wind category 5
Winds more than 252 km/h
Catastrophic damage will occur

A high percentage of frame homes will be destroyed, with total roof failure and wall collapse. Fallen trees and power poles will isolate residential areas. Power outages will last for weeks to possible months. Most of the area will be uninhabitable for weeks to months.

▲ Figure 3.10 Saffir–Simpson hurricane wind scale

Governments, NGOs and charities aim for **sustainable development** after the initial relief effort has saved lives. These are long-term responses (see Chapter 3). They take longer to implement but have a longer lasting impact. Projects range from repairing damage to existing buildings, infrastructure and businesses, to ensuring the country is capable of managing a future hazard by investing in methods of protection and prediction of tropical storms, such as a new early warning system for storm surges or new sea defences. The speed to start and complete long term responses will depend on how much destruction was caused, the wealth of the country to pay for the work, and the help available from other countries, organisations and charities.

→ Activities

1. Define primary and secondary effects.
2. List three ways tropical storms affect (a) people and (b) the environment.
3. Describe the scale used to measure tropical storms.
4. Distinguish between immediate and long-term responses.
5. Suggest why food, clothes and water are needed following a tropical storm.
6. Explain why emergency aid can be slow to arrive after a tropical storm.
7. Why are both immediate and long-term responses necessary? Explain your answer.
8. Suggest three reasons why some tropical storms could cause more damage than others.
9. How would you expect the responses to be different in an HIC to an LIC?

✪ KEY LEARNING

➤ Primary and secondary effects of a typhoon
➤ The immediate and long-term responses

In numbers

- 6,190 people died
- 14.1 million people affected, of which 4.8 million already lived in poverty
- US$12 billion overall damage
- Over 1 million farmers and 600,000 hectares of agricultural land affected
- 1.1 million tonnes of crops destroyed
- 1.1 million houses damaged (half destroyed)
- 4.1 million people made homeless

The trouble with Typhoon Haiyan

What were the primary effects of Typhoon Haiyan?

On 8 November 2013 at 4.40 a.m. local time, a category 5 typhoon struck the Philippines (Figure 3.11). Typhoon Haiyan, known in the Philippines as Typhoon Yolanda, originated in the northwest Pacific Ocean. Typhoon Haiyan was one of the most powerful typhoons to hit the Philippines, with recorded wind speeds of up to 314 kilometres per hour.

The strong winds battered people's homes and even the evacuation centre buildings. Those made homeless were mainly in the Western and Eastern Visayas. Power was interrupted, the airport was badly damaged and roads were blocked by trees and debris. Leyte and Tacloban had a five-metre storm surge, and 400 millimetres of heavy rainfall flooded one kilometre inland. Ninety per cent of the city of Tacloban was destroyed (Figure 3.14).

Coconut, rice and sugarcane production made up 12.7 per cent of the Philippines' GDP before Typhoon Haiyan hit. The harvest season had just ended before Typhoon Haiyan struck, but rice and seed stocks were lost in the storm surges. The damage to rice cost US$53 million. Three-quarters of farmers and fishers lost their income. The UN totalled the recovery costs for agriculture and fishing at US$724 million.

➤ Figure 3.11 Satellite image of Typhoon Haiyan approaching the Philippines

▼ Figure 3.13 Tracking co-ordinates of Typhoon Haiyan

Date	Time	Lat.	Long.	Wind (mph)
05/11	00:	6°N	146°E	75
05/11	12:	7°N	143°E	105
06/11	00:	7°N	140°E	150
06/11	12:	8°N	136°E	160
07/11	00:	9°N	133°E	175
07/11	12:	10°N	129°E	190
08/11	00:	11°N	125°E	185
08/11	12:	12°N	121°E	155
09/11	00:	12°N	116°E	135
09/11	12:	15°N	113°E	115
10/11	00:	16°N	110°E	100
10/11	12:	19°N	108°E	85
11/11	00:	22°N	107°E	70

▲ Figure 3.12 Location map of Typhoon Haiyan

What were the secondary effects?

An oil barge ran aground at Estancia in Iloilo causing an 800,000 litre oil leak. Most of this washed ashore, contaminating 10 hectares of mangroves (a type of tree, see Figure 3.17) that grow 10 kilometres down the coast of Estancia. Fishing at Estancia had to stop due to the contaminated fishing waters. Looting was rife as survivors fought for food and supplies. Eight deaths were reported in a stampede for rice supplies. By 2014, rice prices had risen by 11.9 per cent.

The flooding caused surface and groundwater to be contaminated with seawater, chemicals from industry and agriculture, and sewage systems. The likelihood of infection and diseases spreading increased.

What were the immediate responses?

The president televised a warning. The authorities evacuated 800,000 people. Many sought refuge in an indoor stadium in Tacloban. Although this had a reinforced roof to withstand typhoon winds, they died when it was flooded. The government ensured essential equipment and medical supplies were sent out. In one region these supplies were washed away.

Emergency aid supplies arrived three days later by plane once the main airport was reopened. It was a week before power was restored in some regions and partially in others. Within two weeks, over one million food packs and 250,000 litres of water were distributed. A curfew was imposed two days after Typhoon Haiyan to reduce looting.

Thirty-three countries and international organisations pledged help, with rescue operations and an estimated US$88.871 million. Celebrities such as the Beckhams, *The X Factor* and large multinational companies and organisations, such as Coca-Cola, Walmart, Apple and FIFA donated and used their status and influence to raise awareness and encourage public donations. More than US$1.5 billion was pledged in foreign aid.

What were the long-term responses?

In July 2014, the Philippine government declared it was working towards the country's long-term recovery. 'Build Back Better' is the intention that buildings would not just

▲ Figure 3.14 Tacloban city after typhoon Haiyan, 2013

be rebuilt, but upgraded and therefore protected in the event of future disasters. Additionally, they have:

- a 'no build zone' along the coast in Eastern Visayas
- a new storm surge warning system
- mangroves replanted
- plans to build the Tacloban-Palo-Tanauan Road Dike.

→ Activities

1 Study Figure 3.14. What evidence is there of (a) strong winds, (b) torrential rainfall and (c) storm surges?

2 Describe the following effects of the typhoon: (a) social (b) environmental (c) economic.

3 Suggest reasons why Typhoon Haiyan was so destructive.

4 Do you think people in the Philippines were prepared for the typhoon? Explain your answer.

Geographical skills

1 Use Figure 3.13 to:

a) Plot the path of Typhoon Haiyan on a map using the latitude and longitude co-ordinates.

b) Label the countries and islands Typhoon Haiyan passed through on your map. Give your map a suitable title.

c) Draw a line graph to show the speed of wind over the duration of the typhoon.

d) State a reason why a line graph is more suitable than a bar graph.

Reducing the effects of tropical storms

How are tropical storms monitored?

Monitoring tropical storms allows predictions to be made which can save lives and reduce damage. The following explains some of the ways in which they are monitored.

Satellites

There is a classic cloud pattern associated with tropical storms that satellites monitor. In 1997, it was accidentally discovered that the appearance of rainclouds which reach approximately 16 km in altitude are more likely to indicate that a tropical storm will intensify within 24 hours. Seven years later, the Global Precipitation Measurement satellite was launched. It monitors precipitation every three hours between latitudes 65° north and south of the Equator to identify the high-altitude rainclouds.

Aircraft

A plane first flew purposely into a hurricane in 1943 to make observations. Now, specially equipped aircraft frequently fly through tropical storms at 10,000 feet to collect air pressure, rainfall and wind speed data. They release dropsondes (sensors) which send measurements every second by radio back to the aircraft.

The US National Aeronautics and Space Administration (NASA) monitors weather patterns across the Atlantic using two unmanned aircraft called Global Hawk drones (Figure 3.15). On-board radar and microwaves help scientists to understand more about the formation of tropical storms to improve forecasting models.

▲ Figure 3.15 Global Hawk drone

Can tropical storms be predicted?

All available weather data are fed into supercomputers which run models to predict the path and intensity of tropical storms. In the following cases, this has had some success:

- In 2013, the National Oceanic and Atmospheric Administration (NOAA) developed two new supercomputers. In 1992, the location of a tropical storm, predicted with three days' warning, could be wrong by 480 kilometres. Supercomputers can now give five days' warning and a more accurate location within 400 kilometres.
- The National Hurricane Centre in Florida predicts a tropical storm's path and intensity for up to seven days using a 'track cone' (Figure 3.16). The cone shape allows for error with the unpredictable behaviour of the tropical storm, especially when it hits land. Around 70 per cent of tropical storms occur within the predicted cone.
- In 2013, Cyclone Phailin in India was successfully predicted. As many as 1.2 million people were evacuated. Twenty-one people died. A further 23 died in flash flooding after. Yet in 1999, a similar cyclone hit the same area and more than 10,000 lives were lost.
- National Hurricane Centres around the world issue early warnings so people have time to prepare to evacuate – but some may not bother. Additionally, evacuation is costly and time-consuming, particularly if the path of the tropical storm does not actually pass the area in which they live.

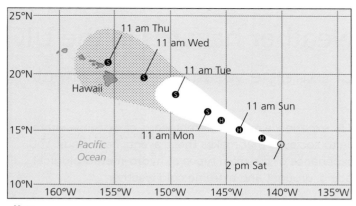

Key

Potential track
☐ Days 1–3
▨ Days 4–5
● Tropical cyclone

Sustained winds
S 39–78 mph
H 79–110 mph

Current information
○ Saturday,
1st August, 2015
13.8°N 140.1°W
Max sustained
wind 105 mph

▲ Figure 3.16 Hurricane Guillermo track forecast cone

Is protection possible?

Buildings have areas of weakness which can be reinforced to reduce damage caused by the forceful winds of tropical storms. This is called mitigation (see Chapter 4, Section 4.5). In the USA, the Federal Emergency Management Agency (FEMA) advises homeowners to:

- install hurricane straps (galvanised metal) between the roof and walls
- install storm shutters on windows
- install an emergency generator
- tie down windborne objects such as garden furniture
- reinforce garage doors
- remove trees close to buildings.

With the fierce storm surges and flooding that occur in tropical storms, salt marshes, wetlands and mangroves can protect against storm surges by reducing the waves' energy. Additionally, trees reduce wind energy and trap debris, which can cause damage – although trees can cause devastation too, if uprooted near buildings and infrastructure. Another way to protect land is to ensure that low-lying areas are not built on. Coastal flood defences (Chapter 10) such as levées and floodwalls reduce the impact of storm surges as they hold back the seawater.

How can planning reduce the risks?

American National Hurricane Preparedness Week in May aims to encourage people to plan what they need to have and do in the event of a tropical storm. Advice included:

- preparing disaster supply kits
- having fuel in vehicles
- knowing where official evacuation shelters are
- storing loose objects
- planning with family what to do.

▲ Figure 3.17 Mangroves

→ Activities

1 Complete this spider diagram to show how the effects of a tropical storm can be reduced. List the different types of (a) monitoring, (b) prediction, (c) protection and (d) planning.

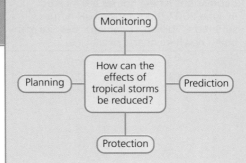

2 Describe how the Global Hawk drone (Figure 3.15) helps to monitor tropical storms.

3 How might track forecast cones such as the one in Figure 3.16 be (a) of benefit to citizens, (b) to the detriment of citizens.

4 Do you think all countries can plan for tropical storms? Explain your answer.

5 Study Figure 3.16.

a) Give the latitude and longitude co-ordinates that Hurricane Guillermo was expected to be on Thursday 6 August at 11 am.

b) Describe the track of Hurricane Guillermo.

Extreme weather hazards in the UK

What kinds of extreme weather events affect the UK?

Communities in the UK experience many different kinds of **extreme weather**. It is the damage done to societies that makes these events hazardous. You may have personal experience of the main types of **hydro-meteorological hazards**: storms, flooding, drought and extreme cold weather.

Storm events

The UK is regularly hit by depressions (page 23) which bring very heavy rain and trigger river floods. They can cause great storm damage, especially to the west coast of the UK (Figure 3.18). A cluster of strong depressions was responsible for widespread wind damage in late 2013. During the St Jude storm of 28 October, 160 km/h winds killed five people, felled trees and toppled lorries. Subsequent storms brought further death and disruption, notably in the week before Christmas, when more people died, thousands lost power and travellers were stranded at Gatwick Airport.

On top of storms, the UK can also experience tornadoes. A tornado is a rotating column of spiralling air whose formation is triggered by strong heating of the ground. In 2005, a tornado caused 19 injuries and damage costing £40 million in Birmingham.

▲ Figure 3.18 Storm damage in the UK, Wales, 2014

Flooding

Figure 3.19 shows four different types of flood hazards that may affect people in the UK. Often, flooding is caused by the heavy rainfall or strong waves brought by a depression.

Flooding may also trigger **landslides** as a secondary hazard. At Ockley in southern England, in 2013, more than 40 metres of a major railway embankment collapsed after heavy rainfall (Figure 3.20). This brought weeks of disruption for London commuters and financial costs for their employers.

▼ Figure 3.19 Different types of flooding associated with extreme weather in the UK

Coastal flooding	• A deep depression brings a storm surge to a major river estuary. • Strong winds funnel coastal water into the mouth of a river. • In 1953, 300 people died when a storm surge hit the Thames Estuary.
River flash flooding	• High-intensity rainfall brings flash flooding, especially on steep slopes. • The village of Boscastle was overwhelmed without warning in 2004, when 185 millimetres of rain fell on the River Valency catchment in just five hours.
Slow-onset river flooding	• A long period of steady rainfall gradually saturates the catchment soil. • The River Thames burst its banks at Henley in 2014, when fresh rain could not soak into the waterlogged ground and instead ran straight into the river.
Surface water flooding	• Intense rainfall collects in hollows and depressions where homes are located. • In June 2007, thousands of UK homes suffered £3 billion of damage, despite many being nowhere near a river.

Drought events

Drought is defined as an extended period of low or absent rainfall relative to the expected average for a region. In the UK, this means fifteen consecutive days with less than 0.2 millimetres of rain on any one day. Beyond this point, there may be insufficient moisture for average crop production, particularly if there are low water reserves in reservoirs.

The longest drought on record in the UK occurred over an 18-month period in 1975 and 1976. In Cheltenham, the temperature exceeded 32 °C for seven successive days (a record that still stands). In recent years, drought conditions struck the UK in 2003, 2006 and 2012. The 2003 drought affected large parts of Europe, where it was linked with 20,000 deaths. It brought the highest temperature ever recorded in the UK: 38 °C in Faversham, Kent.

▲ Figure 3.20 The collapse of Ockley embankment after severe storms in 2013

Extremes of cold weather

The winters of 1946–47, 1962–63 and 1978–79 were exceptionally cold. Other unusually cold winters include 2010–11 and 2014–15. Cold conditions take over if depressions are not passing over the UK as usual. Weather risks include:

- frost – crops and cattle may not survive extremes of around –10 °C.
- freezing conditions – over 17,000 trains were cancelled in January 2014 because of freezing conditions.
- blizzard conditions – transport grinds to a halt, creating costly airline delays.

→ Activities

1. Identify and explain the secondary hazards that can be triggered by (a) a storm and (b) a drought.
2. Suggest how the following could be affected by extreme weather hazards: (a) farming, (b) transport, (c) communities, (d) businesses and (e) people's health.
3. Sketch an annotated map of the UK showing the different extreme weather hazards that different regions are particularly at risk from. Use all of the information provided on these pages to help you.
4. Most parts of the UK are at risk from one or more types of extreme weather. What questions should somebody moving to a new area ask in order to identify possible risks? Who would know the answers?

 ### Fieldwork: Get out there!

'What are the extreme weather hazards in my local area?'

- Think of secondary data sources you could use to investigate the historical record of local weather hazards.
- Think of people you could interview who would know about past events. This counts as primary data.
- What types of weather data could you collect yourself over the course of a year? What equipment would you need? Could technology, such as a smartphone, help you collect these primary data?

⊕ KEY LEARNING

➤ The causes of record rainfall and flooding
➤ The social, economic and environmental impacts for people and places

Record rainfall and flooding in Cumbria (2009)

What caused the record rainfall and flooding in Cumbria in 2009?

Average rainfall in the UK ranges from 700 millimetres to 2,500 millimetres annually, depending on location. The heaviest rainfall ever recorded (at that time) in the UK fell on Cumbria in northern England in November 2009 (Figure 3.22). More than 1,500 homes were affected by river flooding and surface water flooding. The cause of the rainfall was a very deep Atlantic depression moving northeastwards over Scotland and northern England.

Extreme rainfall causes flooding

Cumbria had already received a month's worth of average rainfall before the extreme event of 17–20 November. This meant that the soil was already very wet and new rainfall could not soak in. It flowed straight down the steep slopes of the Lake District into its rivers for 36 hours.

By 20 November, the River Derwent was ten metres wider than during normal conditions. Water was flowing at a rate 25 times higher than the normal average. The River Cocker's flow exceeded anything seen in the previous 30 years (see Figure 3.24).

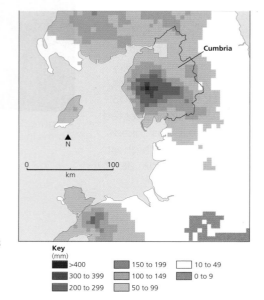

Key
(mm)
■ >400	▨ 150 to 199	☐ 10 to 49
▨ 300 to 399	▨ 100 to 149	▨ 0 to 9
▨ 200 to 299	▨ 50 to 99	

▲ Figure 3.22 Distribution of rainfall in the 72-hour period from 9 a.m. on 17 November to 9 a.m. on 20 November 2009

▼ Figure 3.23 Rainfall at gauging stations in Cumbria, 17–19 November 2009

Location	17 Nov. (mm)	18 Nov. (mm)	19 Nov. (mm)
Seathwaite Farm	60.8	142.6	253.0
High Snab Farm	63.0	110.2	228.2
Honister Pass	73.4	151.8	229.2
Thirlmere	72.6	103.0	151.2

What were the impacts for people and places?

A range of harmful social, economic and environmental impacts were experienced immediately during and after the storm and floods (Figure 3.24).

■ **Social**: Police officer Bill Barker was killed when a bridge in Workington collapsed. Many more people were injured and 1,500 homes were flooded, causing great distress. River water contaminated with sewage brought health risks.

▲ Figure 3.21 The weather chart for 6 am Thursday, 19 November 2009, with a deep Atlantic depression.

- **Economic**: The regional economy was instantly hit. Many businesses closed and did not reopen until long afterwards. Debris transported by the river destroyed six important regional bridges.
- **Environmental**: At its peak flow, water erosion by the River Derwent triggered landslides along its banks. The river tore loose and carried away hundreds of trees, damaging local ecosystems and habitats.

The worst-hit single place was the town of Cockermouth (see Figure 3.24). As time passed, longer-term impacts became apparent too. In total, the floods caused £100m damages (including insurance claims, business losses and the cost of rebuilding roads and bridges). Many businesses, however, took the opportunity to improve their shop fronts. The rebuilt town centre now looks smarter than before.

→ **Activities**

1 Look at Figure 3.21. Identify evidence from the weather chart showing which parts of the UK experienced storms.

2 Look at Figure 3.22. Describe the pattern of rainfall shown. Suggest reasons for the very high figures recorded in some places.

3 Study Figure 3.23. Compare the rainfall data provided. Which location recorded the least rainfall from 17 to 19 November?

4 Look at Figure 3.24. Explain why some places were worse-affected than others by flooding.

5 A whopping 316.4 millimetres of rain fell over 24 hours at Seathwaite. Using Figure 3.23, suggest when this period started and ended.

6 Research the damage done to the area's bridges at www.metoffice.gov.uk/climate/uk/interesting/nov2009.

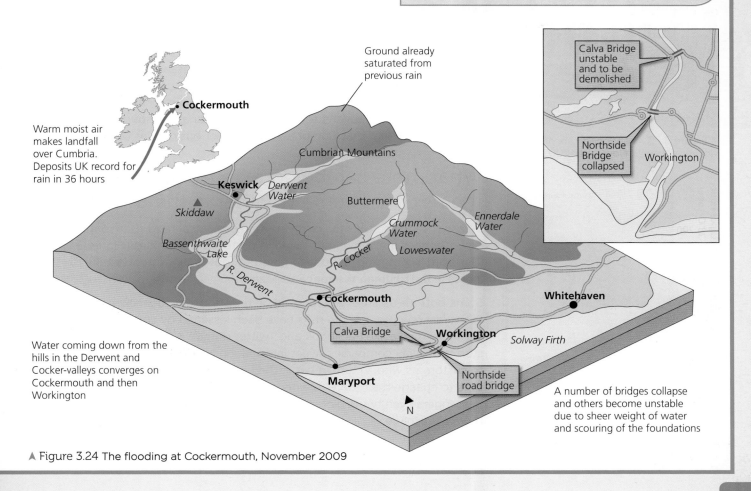

▲ Figure 3.24 The flooding at Cockermouth, November 2009

Responding to the risk of extreme weather

How have management strategies reduced the risk of extreme weather in Cumbria?

The Cumbrian rainfall and floods of 2009 were exceptional (Figure 3.25). Climate change may cause extreme events like these to become more common. In Cockermouth, action has been taken to increase the level of protection.

New flood defences have been built there at a cost of £4.5 million, funded by central government and the local community. While new defences were essential to reduce extreme weather risk, it was important that they did not harm the town's tourist economy. A cleverly engineered mobile wall was built. It rises when needed, yet disappears from view at other times, protecting the river view for cafes and restaurants.

The **Environment Agency (EA)** has played an important role in providing residents with improved flood warning information. This improves safety by giving people more time to evacuate and to protect their own properties. Some people living in Cockermouth have also asked the EA to send flood warning messages directly to their smartphones.

Another important element of risk management in Cumbria is helping local businesses get back on their feet and earning money as soon as possible. After the 2009 floods, the West Cumbria Development Agency paid for adverts to be placed in national newspapers announcing that Cumbria was 'open for business as usual'.

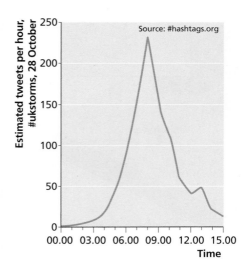

▲ Figure 3.25 Abandoned cars on the main street in Cockermouth, 19 November 2009

▲ Figure 3.26 How Twitter users helped to spread information about the St Jude storm of 28 October 2013

What more can be done to manage the risk of extreme weather in the UK?

In response to the UK's major hydro-meteorological hazards, a range of 'top-down' actions are sometimes taken to protect communities (Figure 3.27). There are 'bottom-up' actions to consider too – actions people take to increase their **human resilience** to hazards.

Some bottom-up steps are focused on reducing individual exposure to harm. In the last few years, some risks have been lessened by communication via social media like Facebook and Twitter (Figure 3.26). During winter storms, railway companies tweet information and pictures of fallen trees to customers who, in turn, retweet them to others. That way, people have time to change their route to work or can choose to return home. Some passengers even tweet pictures to the railway companies in order to alert them to damage.

Additionally, some homeowners in flood-prone areas 'future-proof' their homes by very sensibly having stone tiles rather than carpets in their ground floor and basement rooms. They know that although their insurance policy will cover any damage to carpets, it could take a very long (and stressful) time to get the problem sorted – especially if the insurance companies are dealing with thousands of similar claims.

→ Activities

1 Look at Figure 3.26. In what other situations could social media be used by citizens to help reduce their own exposure, or that of other people, to different types of weather hazard?

2 Look at Figure 3.27. Identify the different types of organisations who try to tackle the risk posed by extreme weather to the public.

3 What other possible top-down actions could be added to Figure 3.27? Suggest two for each type of weather hazard.

→ Going further

Discussion time! Are there limits to how far we can reduce the risk posed by extreme weather events, even if technology keeps improving? As part of a class discussion, you can consider the following points and other points of your own:

■ Some events are still very hard to predict, such as tornadoes.

■ The scale of some events, such as storm surges, could mean damage will always be unavoidable.

■ People do not always act on weather warnings, for a range of reasons (see how many you can think of).

Drought

A hosepipe ban can be put in place in affected regions. During the 1976 drought, the Minister for Drought sent 'hosepipe patrols' in search of breakers of the hosepipe ban. Offenders can be fined.	Water companies can apply to the government for an official Drought Order in an extreme drought. Water supplies to houses are turned off, and members of the public take their turn queuing in the street at standpipes.	Water companies can encourage people to have a water meter fitted (this tends to stop people from leaving taps running) and do more to repair old water pipes (London has 6,000 km of 150-year-old pipes).

Storms

The UK Met Office is constantly improving its ability to make weather predictions. Extreme storm events can be predicted several days in advance.	Severe weather warnings can now be issued using a whole range of media (in the past, only television and radio were available). Smartphones allow people to receive weather warnings while away from home.	Airlines and rail companies cancel their services when very strong winds are forecast in order to minimise the safety risk to their customers.

Flood

The Thames Barrier was completed at great cost in 1982 to protect London from any future storm surges of similar magnitude to the 1953 event.	The Environment Agency constantly monitors ground moisture levels in river basins using its own communications system. This helps the EA to make accurate flood predictions and give timely evacuation orders.	A new agreement called Flood Re was recently reached between UK insurance companies and the government. Any new housing on floodplains faces higher insurance bills. This should deter construction in risky places.

Cold weather

The responsibility for clearing roads of snow and ice lies with local councils. Councils in the UK are legally responsible for safety along nearly 500,000 km of road.	National organisations like Public Health England make announcements in the media warning people to take great care while out and about in blizzard conditions.	Charities for the elderly raise awareness about the heightened health risks for old people during cold conditions. They work to raise public awareness of this issue.

▲ Figure 3.27 Top-down actions to manage extreme weather risks

KEY LEARNING

➤ Rainfall record and changes in storm frequency

➤ Temperature record and changes in drought frequency

➤ Future extreme weather predictions

Extreme weather on the rise in the UK

What does the rainfall record tell us about changes in storm frequency?

A range of data suggests that the Earth's climate is currently warming and changing. The rate of change is unprecedented in historical terms, which is why the majority scientific viewpoint is that humans are to blame (see Chapter 4). The Intergovernmental Panel on Climate Change (IPCC) of the United Nations has warned that temperatures could increase by several degrees during this century. A warmer world will be one where more evaporation takes place over the oceans – and what goes up must ultimately come down. Climate change scientists therefore believe that rainfall patterns are likely to change in the UK as the world's oceans warm, and may in fact be doing so already.

Figure 3.28 shows the UK winter rainfall figure (millimetres) recorded between 1910 and 2015. There is great natural variability in how much rain falls in June from year to year. One interpretation of these data is that there has been an increase in extreme winter rainfall since the 1980s. The suggested reason for this is increased warming of the Atlantic ocean during the same period. As a result, rain-bearing depressions will be gaining more energy and moisture.

Although winter rainfall was not exceptional in 2013, storms were accompanied by very strong winds. Some weather experts at the UK's Hadley Centre for Climate Prediction and Research say the increased incidence of rain-bearing storms since the 1980s is in line with global climate change predictions. Recent storms and floods may already have 'human fingerprints' on them.

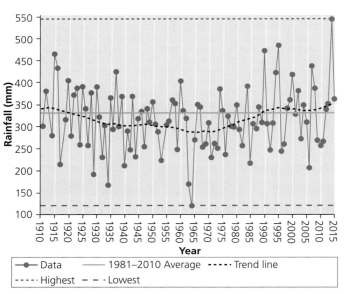

▲ Figure 3.28 Winter rainfall in the UK, 1910–2015

What does the temperature record tell us about changes in drought frequency?

UK temperatures have increased by about 1°C since 1980. Figure 3.29 shows the average July temperature between 1910 and 2015. There certainly appears to be an upward trend beginning with 1976, the year of the famous drought (page 37). However, high temperatures alone do not cause drought; there must also be rainfall deficiency. Not all of the temperature spikes shown in Figure 3.29 correspond with a period of drought. IPCC scientists admit that future rainfall trends for a warming world are hard to predict, as there are so many variables to consider.

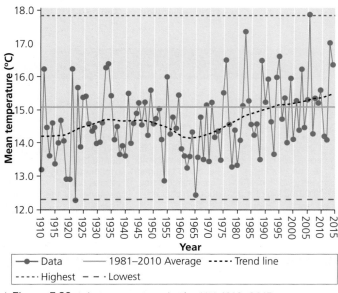

▲ Figure 3.29 July temperature in the UK, 1910–2015

What are the extreme weather predictions for the future?

Some scenarios suggest a global average temperature rise of 2–3°C in the twenty-first century (Figure 3.30). Several agencies, including the Environment Agency and the Met Office, believe that if this happens, the UK could be faced with warmer and wetter winters (Figure 3.31). There will most likely be more rain-bearing depressions affecting all parts of the UK, not just northern regions. The risk of extreme flooding and high wind speeds will probably increase.

Yet there are long-term cyclical changes also taking place in the temperature of the Atlantic ocean and the position of the jet stream. Researchers have shown that the Atlantic ocean was relatively warm between 1931 and 1960, before cooling from 1961 and 1990. Since then it has warmed again. These changes may be part of a cycle that is happening independently, whether or not the world is warming.

In the highest scenario shown in Figure 3.30, even more variables come into play. A global average temperature rise of 4–5°C would cause widespread melting of land and sea ice in the Arctic. Colder water would pour into the north Atlantic, with unknown effects for the movement of the air masses and ocean currents that regulate the UK's climate. In a warmer world, the UK could be left facing more extremes of cold weather!

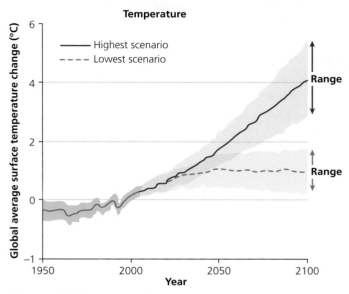

▲ Figure 3.30 The range of projected global temperature changes up to 2100

	Precipitation ↓	River flow →	Evaporation ↑
What has already happened? (actual change)	No change in annual UK total. But more winter rain has fallen in heavy events since the 1980s	The frequency and magnitude of winter river flooding has increased since the 1980s	We are not sure if evaporation has increased or not. But we know temperature has risen by 1°C
What will happen next? (predicted change)	Precipitation will become even more seasonal. But the annual UK total will stay the same	Some scientists predict that certain UK rivers will flood more in future winters	Evaporation will increase due to higher air temperatures, causing more drought

▲ Figure 3.31 Recorded and projected UK weather changes

→ Going further

Research the most recent 2015 floods in Cumbria. To what extent do they conform to the trends outlined in Section 3.11?

- The Met Office and Hadley Centre have websites full of information about climate change and the UK's weather.
- The LWEC Climate Change Impact Report Cards are a quick, easy-to-use, pull-together of the latest evidence on climate impacts: www.lwec.org.uk/resources/report-cards.
- This article looks at new jet stream research: www.bbc.co.uk/news/science-environment-19848112.

→ Activities

1. Look at Figure 3.28. Identify the years when exceptionally high winter rainfall was recorded.

2. Look at Figure 3.29. Describe the trend shown in July temperatures. How does this information help explain the occurrence of some serious droughts you have learned about?

3. Look at Figure 3.30. Compare the two scenarios shown for 2100.

4. Look at Figure 3.31. Suggest how a warming climate could be responsible for the actual and predicted changes shown.

5. Use the data in Figure 3.28 to argue for and against the proposition that winter rainfall is increasing in the UK.

4 Climate change

KEY LEARNING

- What the Quaternary period is
- Changes in climate through time
- Evidence of climate change

Climate change? Prove it!

What is the Quaternary period?

The Earth is believed to be 4.55 billion years old. Studying the Earth has led us to devise a geological timeline that divides its history into a series of eras, periods and epochs. The period of time that stretches from 2.6 million years ago to the present day is called the **Quaternary period**, which is in the Cenozoic era. This period marks a time when there was a global drop in temperature and the most recent ice age began. (It is thought that the Earth has experienced five ice ages in its history.) The Quaternary period is split into two epochs, the **Pleistocene epoch** and the **Holocene epoch**.

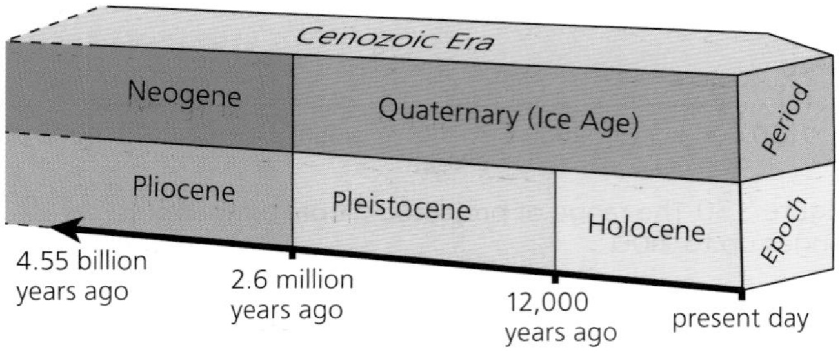

▲ Figure 4.1 Geological timeline of the Quaternary period

How has climate changed?

The entire Quaternary period is often called an ice age due to the presence of a permanent ice sheet on Antarctica. During the Pleistocene epoch there were cold **glacial episodes** lasting approximately 100,000 years. Thick ice would expand, covering vast areas of continents, but then retreat, as each glacial episode was followed by a warmer **interglacial episode**. The warmer intervals were much shorter, lasting for approximately 10,000 years. The Holocene epoch began when the last glacial expansion ended and the current interglacial episode started. This is what we live in today. There are still sheets of ice covering Greenland and Antarctica, but our climate has remained relatively stable.

What is the evidence for climate change?

Climate change is the long-term change in the weather. Global climate change occurs very slowly over thousands of years. Since 1914 the Met Office has recorded reliable climate change data using weather stations, satellites, weather balloons, radar and ocean buoys. The Earth's average surface air temperature has increased by approximately 1°C over the last 100 years. In addition:

- sea levels have risen by 19 centimetres since 1900 and are expected to continue to rise – this is due to thermal expansion and ice sheets melting
- ocean temperatures are the warmest they have been since 1850, and the world's glaciers and ice sheets are decreasing in size
- NASA data show that since 2002, the volume of ice lost in Antarctica is 134 billion tonnes per year, and 287 billion tonnes per year in Greenland.

For the era before there were reliable data records, we need to take clues from **proxy data** (natural recorders), such as tree rings, fossil pollen, ice cores and ocean sediments to estimate what the climate was like. However, these records are not as reliable, because these only indicate climate change rather than providing direct evidence of accurate temperatures.

Ice cores

Antarctic **ice cores** are crucial in understanding long-term climate change. Antarctica is a wilderness with no permanent residents, so the layers of snow remain unaltered. They act like time capsules, holding information about climate change as different layers of snow build up over thousands of years. The ice cores can be drilled (see Figure 4.2) so that the information about what the climate was like when the snow fell can be analysed. The deeper the snow that is drilled, the older the snow. Records go back to about 800,000 years ago.

Oxygen isotopes in the ice cores are commonly used to estimate what the temperatures would have been. The isotopes are atoms with different numbers of neutrons. There are three different oxygen isotopes. The ratio of two types of oxygen isotopes are measured to work out what the climate was like (see Figure 4.3). Additionally, when the ice cores are melted, trapped carbon dioxide and methane are released, which can be compared to present levels to see the differences between climate then and now.

⬆ Figure 4.2 Measuring the data held in ice cores

Ocean sediments

As with ice, the deeper the sediment, the older the sediment. The billions of tonnes of sediment deposited at the bottom of the sea also act as a timeline for providing evidence of climate change. Organisms and remains of plankton in the sediment reveal information such as past surface water temperatures, and levels of oxygen and nutrients. Figure 4.3 shows the oxygen isotopes found in ocean sediments through the Quaternary period. The spikes represent interglacials (warmer times with less ice) and the troughs show glacials (colder times with more ice).

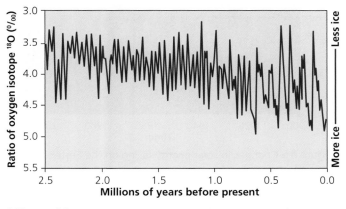

⬆ Figure 4.3 Climate change evidence from ocean sediments

→ Activities

1 Describe the Quaternary period timescale using Figure 4.1.

2 What do glacial and interglacial episodes mean?

3 Contrast the Holocene epoch with the Pleistocene epoch.

4 If there was no human effect on climate, how would you expect the climate to change over the next several thousand years?

5 List the different ways that evidence is found for climate change.

6 Why do natural recorders (or proxy data) have to be used to show evidence of climate change?

7 Use Figure 4.3 to describe how ocean sediments have provided evidence of climate change.

8 Why might people argue that some evidence for climate change is better than others?

9 Research how tree rings and fossil pollen provide evidence of climate change.

Climate change as a natural phenomenon

Is climate change a natural phenomenon?

The geological evidence we have found suggests that climate change has been happening throughout the Quaternary period, before humans were present on the planet. This clearly suggests that long-term climate change is a result of natural causes.

Solar output

The output of the Sun is measured by observing **sunspots** on the Sun's surface. Sunspots are caused by magnetic activity inside the Sun, which results in dark patches on the surface of the Sun (see Figure 4.4). The output of the Sun increased slightly from 1900 to 1940. Satellites have recorded the intensity of solar energy output using **radiometers**, since 1978. These data show that overall solar output, from the Sun has barely changed in the last 50 years; in fact, it has decreased slightly (see Figure 4.5). Therefore solar output cannot be responsible for the cause of the climate change seen from the 1970s.

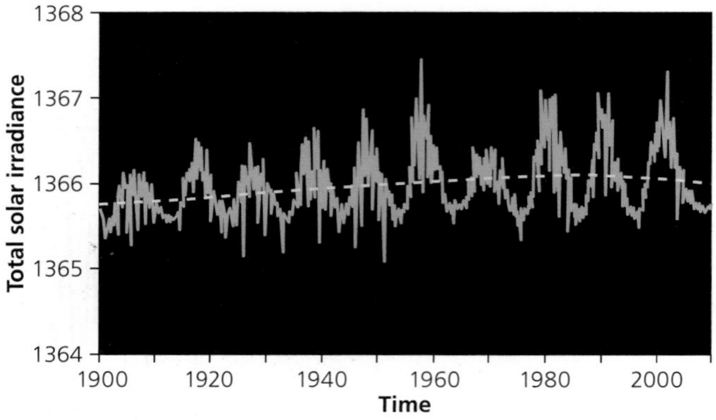

▲ Figure 4.4 Sunspots on the Sun's surface

▲ Figure 4.5 Changes in solar energy falling on the Earth's surface

Orbital changes

The distribution of the Sun's energy on the Earth changes due to the Earth's orbit:

- The Earth's orbit is an ellipse. The Sun is not perfectly in the centre of the ellipse and the ellipse changes shape every 100,000 years. This means the distance between the Earth and the Sun changes as the Earth orbits. As the Earth orbits closer to the Sun, the climate becomes warmer, and the opposite happens as it orbits away.
- The Earth's axis is tilted on an angle. The angle of the tilt changes due to the gravitational pull of the Moon. When the angle of the tilt increases, this can exaggerate the climate, so summers get warmer and winters get colder. The angle of the tilt moves back and forth every 41,000 years.
- The Earth is not a perfect sphere, so as the Earth spins, it wobbles on its axis in a 20,000-year cycle.

Together, these three **orbital changes** vary the distribution of the Sun's energy on the Earth. This can mean a significant impact on climate change. However, scientists suggest that orbital changes would not cause an ice age for at least 30,000 years.

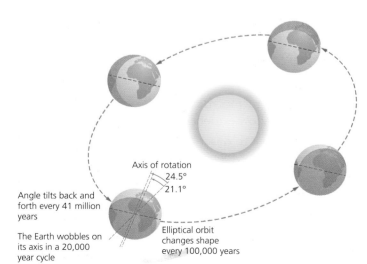

Axis of rotation
24.5°
21.1°

Angle tilts back and forth every 41 million years

The Earth wobbles on its axis in a 20,000 year cycle

Elliptical orbit changes shape every 100,000 years

▲ Figure 4.6 Orbital changes of the Earth

Volcanic activity

Volcanic eruptions can temporarily cause climate change. In June 1991, Mount Pinatubo in the Philippines erupted. An ash cloud was thrown vertically 40 kilometres into the stratosphere and carried around the world for about three weeks. This is extremely important in understanding the impact the volcanic eruption had on climate change. Approximately 20 million tonnes of sulphur dioxide (SO_2) was released by Mount Pinatubo (see Figure 4.8). When SO_2 mixes with water vapour, it becomes a volcanic (sulphate) aerosol. Volcanic aerosols reflect the sunlight away and reduce the Sun's heat energy entering the Earth's atmosphere. Following Mount Pinatubo's eruption, global temperatures dropped by approximately 0.5°C.

Carbon dioxide (a greenhouse gas) also erupted from Mount Pinatubo. Carbon dioxide should help to trap the Sun's heat in the Earth's atmosphere. Instead, the temperature dropped, as the cooling effect of the sulphate aerosols was greater.

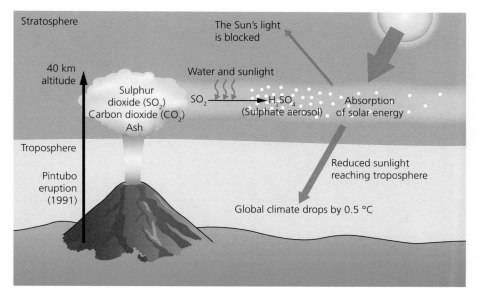

▲ Figure 4.7 The Mount Pinatubo eruption in the Philippines, 1991

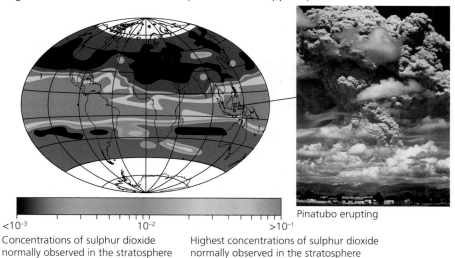

Pinatubo erupting

<10⁻³ 10⁻² >10⁻¹

Concentrations of sulphur dioxide normally observed in the stratosphere

Highest concentrations of sulphur dioxide normally observed in the stratosphere

▲ Figure 4.8 Sulphur dioxide levels during the 40 days after Mount Pinatubo's eruption

→ Activities

1 Why are orbital changes, volcanic eruptions and solar output categorised as natural causes of climate change?

2 'Solar output is responsible for climate change.' True or false? Explain your answer.

3 Explain how shifts in the Earth's orbit can cause climate change.

4 Use Figure 4.7 to describe how volcanic eruptions cause climate change.

5 Use Figure 4.8 to describe the location of sulphur dioxide across the world following the Mount Pinatubo eruption.

6 How important do you think natural causes are in explaining climate change? Explain your opinion.

➤ How the greenhouse effect works

➤ How humans have contributed to climate change

Climate change: our fault?

What is the greenhouse effect?

The greenhouse effect is a naturally occurring phenomenon that keeps the Earth warm enough for life to exist. It is thought that without the greenhouse effect, the Earth would be approximately 33 °C colder and therefore life would not exist as we know it today. The Sun's infrared heat rays enter the Earth's atmosphere. The heat is reflected from the Earth's surface. The natural layer of greenhouse gases allows some heat to be reflected out of the Earth's atmosphere, but some of the Sun's infrared heat is trapped, which keeps the Earth warm enough.

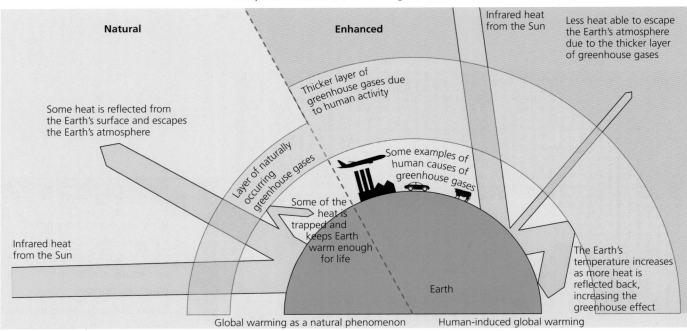

Natural

Enhanced

Infrared heat from the Sun

Less heat able to escape the Earth's atmosphere due to the thicker layer of greenhouse gases

Thicker layer of greenhouse gases due to human activity

Some heat is reflected from the Earth's surface and escapes the Earth's atmosphere

Layer of naturally occurring greenhouse gases

Some examples of human causes of greenhouse gases

Some of the heat is trapped and keeps Earth warm enough for life

Infrared heat from the Sun

The Earth's temperature increases as more heat is reflected back, increasing the greenhouse effect

Earth

Global warming as a natural phenomenon   Human-induced global warming

▲ Figure 4.9 The greenhouse effect: natural and enhanced

However, there is an **enhanced greenhouse effect** whereby human activity has increased the layer of greenhouse gases which naturally exists. Activities which generate more greenhouse gases include burning fossil fuels in industry, agriculture, transport, heating and deforestation. Less heat escapes from the Earth and more is trapped by the thicker layer of greenhouse gases, which means the Earth warms up even more.

Are humans causing climate change?

Scientists have measured and proved that natural causes are responsible for climate change, yet natural causes cannot account for the increases in temperature since the 1970s (see Figure 3.29 in Section 3.11). The link between increasing carbon dioxide levels and increasing global temperatures can be seen in Figure 4.10. Carbon dioxide emissions have increased since the Industrial Revolution, and especially since the 1970s. The Intergovernmental Panel on Climate Change (IPCC) reports that it is very likely that rising levels of carbon dioxide are the main cause of climate change.

Despite volcanoes naturally releasing carbon dioxide, it is thought that humans generate more than 130 times the volume of carbon dioxide than volcanoes. Greenhouse gases consist of 77 per cent carbon dioxide, 14 per cent methane, 8 per cent nitrous oxide and 1 per cent chlorofluorocarbons (CFCs). Although carbon dioxide makes up the majority of greenhouse gases, each molecule of methane has 25 times and each molecule of nitrous oxide has 125 times the global warming potential over 100 years, when compared to carbon dioxide.

How do humans cause climate change?

Fossil fuels

Fossil fuels account for the majority of global greenhouse gas emissions – over 50 per cent. Burning these releases carbon dioxide into the atmosphere. Fossil fuels are used in transportation, building, heating homes, and the manufacturing industry. Additionally, they are burnt in power stations to generate electricity. As the world's population grows and wealth increases, people are demanding more and more energy, which increases the level of fossil fuels and carbon dioxide.

Agriculture

Agriculture contributes to approximately 20 per cent of global greenhouse gas emissions. It also produces large volumes of methane: cattle produce it during digestion, and microbes produce it as they decay organic matter under the water of flooded rice paddy fields.

As the world's population increases, more food is required, especially in areas such as Asia where rice is the staple diet. When countries increase their standard of life, there is almost always an increasing demand for meat (Section 23.1). If current population rates continue, it is inevitable that large-scale agriculture's contribution to climate change will continue to grow.

Deforestation

Deforestation is the clearing of forests on a huge scale. If deforestation continues at the current rate, the world's forests could disappear completely within a hundred years.

There are several reasons why forests are cut down:

- clearing land for agriculture so that farmers have space to plant crops and graze livestock
- logging for wood and paper products
- building roads to access remote areas
- making room for the expansion of urban areas.

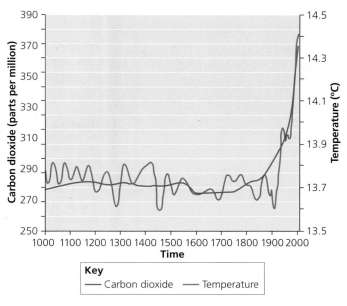

▲ Figure 4.10 Carbon dioxide and global temperature change

▼ Figure 4.11 World agriculture emissions

Continent	% of world emissions from agriculture
Africa	15%
Americas	25%
Asia	44%
Europe	12%
Oceania	4%

During the process of **photosynthesis**, trees absorb carbon dioxide, which reduces the amount of carbon dioxide in the atmosphere. The process of deforestation leaves fewer trees to absorb carbon dioxide. Therefore the enhanced greenhouse gases contribute to rapid climate change. When trees are burnt to clear an area, such as with **slash and burn**, the carbon dioxide that has been stored is also released, which again contributes to climate change.

→ **Activities**

1 Describe the relationship between temperature and atmospheric carbon dioxide in Figure 4.10.

2 Write a sequence of numbered statements to explain the greenhouse effect.

3 How do (a) fossil fuels, (b) agriculture and (c) deforestation each contribute to climate change?

4 'Humans are to blame for climate change.' To what extent do you think this statement is true?

● KEY LEARNING

➤ The likely effects of climate change

➤ How people and the environment may be affected by global climate change

The effects of climate change

What are the likely effects of climate change?

The IPCC states that the cost of damage caused by climate change is likely to be 'significant and to increase over time'. The effects of climate change are not certain. The likely effects will vary and be uneven globally and regionally.

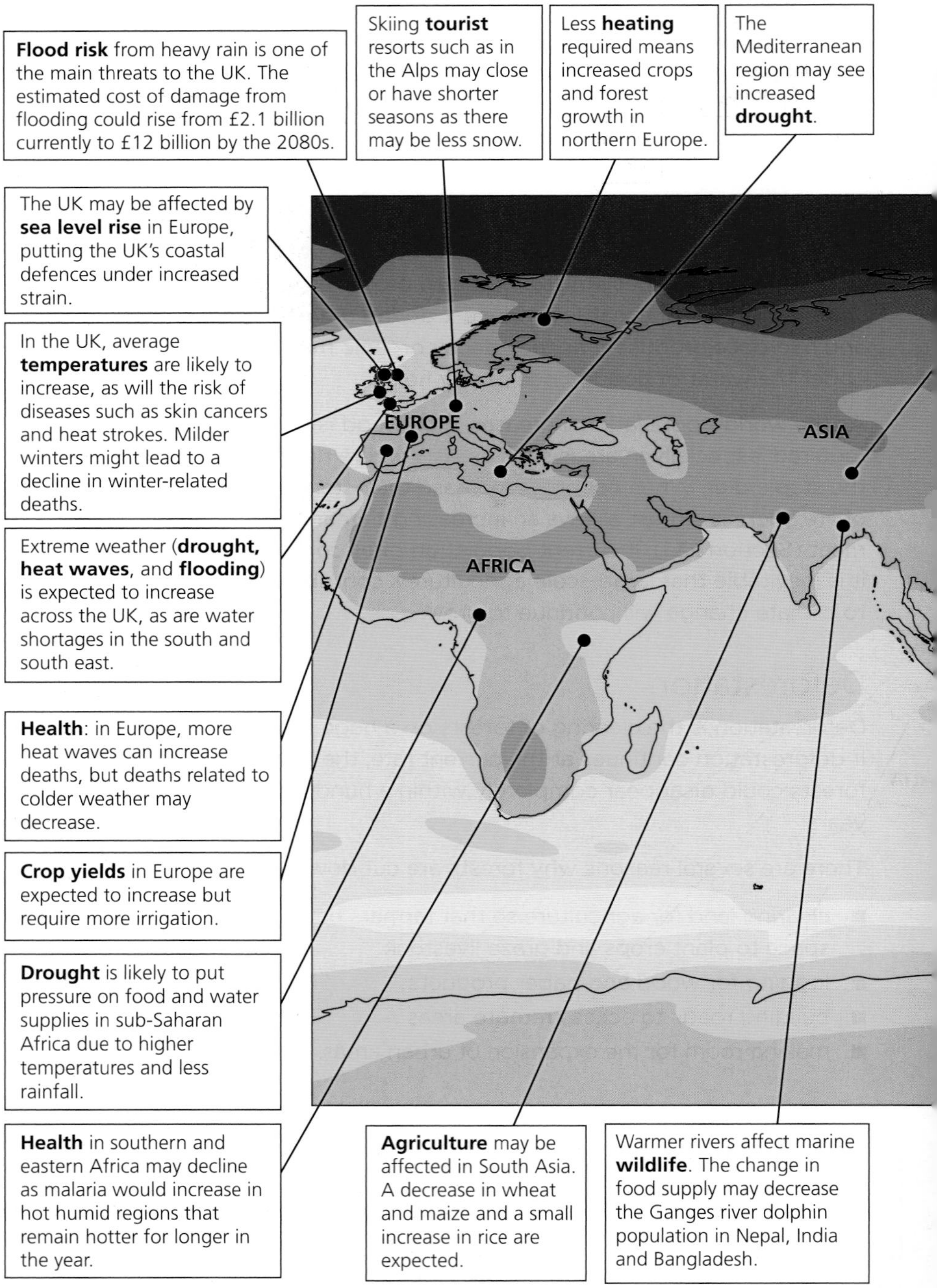

Flood risk from heavy rain is one of the main threats to the UK. The estimated cost of damage from flooding could rise from £2.1 billion currently to £12 billion by the 2080s.

Skiing **tourist** resorts such as in the Alps may close or have shorter seasons as there may be less snow.

Less **heating** required means increased crops and forest growth in northern Europe.

The Mediterranean region may see increased **drought**.

The UK may be affected by **sea level rise** in Europe, putting the UK's coastal defences under increased strain.

In the UK, average **temperatures** are likely to increase, as will the risk of diseases such as skin cancers and heat strokes. Milder winters might lead to a decline in winter-related deaths.

Extreme weather (**drought, heat waves,** and **flooding**) is expected to increase across the UK, as are water shortages in the south and south east.

Health: in Europe, more heat waves can increase deaths, but deaths related to colder weather may decrease.

Crop yields in Europe are expected to increase but require more irrigation.

Drought is likely to put pressure on food and water supplies in sub-Saharan Africa due to higher temperatures and less rainfall.

Health in southern and eastern Africa may decline as malaria would increase in hot humid regions that remain hotter for longer in the year.

Agriculture may be affected in South Asia. A decrease in wheat and maize and a small increase in rice are expected.

Warmer rivers affect marine **wildlife**. The change in food supply may decrease the Ganges river dolphin population in Nepal, India and Bangladesh.

EUROPE

ASIA

AFRICA

▲ Figure 4.12 Global effects of climate change

➔ Activities

1   State two positive and two negative effects of climate change, using Figure 4.12.

2   Complete the following table (notice the impacts are plural):

Social impacts	
Environmental impacts	

3   Rank in order which parts of the world you think may be affected, most to least.

4   How might the effects in one part of the world impact on another? Give at least one example.

How might people and the environment be affected by climate change around the world?

Expected global effects of climate change can be seen on the map in Figure 4.12, which is centred on the Pacific Ocean.

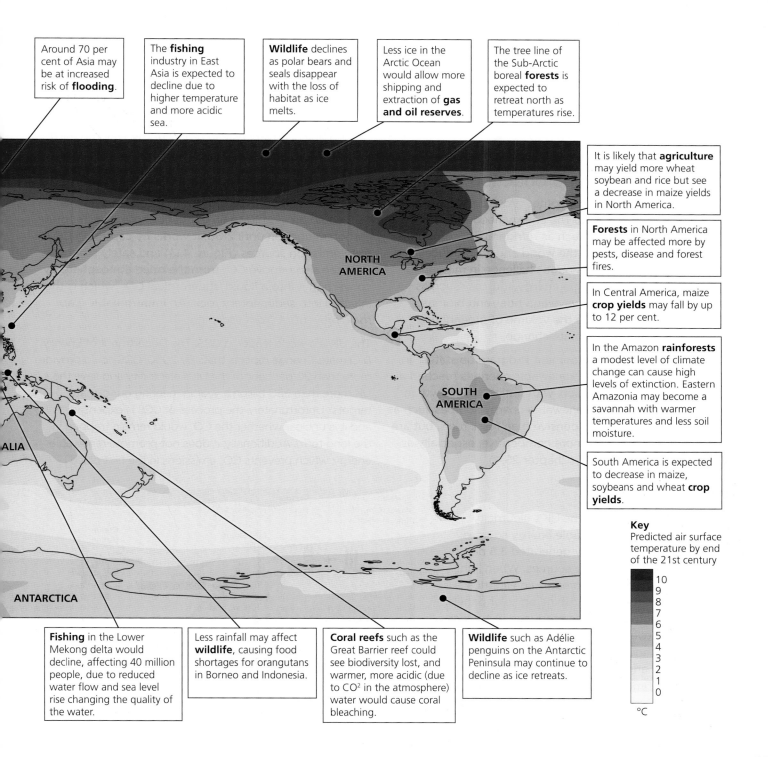

Around 70 per cent of Asia may be at increased risk of **flooding**.

The **fishing** industry in East Asia is expected to decline due to higher temperature and more acidic sea.

Wildlife declines as polar bears and seals disappear with the loss of habitat as ice melts.

Less ice in the Arctic Ocean would allow more shipping and extraction of **gas and oil reserves**.

The tree line of the Sub-Arctic boreal **forests** is expected to retreat north as temperatures rise.

It is likely that **agriculture** may yield more wheat soybean and rice but see a decrease in maize yields in North America.

Forests in North America may be affected more by pests, disease and forest fires.

In Central America, maize **crop yields** may fall by up to 12 per cent.

In the Amazon **rainforests** a modest level of climate change can cause high levels of extinction. Eastern Amazonia may become a savannah with warmer temperatures and less soil moisture.

South America is expected to decrease in maize, soybeans and wheat **crop yields**.

NORTH AMERICA

SOUTH AMERICA

ALIA

ANTARCTICA

Key
Predicted air surface temperature by end of the 21st century

10
9
8
7
6
5
4
3
2
1
0

°C

Fishing in the Lower Mekong delta would decline, affecting 40 million people, due to reduced water flow and sea level rise changing the quality of the water.

Less rainfall may affect **wildlife**, causing food shortages for orangutans in Borneo and Indonesia.

Coral reefs such as the Great Barrier reef could see biodiversity lost, and warmer, more acidic (due to CO_2 in the atmosphere) water would cause coral bleaching.

Wildlife such as Adélie penguins on the Antarctic Peninsula may continue to decline as ice retreats.

Managing climate change: mitigation

What is mitigation?

The challenge of reducing the impact of climate change can be managed by two main approaches: mitigation and **adaptation**. (See Section 4.6 for information about adaptation.) Mitigation strategies, whether local or global, deal with the cause of the problem. They reduce or prevent the greenhouse gases which cause climate change and protect carbon sinks, such as forests and oceans.

How can the causes of climate change be reduced?

Alternative energy production

As world population and incomes grow, the demand for energy also increases. The energy needed to power more consumer goods, such as refrigeration, computers and car, to travel around the world and, to produce food, especially meat, causes a huge challenge in mitigating climate change. **Renewable energy sources** (such as wind, solar, geothermal, wave and tidal, and biomass) offer a solution to reduce the volume of greenhouse gases contributing to climate change.

The United Nations Environment Programme states: 'In 2010, new investments in renewable energies reached a record high of US$211 billion, with noticeable growth in emerging economies'. Renewable energy sources such as solar energy are more expensive than fossil fuels, but are becoming cheaper and more competitive, especially as they do not produce CO_2 (Chapter 25.4).

Solar energy

In 2013, 14.9 per cent of the UK's electricity was generated by renewable energy sources. Photovoltaic solar energy generated 3.8 per cent of renewable energy sources. When light shines on solar panels it creates an electrical field. The stronger the sunshine on solar panels, the more electricity that is produced. A typical home saves over a tonne of CO_2 per year as there are no greenhouse gas emissions to contribute to climate change (Energy Saving Trust, 2014). However, at times when there is no sunshine, such as night, solar energy cannot be relied on to generate electricity.

Carbon capture

Technological advances can replicate the way the Earth stores carbon dioxide (underground in rock formations and the oceans) in a process known as **carbon capture and storage (CCS)**. CCS can be used with existing and new power plants. The IPCC (page 42) estimates that CCS could provide 10 to 55 per cent of the world's total carbon mitigation until 2100. It works by capturing CO_2 from emission sources (Figure 4.13) and safely storing it. CCS can also remove CO_2 from the open atmosphere by converting it into a liquid 'supercritical CO_2' which is then injected into sedimentary rock. An **impermeable** 'cap rock' prevents it from escaping.

The UK is a world leader in CCS. The Department of Energy & Climate Change reports that: 'By 2050, CCS could provide more than 20 per cent of the UK's electricity and save the UK more than £30 billion a year in meeting our climate targets'. Unfortunately, the process of CCS is expensive, and it is unclear whether the CO_2 would remain trapped in the long term. Additionally, it does not promote renewable energy, which prevents CO_2 emissions in the first place.

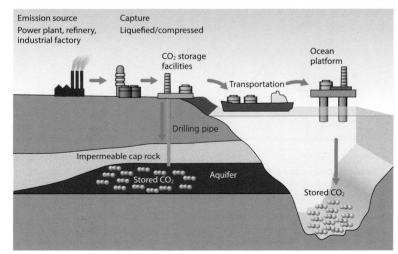

▲ Figure 4.13 Carbon capture and storage

Planting trees

Deforestation is a global problem as it is a major driver of climate change (see Chapter 6). According to the United Nations Environment Programme, deforestation and forest degradation occurs at a rate of 13 million hectares per year. A US$40 billion investment in **reforestation**, and payments to landholders for conservation each year from 2010 to 2050, could increase forest carbon storage by 28 per cent.

The UK has a £24.9 million project (funded by the Department for Environment, Food and Rural Affairs) to reduce deforestation and increase forest and land restoration in Brazil (see Chapter 6). It aims to tackle climate change by reducing 10.71 million tonnes of CO_2 emissions over 20 years by recovering 41,560 hectares of degraded forests.

International agreements

The UN negotiated a new international climate change agreement for all countries at the 2015 Paris climate conference. It will be implemented from 2020. The European Commission has set the EU's vision for a new agreement that will reduce global emissions by at least 40 per cent below 2010 levels by 2030, and by 60 per cent by 2050. It was a challenge for countries to agree on targets that will go far enough to manage climate change. Some countries can afford to mitigate climate change more than others, and some are considered more responsible for causing climate change than others.

▲ Figure 4.14 'Wind farm where a forest once stood.'

➜ Activities

1 Define mitigation.

2 Use Figure 4.13 to describe the process of carbon capture and storage.

3 Complete the following table:

Mitigating climate change			
Method	How does it reduce CO_2?	Advantage of the method	Disadvantage of the method
Alternative energy production			
Carbon capture and storage			
Planting trees			
International agreements			

4 Explain the message in the cartoon (Figure 4.14). Include the following words in your answer: wind farm, renewable energy, deforestation, carbon dioxide.

Managing climate change: adaptation

What is adaptation?

Section 4.5 considered how mitigation can manage the challenges of climate change. Another approach, adaptation, responds to the impacts of climate change and tries to make populations less vulnerable. Adaptation strategies are local rather than global, to respond to localised impacts. If mitigation stopped all carbon emissions from human activity, adaptation would still be required to manage the impacts of climate change that are naturally occurring and those that have already occurred.

How can climate change be managed through adaptation?

★ KEY LEARNING

➤ What adaptation is
➤ Managing climate change through adaptation
➤ The costs and benefits of methods of adaptation

Potato Park in Peru

Peru's Potato Park is a 12,000 hectare reserve high in the Andes, near Cusco. It was established to conserve the region's potato biodiversity (more than 1,345 varieties). Warmer climates have altered the growing patterns of some local potato varieties.

The 8,000 residents, from six indigenous Quechua communities, own the land and control access to local resources. The organisation Papa Arariwa Collective ('guardian of native potatoes') helps manage all the land so everyone can benefit.

A typical family farm grows 20 to 80 varieties of potato. Most are grown for local consumption. As the climate becomes warmer farmers have begun experimenting with different varieties at higher altitudes where temperatures are lower.

Varieties which are disappearing have been conserved in the gene bank of the International Potato Centre. The disease-free varieties have helped increase crop yields.

Conservation of potato varieties will provide invaluable help to local communities adapting to climate change.

Change in agricultural systems

Although the effects of climate change are uncertain, agriculture will need to adapt to them. Some ways of adapting include:

- moving production to another location due to changing temperatures and extreme weather
- increasing irrigation in areas due to changing precipitation
- changing the crops and varieties grown and the time of year they are planted, such as drought-resistant crops or switching to livestock production which tends to have more guaranteed returns.

The cost of adapting to climate change is more difficult for poorer subsistence farmers.

The United Nation's Food and Agriculture Organization states that agriculture needs to be 'climate-smart' if it is to feed the world (Figure 4.15).

▲ Figure 4.15 An example of climate-smart agriculture

Managing water supply

In the UK, Londoners consume 167 litres of water each day compared to the national average of 146 litres. It is the driest part of England, contains 13 per cent of the population of the UK and faces the challenge of climate change: summers will get drier and winters will get wetter (Chapter 15).

There are two ways water can be managed:

- Reducing demand. As Mayor of London, Boris Johnson developed a Water Strategy to reduce London's water demand. By 2030, all London homes should have been offered a free **retrofit** package of water-efficient devices (Figure 4.16), including aerators.
- Increasing supply. Thames Water opened a **desalination** plant in Beckton in 2010 to increase water supply. Water is taken from the River Thames at low tide (when it is least salty). A process called reverse osmosis is used to produce drinking water for 400,000 homes. The plant requires a lot of energy (enough to power 8,000 homes), so carbon emissions need to be offset by a **biodiesel** electricity plant.

Reducing risk from rising sea levels

London is currently well protected against rising sea levels. The Thames Barrier has been closed over 100 times since it was built, in 1982, to stop tidal surges entering central London. The Barrier was designed with an expectation it would be breached once every 1,000 years, but a 50 centimetre rise in sea level would increase the risk to once every 100 years.

Fieldwork: Get out there!

Investigate the impact that mitigation (such as alternative energy) and adaptation (such as flood prevention) strategies have on the physical environment in your local area.

a) Decide what evidence of mitigation and/or adaptation you could survey.

b) Suggest where you could carry out your fieldwork.

c) i) Devise an environmental survey to assess its impact on the physical environment.
   ii) Explain why an environmental survey is **qualitative data**.

▼ Figure 4.16 Impact of water-efficient devices installed in London homes

Product	Energy saved/year	Water saved (L/year)	CO$_2$ saved/year	Water bill savings (£/year)	Energy bill savings (£/year)
Tap aerators (saving for whole house)	199.67	6,570	42.35	11.12	8.84
Dual flush	0	17,155	0	29	N/A
Shower timers	46	12,78	14.1	2.16	2.04
Aerator showerheads	440.36	10,950	93.41	18.53	19.5
Total	686.03	35,953	149.86	60.81	30.38

Geographical skills

What type of graph would be appropriate to show 'water bill savings' in Figure 4.16? Construct your chosen graph.

→ Activities

1 Distinguish between mitigation and adaptation.

2 Why will agricultural systems need to adapt in the future?

3 Use Figure 4.15 to describe how agricultural systems can change to cope with climate change.

4 Why has the 'Water Strategy' been established in London?

5 Suggest at least three other ways in which water consumption in the home can be reduced.

6 Describe how water supply can be managed to cope with changes caused by climate change.

7 Which water efficiency measure in Figure 4.16 has the greatest impact on London's water supply? Explain your choice.

8 Suggest how existing flood defence schemes such as the Thames Barrier may need to change due to climate change.

9 'The best of both worlds'. Explain why both mitigation and adaptation are needed to manage climate change.

Unit 1 Section A

1 Complete the sentence using Figure 1.3.

The continent that experienced the biggest natural disasters caused by tornados in 2010 was _____. Countries in Europe were mostly at _____ risk of natural hazards.

> You should aim to allow yourself about one minute per mark.

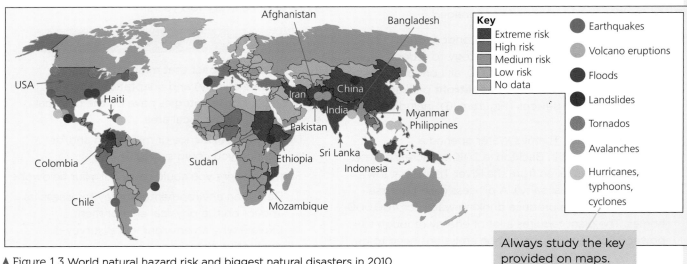

▲ Figure 1.3 World natural hazard risk and biggest natural disasters in 2010

> Always study the key provided on maps.

[2 marks]

2 Describe the distribution of volcanoes shown in Figure 2.3.

> Avoid naming the location of individual volcanoes. Use words such as a 'cluster' and 'chain'. Use directions such as 'along the east/west coast' and 'central'.

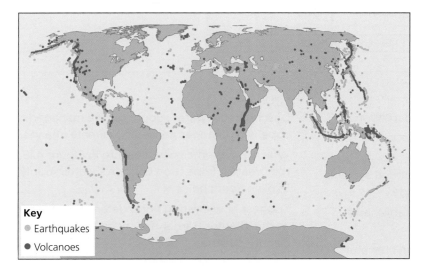

▲ Figure 2.3 Tectonic hazards around the world

[2 marks]

3 Which two of the following statements describes a constructive plate margin?

Read all multiple choice answers first.

   A Tectonic plates move towards each other.
   B Tectonic plates slide past each other side by side.
   C Tectonic plates move away from each other.
   D Magma rises to fill the gap between tectonic plates to form new land.
   E The continental plate is pushed under the oceanic plate.
   F The oceanic plate is pushed under the continental plate.

   [2 marks]

4 'It is **not** worth continuing to live at risk from a tectonic hazard.' Do you agree with this statement? Justify your decision.

   The command 'justify' means you should reach a conclusion and give reasons for the choice you made. You should aim to write a response, however short, to gain at least one SPaG mark. It must relate to the question.

   [9 marks]
   [+ 3 SPaG marks]

5 'Human activity is the main cause of climate change.' Use evidence to support this statement.

   Remember to develop your sentences.

   [6 marks]

6 Define the term 'adaptation'. [1 mark]

7 Suggest how climate change might affect tropical storms. [2 marks]

8 Outline **one** weather hazard that southern England sometimes experiences during summer.

   'Outline' means providing information that progresses beyond a simple or basic statement, but not too far. Where two marks are on offer, aim to write two short sentences, or one longer sentence. You should aim not to write a lengthy answer that may leave you short of time to answer other exam questions.

   [2 marks]

9 Suggest how weather hazards, such as those shown in Figure 3.20 (page 37) and Figure 3.28 (page 42), may support the view that UK weather is becoming more extreme.

   This question asks you to make use of photographs and data in order to support understanding of an important geographical issue.

   [4 marks]

5 Ecosystems

How ecosystems operate

How are the different parts of an ecosystem linked?

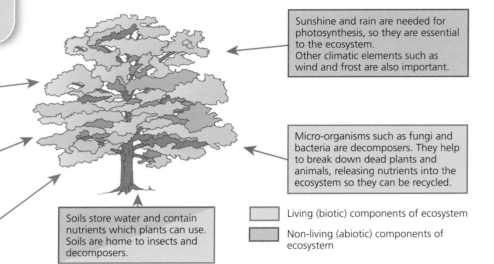

Animals found in a woodland include many species of insects and birds, and mammals such as rabbits, squirrels and foxes.

Plants include trees, wild flowers, grasses, mosses and algae. They provide food and shelter for many animals.

Rocks help in the formation of soils and rock type is important. Weathering releases nutrients stored in rocks into the ecosystem.

Soils store water and contain nutrients which plants can use. Soils are home to insects and decomposers.

Sunshine and rain are needed for photosynthesis, so they are essential to the ecosystem.
Other climatic elements such as wind and frost are also important.

Micro-organisms such as fungi and bacteria are decomposers. They help to break down dead plants and animals, releasing nutrients into the ecosystem so they can be recycled.

☐ Living (biotic) components of ecosystem

☐ Non-living (abiotic) components of ecosystem

▲ Figure 5.1 The biotic and abiotic components of an ecosystem

An ecosystem is made up of plants, animals and their surrounding physical environment, including soil, rainwater and sunlight (Figure 5.1). Important interrelationships link together the **biotic** (living) and **abiotic** (non-living) parts of the ecosystem. These interrelationships consist of:

■ physical linkages between different parts of the ecosystem (animals eating the plants, for example)
■ chemical linkages (mild acids in rainwater speed up the decay of dead leaves, for example).

In ecosystems, plants and animals can migrate from one place to another, bringing change. There are also inputs and outputs from the ecosystem to other places (Figure 5.2). Most importantly of all, ecosystems depend on a constant input of light from the Sun, as well as rain from the atmosphere. In turn, rainwater leaves the ecosystem when it evaporates and returns to the atmosphere, or runs into a river.

Ecosystems can be any size:

■ local (a small-scale ecosystem is also called a habitat)
■ regional (England's Lake District moorland)
■ global **biomes** (South America's tropical rainforest)
■ Earth (some scientists argue that all of the planet's organisms are linked together).

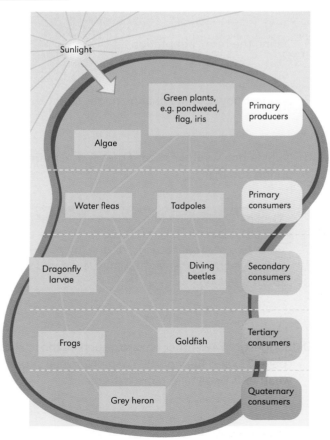

Sunlight

Green plants, e.g. pondweed, flag, iris	Primary producers
Algae	
Water fleas	Tadpoles	Primary consumers
Dragonfly larvae	Diving beetles	Secondary consumers
Frogs	Goldfish	Tertiary consumers
Grey heron	Quaternary consumers

▲ Figure 5.2 Linkages, inputs and outputs for a garden pond ecosystem

How do food chains and nutrient cycles work?

The biotic community of an ecosystem consists of different species of plants and animals in different feeding groups:

- plants or primary producers: green plants that use photosynthesis and take nutrients from the soil using their roots
- herbivores or primary **consumers**: plant-eating animals (cows, or rabbits)
- carnivores or secondary consumers: these animals feed on herbivores (foxes, or cats)
- top carnivores: these animals will hunt and eat other carnivores in the ecosystem, as well as the herbivores. They include the largest and fastest hunters, like lions and wolves.

The interrelationships between these feeding groups are shown in the **food chain** in Figure 5.3. This diagram shows the weight of **biomass** getting smaller at each level. For instance, in the tropical rainforests of Brazil (see Chapter 6), there are only five kilograms of animal biomass per 40 kilograms of plant biomass. There are two important reasons for this reduction:

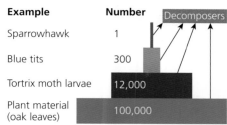

Example	Number		Trophic level
Sparrowhawk	1	Decomposers	Tertiary consumers (top predators)
Blue tits	300		Secondary consumers (carnivores)
Tortrix moth larvae	12,000		Primary consumers (herbivores)
Plant material (oak leaves)	100,000		Primary producers (plants)

▲ Figure 5.3 An ecosystem food chain

- Many parts of plants are simply not eaten by animals, and carnivores do not eat all of their prey (such as the bones). Also, much of what the animals do eat is excreted.
- Energy is last at each level. Hunters use a lot of kinetic energy: chasing prey can be time-consuming and exhausting. Some herbivores search around a lot for plants to eat. In addition, energy is constantly being used up in respiration. Much of an animal's daily calorie intake is used simply to stay alive, rather than to build new biomass.

The **decomposers** in Figure 5.3 are the organisms that, over time, break down dead organic matter and animal excretions. They include a mixture of: scavengers (such as insects that eat dead wood) and detritivores (such as bacteria). Decomposers help to return nutrients to the soil in the form of an organic substance called humus.

The importance of nutrient cycling

All plants and animals depend on nutrients in food for their health and vitality. Nutrients occur naturally in the environment and are constantly recycled in every ecosystem. Figure 5.4 shows these important pathways.

→ Activities

1. Look at Figure 5.1. Identify three interrelationships between the biotic and abiotic components of the ecosystem.

2. Look at Figure 5.2.
   a) Identify plant and animal species that are part of the pond ecosystem.
   b) Suggest why the input shown is important for the health of the ecosystem.

3. Look at Figure 5.3. Use the information provided to help you describe and explain the difference in size between the primary producers and the top predators.

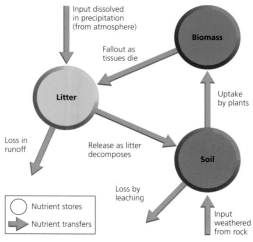

▲ Figure 5.4 The nutrient cycle

⊛ KEY LEARNING

➤ The characteristics of Epping Forest's food web
➤ The interdependence of the ecosystem
➤ The characteristics of Epping Forest's nutrient cycle

Epping Forest ecosystem, UK

Key facts

- Located east of London (Figure 5.5), Epping Forest is all that remains of a larger forest that colonised England at the end of the last Ice Age.
- Bogs and ponds in the forest have their own unique species, including 20 kinds of dragonfly.
- For 1,000 years, Epping Forest has been managed in a variety of ways: as hunting grounds for royalty, a timber resource and, nowadays, recreation (as Figure 5.5 shows, it is easily accessible).

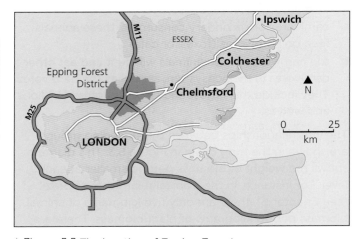

▲ Figure 5.5 The location of Epping Forest

What are the characteristics of Epping Forest's food web?

Biodiversity in the forest has remained naturally high, thanks to careful management, so there is a complex **food web** composed of thousands of species (Figure 5.6 shows this in a simplified form). Epping Forest is home to:

- a large number of native tree species, including oak, elm, ash and beech
- a lower shrub layer of holly and hazel at five metres, overlying a field layer of grasses, brambles, bracken, fern and flowering plants; 177 species of moss and lichen grow here. Altogether, there is great diversity of producer species (Figure 5.7)
- many insect, mammal and bird consumer species are supported, including nine amphibian and reptile species and 38 bird species
- studies have found 700 species of fungi, which are important decomposers.

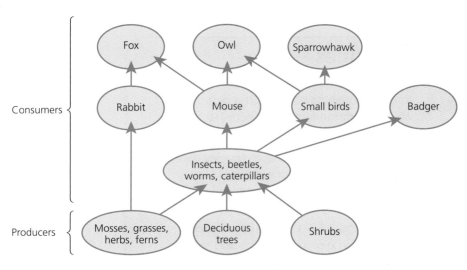

▲ Figure 5.6 Epping Forest's food web

How is the ecosystem interdependent?

The forest's **producers**, consumers and decomposers are all interdependent. This is most clearly shown by the annual life cycle of the trees.

Most of the trees are deciduous, meaning that they lose their leaves in winter. This is an adaptation to the UK's seasonal climate. Winters are darker and cooler than summers (the mean monthly temperature is 18°C in July but just 5°C in January). As a result, the trees grow broad green leaves in spring. This allows them to maximise photosynthesis during the summer. They shed their leaves in the autumn, and so conserve their energy during winter.

By mid-autumn, the forest floor is covered with a thick layer of leaves. Remarkably, by spring, the leaf litter has all but disappeared: the decomposers and detritivores' work (page 59) is now complete. Nutrients stored in the leaves are converted to humus in the soil, ready to support the new season's plant growth. This will ultimately include the fruits and berries that, in turn, support many primary consumers.

Nutrient cycling demonstrates clearly the interdependence of plants, animals and soil. People and ecosystem components are interdependent too. In the past, coppicing was common (cutting back trees to encourage new growth of wood). Today, visitors pick berries and flowers. In turn, this helps spread the seeds, which stick to their clothing.

▲ Figure 5.7 Leaves on the forest floor and Epping Forest's tree canopy, which in places reaches 30 metres

What explains the characteristics of Epping Forest's nutrient cycle?

In Figure 5.8, which reflects Epping Forest, the biomass store is large because of the great height of the trees, and the dense undergrowth beneath them. The soil store is large too because there is always plenty of humus.

The high flow rates between the litter, soil and biomass stores reflect the vigorous cycle of new growth that takes place each year. The forest also loses a lot of nutrients each year, via leaching, during episodes of heavy rainfall.

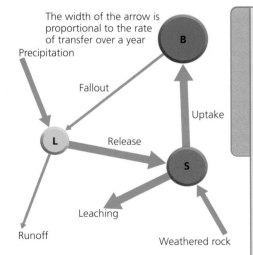

▲ Figure 5.8 The nutrient cycle of a deciduous forest

The width of the arrow is proportional to the rate of transfer over a year

Precipitation
Fallout
Uptake
Release
B
L
S
Leaching
Runoff
Weathered rock

→ Activities

1   Look at Figure 5.6. Suggest how the populations of the different food web species might be affected by the removal of foxes from the ecosystem.

2   Birds, mammals, insects, amphibians and reptiles are all present in Epping Forest. Some are herbivores (primary consumers), whereas others are carnivores (secondary consumers and top carnivores). Draw a table with two columns and add as many named examples as you can of both categories (for instance, the tawny owl is a carnivore).

3   Study Figure 5.7. Describe as many characteristics as you can of the plants shown. What kinds of processes are taking place?

4   Using all the information, explain two ways in which studying Epping Forest has helped you understand the term 'interdependence'.

- How physical and human forces disturb ecosystem balance
- How the loss or gain of one species affects a food web
- How ecosystem balance can be restored through management

Changes affecting ecosystem balance

How do physical and human forces disturb ecosystem balance?

Periods of extreme weather or climate change can disturb the balance of ecosystems. In the years 1976–77, southern England experienced an 18-month drought that killed many trees. A further 15 million English trees were felled by a great storm in 1987. As a result, population numbers declined for many consumer species in the food chain. Secondary forest growth has since taken place, however, and consumer species have migrated back. The recent recovery of English woodland is an example of ecosystem resilience.

Ecosystems are sometimes damaged in permanent ways, especially when human forces are involved, for instance by deforestation (page 49). The removal of the forest exposes the soil beneath to rainfall, and so it can be washed away, making it impossible for the ecosystem to recover. This is especially true in tropical rainforest regions, where heavy rain falls most days (Chapter 6). In the longer-term, human-induced climate change could threaten the ecosystem balance of many places. For instance, changes in temperature and precipitation patterns for southern England might make it harder for ecosystems like Epping Forest to survive in their current form. In some places, grass (rather than trees) may dominate in the future, if climate change predictions are correct.

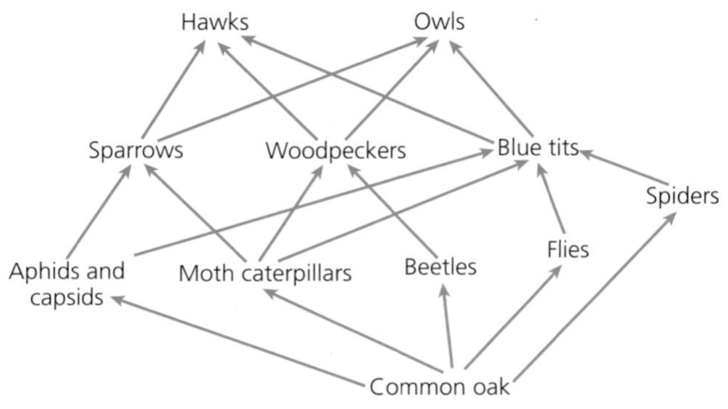

▲ Figure 5.9 A food web supported by an oak tree

How does the loss or gain of one species affect a food web?

Figure 5.9 shows a food web supported by oak woodland. Suppose that the population of beetles is reduced by disease. This would directly impact on the numbers of woodpeckers. With fewer beetles to eat, their numbers may decline. In contrast, we may expect to see more oak tree growth now fewer beetles are feeding on them. In addition to these direct impacts, there are indirect impacts to consider:

- Owl and hawk numbers may also fall because they eat woodpeckers.
- Woodpeckers are carnivorous and have multiple food sources. They may just eat more caterpillars instead. However, this could now impact on blue tit numbers. How would this happen? Can you identify more possible food web impacts that could follow?

How can ecosystem balance be restored through management?

Many species have been hunted to extinction, without a full understanding of how this could affect ecosystem balance. In Europe and the USA, killing wolves and bears removed danger to people and their cattle. But fewer carnivores meant that rabbit and deer populations quickly multiplied and began to eat all available vegetation, stripping the land bare, leading to soil erosion. The ecosystem lacks balance.

Many scientists believe that 'rewilding' or 'ecosystem restoration' is the best way to restore ecosystem balance. Grey wolves were recently reintroduced into Yellowstone National Park in the USA, which has resulted in numerous impacts (Figures 5.11 and 5.12). The wolves have restored balance to the ecosystem and landscape (Figure 5.10).

▲ Figure 5.10 Yellowstone National Park

▲ Figure 5.11 A grey wolf

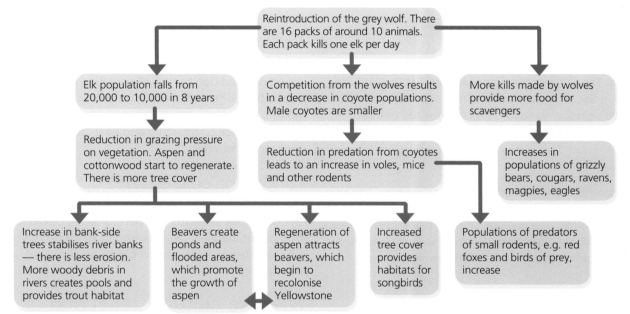

Reintroduction of the grey wolf. There are 16 packs of around 10 animals. Each pack kills one elk per day

Elk population falls from 20,000 to 10,000 in 8 years

Competition from the wolves results in a decrease in coyote populations. Male coyotes are smaller

More kills made by wolves provide more food for scavengers

Reduction in grazing pressure on vegetation. Aspen and cottonwood start to regenerate. There is more tree cover

Reduction in predation from coyotes leads to an increase in voles, mice and other rodents

Increases in populations of grizzly bears, cougars, ravens, magpies, eagles

Increase in bank-side trees stabilises river banks — there is less erosion. More woody debris in rivers creates pools and provides trout habitat

Beavers create ponds and flooded areas, which promote the growth of aspen

Regeneration of aspen attracts beavers, which begin to recolonise Yellowstone

Increased tree cover provides habitats for songbirds

Populations of predators of small rodents, e.g. red foxes and birds of prey, increase

▲ Figure 5.12 Impacts of the reintroduction of the grey wolf to the Yellowstone ecosystem since 1995

→ Activities

1   Look at Figure 5.9.

   a) How many levels are there in the food web?
   b) Name four primary consumers and three secondary consumers.

2   State one human and one physical cause of ecosystem disturbance. Compare their impacts.

3   Humans have deliberately or accidentally helped plant and animal species to migrate to new places. What examples can you think of, other than the ones included on these pages? Did any of these movements impact negatively on ecosystems?

4   Look at Figure 5.12. Make a list of (a) species whose numbers increased and (b) species whose numbers were reduced, after the reintroduction of wolves.

5   Explain how changes to the ecosystem in turn changed the physical environment of Yellowstone.

→ Going further

Discuss whether potentially dangerous wild animals like wolves and bears really can be brought back to the countryside in the UK. How might such a move cause conflict with other uses of the countryside like tourism?

Fieldwork: Get out there!

Look at Figure 5.9. Plan a fieldwork investigation of this small-scale ecosystem. Possible themes to investigate could include the number of different plant and animal species, or evidence of interrelationships and nutrient cycling.

The distribution and characteristics of global ecosystems

How does climate explain the distribution and characteristics of global ecosystems?

Figure 5.13 shows the distribution of the world's large-scale **global ecosystem** in biomes. Figure 5.13 and the this page explain how global-scale variations influence the distribution of ecosystems.

Tropical rainforests

These lie along the Equator in Asia, Africa and South America. The Sun's rays are concentrated at this latitude, heating moist air which rises and leads to heavy rainfall, with little seasonal variation. This creates the perfect conditions for evergreen rainforest.

Deserts

Found close to the Tropics of Cancer and Capricorn. The air that rises over the Equator heads polewards after shedding its moisture as rain. The Sun's rays are still highly concentrated at this low latitude. Combined with the dry air, this brings arid desert conditions to places like the Sahara and Australia.

Tropical grasslands

Sandwiched between the two extremes of tropical rainforest and desert. Conditions are dry for half of the year, due to the seasonal movement of the Hadley cell (page 23).

Temperate grasslands

Short tussock and feather grasses dominate the landscape between 40° and 60° north of the equator, but only in the centre of continents away from the sea.

Mediterranean

Drought-resistant small trees and evergreen shrubs grow between 30° and 40° north and south of the equator, but only on the west coasts of continents.

Deciduous forests

These grow in many places at higher latitudes. Found in western Europe, where rain-bearing storms arrive regularly thanks to the jet stream, and the east coasts of Asia, North America and New Zealand. The Sun's rays are weaker at this latitude.

Coniferous forests

Found at 60° north, where winter temperatures are extremely cold due to lack of insolation (page 22). Due to the Earth's tilt, there is no sunlight for some months of the year. Coniferous trees have evolved needle leaves that reduce moisture and heat loss during the cold, dark winter months.

Tundra (or 'cold desert')

These areas are found at the Arctic Circle, where the Sun's rays have little strength. Temperatures are below freezing for most of the year. Only tough, short grasses can survive, often in waterlogged conditions (due to surface ice thawing).

Why are altitude, relief and ocean currents also important?

Although latitude and distance from the sea are the main factors affecting distribution, the following are also important:

- altitude: temperatures fall by about half a degree for every 100-metre increase in altitude, and tough grasses replace trees on steep mountainsides
- mountain ranges: in the USA and Asia, inland areas isolated from the sea suffer from low rainfall. This is because winds blowing off the oceans quickly lose their moisture when air is forced to rise upwards over a high mountain range. The drier lands found east of the USA's Rocky Mountains are said to be in a **rain shadow**
- ocean currents: a cold ocean current flowing along South America's coast helps to create arid conditions in Chile's Atacama Desert because little evaporation takes place over the cold water. In contrast, the warm Gulf Stream ocean current affects the climate of western Europe.

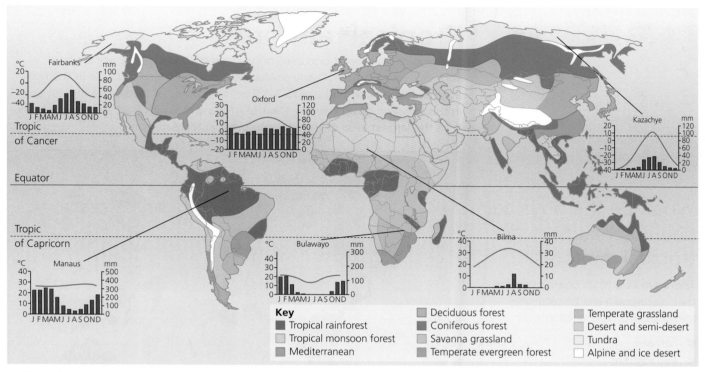

▲ Figure 5.13 World distribution of global ecosystems

Geographical skills

Each global-scale ecosystem in Figure 5.13 has several distributional features. Choose one ecosystem and take the following steps:

1   What is the overall pattern? Does the ecosystem circle the Earth at a particular latitude? Or is it found only in particular regions?

2   Can you name the continents or any particular places where it is found?

3   Are there any unusual features or anomalies? Perhaps the ecosystem is found at an unusually high or low latitude in some places.

4   Can you support your answer with data? Is it possible to estimate the width of the band of vegetation in thousands of kilometres, or the distance it lies from the Equator?

→ Activities

1   Look at Figure 5.13. From the climate graphs:
a)   Identify the annual rainfall that deciduous forest needs.
b)   State the minimum and maximum annual temperatures for the tundra ecosystem.

2   Draw a table showing the three types of forest. In one column, describe the associated climate. In another column, briefly describe the characteristics of the trees.

Type of forest	Climate	Characteristics

3   Kilimanjaro is a six kilometre high mountain located close to the Equator in Africa. At its base it is surrounded by tropical rainforest. Describe how you think the vegetation on Kilimanjaro changes with increasing altitude.

4   Look at Figure 5.13. Is temperature or rainfall the most important influence on whether a place is a desert? Explain your answer.

6 Tropical rainforests

Tropical rainforest

The tropical rainforest occupies only seven per cent of the world's land surface, but it contains many useful resources. It is also valuable in the fight against global warming (page 74). The main areas of tropical rainforest are in the Amazon basin (Brazil), Central Africa and South East Asia (Figure 6.1), with the largest area of tropical rainforest in the Amazon.

What are the physical characteristics of the tropical rainforest?

There are two main characteristics that distinguish the tropical rainforest from other biomes:

- climate
- vegetation.

Climate

Because tropical rainforests occur on or close to the Equator, the climate is typically warm and wet. Annual temperatures average around 26°C and show little variation from day to day or month to month (Figure 6.2). Annual rainfall usually exceeds 2,500 millimetres. This abundant supply of water feeds the huge rivers, such as the Amazon in Brazil and the Congo in Central Africa, which are an impressive physical feature of tropical rainforests.

Vegetation

Tropical rainforests are renowned for their rich vegetation cover. Particularly spectacular are their very tall trees, typically 30 to 45 metres in height. (For more about the vegetation, see Section 6.2.)

Soils

The soils of the tropical rainforest are mainly thin and poor. So, how is there so much luxuriant vegetation? The answer lies in the rapid recycling of nutrients known as the nutrient cycle (Figure 6.3).

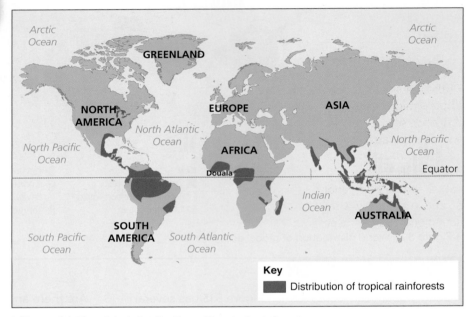

▲ Figure 6.1 The global distribution of tropical rainforests

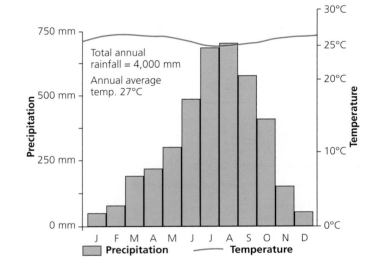

▲ Figure 6.2 The climate at Douala (Cameroon)

Most of the forests' vital nutrients are locked up in:

- the biomass – living vegetation and animals
- the litter – dead wood and leaves, and animal remains on the ground.

The warm, humid conditions cause the litter to decompose very quickly. The little rain that reaches the forest floor often washes away litter nutrients before they become part of the soil. It is not surprising that rainforest soils are rather infertile. Nonetheless, plants can pick up enough nutrients from the soil to survive. Many nutrients are stored in large, thick trees.

How is the tropical rainforest interdependent?

Figure 6.4 shows the main components of the tropical rainforest ecosystem: climate (rain and sunlight), soil, vegetation (trees and plants) and animals. The arrows show how they interact to create an interdependence (Section 6.2). The diagram also shows people. It is possible for native people (indigenous tribes) to live as part of, and in harmony with, the ecosystem. Equally, human activity can badly upset the ecosystem's balance.

Why is there so much biodiversity in tropical rainforests?

Tropical rainforests are renowned for their high level of biodiversity. It is higher than in any other biome. More than two-thirds of the world's plant species are found in these forests. The forests contain around half of the world's known animal species, ranging from mammals (such as monkeys and sloths) and birds, to reptiles (such as snakes and frogs) and insects.

Human exploitation of the rainforest's resources is reducing this rich biodiversity. For more on the individual threats to biodiversity, see Sections 6.3 and 6.4. Many species are becoming endangered; many others have already become extinct. A loss of biodiversity means a decline in ecosystem productivity (Sections 6.5 and 6.6).

The challenge is this: can the tropical rainforest be used in a sustainable way that does not threaten biodiversity (see Sections 6.7 and 6.8)?

Biomass
All living things in the ecosystem, both plants and animals

Nutrients fall to the ground when plants and animals die

Plants take up nutrients which are dissolved in soil

Litter
Dead organic material such as fallen leaves or tree trunks and dead animals

Nutrients become part of the soil when dead matter decomposes

Soil
Developed by the mixing of dead organic material with weathered bedrock

⌃ Figure 6.3 The nutrient cycle

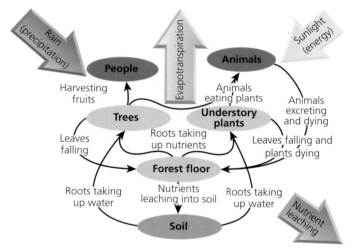

⌃ Figure 6.4 The tropical rainforest ecosystem

Geographical skills

Study Figure 6.2.

  a) Describe the features of Douala's climate.

  b) Calculate the mean annual temperature range.

→ Activities

1   What is the difference between an ecosystem and a biome? (Look back at Chapter 5.)

2   Look at Figure 6.1. Describe the global distribution of tropical rainforests. In which hemisphere is there more tropical rainforest?

3   Suggest reasons why biodiversity is so high in the tropical rainforests.

⭐ KEY LEARNING

➤ How plants have adapted to survive

➤ How animals have adapted to survive

Adaptations to the tropical rainforest environment

Let us now look more closely at the adaptations of plants and animals within the tropical rainforest. They provide examples of the interdependence that exists within its ecosystem (Figure 6.4 on page 67).

How have plants adapted to survive?

Factors that help poor soils

In Section 6.1, the soils of the tropical rainforest were described as being very poor. Not good news for plants! So how do they manage to survive and prosper? The answer lies mainly in four factors:

■ a rapid cycling of nutrients through the ecosystem (see Figure 6.3) – a sort of fast-food delivery

■ the absorption of sunlight, leading to photosynthesis

■ the warm, humid climate, which is ideal for plant growth throughout the year

■ the ability of plants to adapt as they compete for sunlight and nutrients.

Heat and humidity

The nutrient cycle is one way in which the three components of an ecosystem work together. Another is the water cycle (Figure 6.5). This constant recycling of water occurs every day.

The leaves of many trees are waxy and have tips that allow water to run off them. Leaf stems are also flexible to allow leaves to move with the Sun to maximise photosynthesis. Therefore, vegetation copes with both heat and heavy rainfall by:

■ using the circulating water as a sort of cooling system

■ passing water to the soil or returning it to the atmosphere

■ having leaves that can cope with the large amounts of water falling on them.

Competition for sunlight

Although photosynthesis is important for plant growth, plants still need minerals and these come mainly from the soil. The dense vegetation of the tropical rainforest shows four distinct layers (Figure 6.6). In each layer, the plants have adjusted to the physical conditions, particularly to available sunlight. Most sunlight is received by the tops of tall trees and, due to a shading effect, the least sunlight is received close to the forest floor.

In the lowest two layers there is little photosynthesis to convert the small amount of sunlight into plant food. So plants have to rely on other ways of getting their food supply. In most cases, this means from the soil. In some cases, a different strategy is used. For instance, parasitic plants have developed a way of attaching themselves to a host tree or shrub and sharing its supply of food and water.

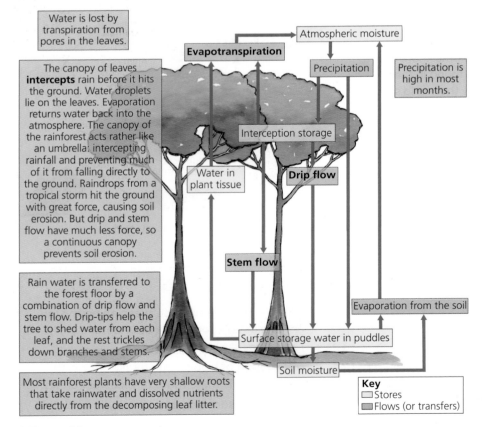

Water is lost by transpiration from pores in the leaves.

Evapotranspiration

Atmospheric moisture

Precipitation

Precipitation is high in most months.

Interception storage

Water in plant tissue

Drip flow

Stem flow

Evaporation from the soil

Surface storage water in puddles

Soil moisture

The canopy of leaves **intercepts** rain before it hits the ground. Water droplets lie on the leaves. Evaporation returns water back into the atmosphere. The canopy of the rainforest acts rather like an umbrella: intercepting rainfall and preventing much of it from falling directly to the ground. Raindrops from a tropical storm hit the ground with great force, causing soil erosion. But drip and stem flow have much less force, so a continuous canopy prevents soil erosion.

Rain water is transferred to the forest floor by a combination of drip flow and stem flow. Drip-tips help the tree to shed water from each leaf, and the rest trickles down branches and stems.

Most rainforest plants have very shallow roots that take rainwater and dissolved nutrients directly from the decomposing leaf litter.

Key
☐ Stores
▨ Flows (or transfers)

▲ Figure 6.5 The water cycle

How have animals adapted to survive?

The climate, vegetation and food chains of the tropical rainforest are ideal for animal life of all sorts. There is plenty of food and water throughout the year.

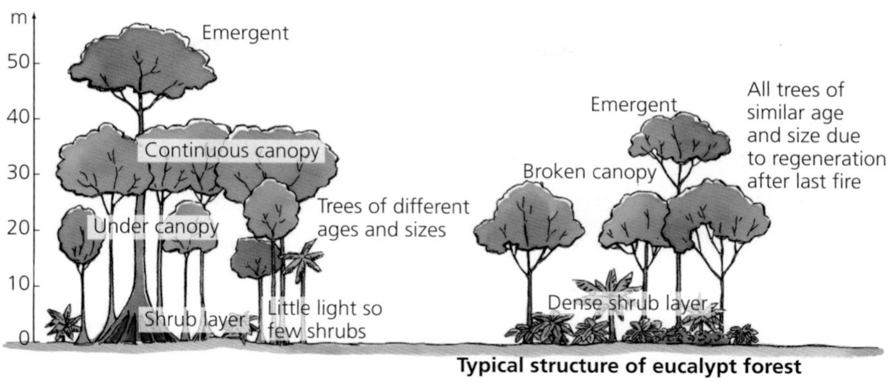

Competition for food

Because there are so many animals, there is a great deal of competition for food. Some animals are very specialised and live off a specific plant or animal that few others eat. For example, parrots and toucans have developed big strong beaks to crack open hard nuts (Figure 6.7).

There are relationships between animals and plants that benefit both. Some trees depend on animals to spread the seeds of their fruit. Birds, and mammals such as bats, eat the fruit and travel some distance before the seeds pass through their digestive systems in another part of the forest.

Other survival strategies

Many animals use camouflage to escape becoming prey, and predators use it to help them catch their prey. Some animals are poisonous and use bright colours to warn predators to leave them alone. There are several species of brightly-coloured, poisonous arrow frogs. Native Central and South American tribes used to wipe the ends of their arrows on the frogs' skin to transfer the deadly poison.

From all this, it can be seen that the plants and animals of the tropical rainforest are finely tuned to the environment and to each other. Everything works well, but only as long as it is not disturbed by people.

Typical structure of eucalypt forest

▲ Figure 6.6 The layers of the tropical rainforest

Emergent

Continuous canopy

Under canopy

Trees of different ages and sizes

Shrub layer

Little light so few shrubs

Emergent

Broken canopy

All trees of similar age and size due to regeneration after last fire

Dense shrub layer

▲ Figure 6.7 A keel-billed toucan with its colourful but strong nut-cracking bill

→ Activities

1 What are the main sources of food for plants?

2 Make a copy of Figure 6.6. Add the following labels to the diagram to match the layers (emergent, continuous canopy, under canopy, and shrub):

- This layer is dark and gloomy. There is little vegetation between the tree trunks.
- Sunlight is limited in this layer. Saplings and seedlings wait here for larger trees to die and leave gaps into which they can grow.
- This is the layer in which the upper parts of most trees are found. The leaf environment is home to insects, birds and mammals.
- The tops of the tallest trees are in this layer. Because of their height, these trees are able to get more light than the tree with their tops in the canopy.

3 Suggest why parasitic plants are found in the canopy layer, close to where sunlight first reaches the trees.

4 Explain how animals contribute to the nutrient and water cycles.

→ Going further

Research ways in which each of the following have adapted to the environment:

- three-toed sloths
- leaf-cutter ants
- boa constrictors
- toucans.

Deforestation of tropical rainforests

Deforestation (page 49) has a very long history. But it is only over the last 100 years or so that it has begun to have a serious impact on the tropical rainforest.

What is the scale of deforestation?

There are 62 countries with a tropical rainforest within their borders. Figure 6.8 shows the top 30 countries. Few, if any, early records were kept of the original extent of tropical rainforest. The UN Food and Agriculture Organization estimates that about half the world's tropical rainforest has now been cleared. The scale and accelerating rate of the deforestation are truly worrying.

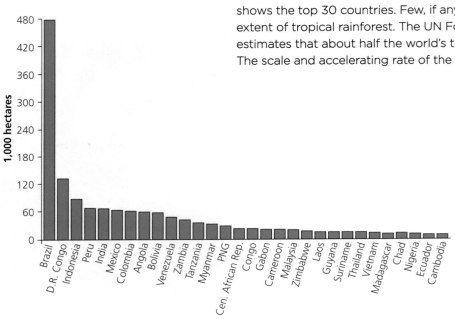

▲ Figure 6.8 The leading 'owners' of tropical rainforest

Is the rate of deforestation changing?

Figure 6.9 shows that during the first decade of the twenty-first century, rates of deforestation increased in all three continents: Asia (Indonesia, Malaysia and Thailand), Africa (Madagascar and Mali) and South America (Peru, Guatemala and Bolivia). In Indonesia and Peru the increases were particularly alarming. In Indonesia, the rate of deforestation between 2005 and 2010 was double the rate between 2000 and 2005. In Peru, the rate almost doubled.

Figure 6.9 also shows that the rate of deforestation decreased in seven countries. Most encouragingly, this happened in Brazil where the rate of deforestation has fallen to a record low. It is estimated that about half of Brazil's remaining rainforest now has some form of protected status. The bad news is that 20 per cent of the Amazon rainforest has been cleared since 1970: an area of 761,000 square kilometres (roughly three times the size of the UK).

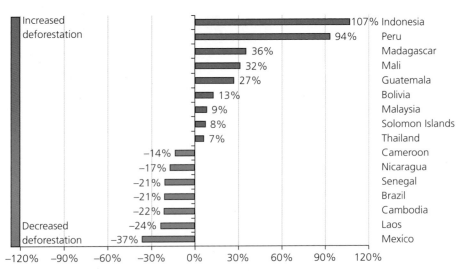

▲ Figure 6.9 Change in the annual deforestation rate (2000–2010)

The rate of reductions elsewhere may also reflect that other countries have already put in place measures to protect their rainforests. In some cases, for example in Mexico, strenuous efforts are being made to save what little is left before it disappears. But it is only the rate that has decreased. Deforestation continues in all the countries shown in Figure 6.8. Indeed, it is still happening in all 62 countries.

Today, the global rate at which the tropical rainforest is being cleared is estimated to be:

- 1 hectare per second
- 60 hectares per minute
- 86,000 hectares per day (an area larger than New York City)
- 31 million hectares per year (an area larger than Poland).

What has happened in Brazil?

The following is an introduction to two important aspects of deforestation: its causes (Section 6.4) and impacts (Section 6.5).

The Brazilian rainforest occupies the huge lowland basin drained by the Amazon and its tributaries. Figure 6.10 clearly shows how much of the tropical rainforest cover

was lost up to 2006. It is interesting that the clearance has been to the south of the Amazon. This is the part of the rainforest most accessible from Brazil's main cities, such as Rio de Janeiro, São Paulo and Brasilia.

An important point here is that for centuries the rainforest has been lived in and used by indigenous (native) tribes. They have:

- harvested fruits and nuts
- cut wood for fuel
- used timber to build their dwellings
- discovered cures for various illnesses (Section 6.6)
- cleared small areas, by a technique known as slash and burn (page 49), a type of **subsistence farming**.

Slash and burn has done little lasting damage to the forest. When the soil in one small area becomes exhausted, the tribe moves on and clears another. This is why it is sometimes referred to as 'shifting cultivation'. Once abandoned, the forest is able to regenerate.

It is important to understand that human use of the rainforest does not always lead to deforestation. In many cases, it leads to **forest degradation**, where the forest ecosystem is changed in a negative way and its supply of resources declines (Section 6.6).

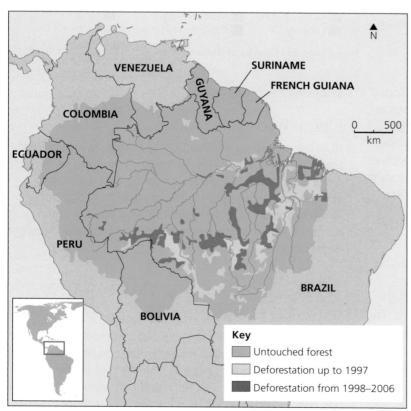

Key

- Untouched forest
- Deforestation up to 1997
- Deforestation from 1998–2006

▲ Figure 6.10 Deforestation in the Amazon basin

→ Activities

1 Identify the differences between deforestation and forest degradation.

2 What conclusions might be made about the tropical rainforest countries not shown in Figure 6.9?

3 Explain why we cannot be sure how much of the world's tropical rainforest has been cleared.

4 Explain why the decreasing rates of deforestation shown in Figure 6.9 are not completely good news.

Geographical skills

Taking data about forest cover from Figure 6.8 and about the changing rate of deforestation from Figure 6.9, plot the following countries on a graph: Bolivia, Cameroon, Indonesia, Laos, Peru and Thailand, to show the area of rainforest lost in each country.

⭐ KEY LEARNING

➤ How people exploit rainforest resources

➤ The activities that cause deforestation

The tropical rainforest in Brazil: causes of deforestation

Brazil is located in South America. It is the fifth largest country in the world and contains the largest area of tropical rainforest. The tropical rainforest of Brazil, as in other countries, is being exploited in two ways:

- by using its resources, such as timber, water and minerals
- by clearing the forest to make way for other activities, such as growing crops and rearing livestock.

What are the main resource-exploiting activities?

Logging

Figure 6.11 shows the main causes of deforestation in Brazil. You may be surprised that the figure for **logging** is so small. Logging is the first step in the conversion of forest land to other uses So it is the eventual use the cleared land is put to that is recorded in the pie chart. Timber companies are most interested in trees such as mahogany and teak, and sell them to other countries to make furniture (**selective logging**). Smaller trees are often used as wood for fuel or made into pulp or charcoal. Vast areas of rainforest are cleared in one go (clear felling).

Mineral extraction

It so happens that some of the minerals that developed countries need are found beneath stretches of tropical rainforest. In the Amazon, mining is mainly about gold. In 1999, there were 10,000 hectares of land being used for gold mining. Today, the area is over 50,000 hectares. The rainforest suffers badly as it is clear-felled. The same applies to the **extraction** of another mineral, bauxite, from which aluminium is made.

Energy development

An unlimited supply of water and ideal river conditions have encouraged dams to be built to generate hydroelectric power (HEP). This involves flooding vast areas of rainforest. Often the dams have a short life. The submerged forest gradually rots, making the water very acidic. This then corrodes the HEP turbines. The dams also become blocked with soil washed down deforested

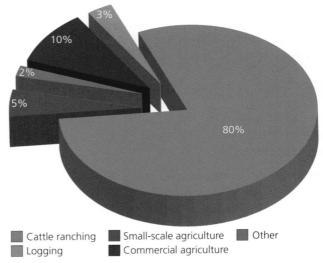

■ Cattle ranching   ■ Small-scale agriculture   ■ Other
■ Logging   ■ Commercial agriculture

▲ Figure 6.11 Causes of deforestation in the Brazilian rainforest

slopes by the heavy rain.

Illegal trade in wildlife

Hunting, poaching and trafficking in wildlife and animal parts are still big business in Brazil. Although this is not a direct cause of deforestation, it is endangering species such as the jaguar, the golden-bellied capuchin and the golden lion tamarind. It is also upsetting the natural balance of the rainforest ecosystem and therefore degrading it.

What activities are causing the forest to be cleared?

Section 6.3 mentions the traditional method of subsistence farming and its temporary clearance of small patches of rainforest (page 71). This is one of the main causes of deforestation – although subsistence farming is not nearly as damaging as **commercial farming**.

Commercial farming: cattle

Large areas of the Amazon rainforest have been cleared to make way for livestock rearing (Figure 6.12). The rearing of cattle is believed to account for 80 per cent of the tropical rainforest destruction in Brazil (Figure 6.11). However, the land cannot be used for long. The quality of the pasture quickly declines (Section 6.5). The cattle farmers then have to move on and destroy more rainforest to create new cattle pastures.

Commercial farming: crops

The forest is being cleared to make way for vast plantations, where crops such as bananas, palm oil, pineapple, sugar cane, tea and coffee are grown. The cultivation of soybean has also caused much forest clearance in Amazonia. The amount of rainforest cleared for this crop doubled between 1990 and 2010 (Figure 6.13). As with cattle ranching, the soil will not sustain crops for long. After a few years, the farmers have to cut down more rainforest for new plantations. Growing sugar cane for biofuel is beginning to become a major crop.

Road building

Roads are needed to bring in equipment and transport products to markets, but road building means cutting great swathes through the rainforest. Additionally, a road built for one particular commercial activity makes the forest accessible to other exploiters of the tropical rainforest resources. The Trans-Amazonian Highway began construction in 1972 and is 4,000 kilometres long. Although only a small part of it is paved, it has played an important part in opening up remote areas of the Amazon rainforest.

Settlement and population growth

All the above activities have a common knock-on effect. They need workers, and workers and their families need homes and services. That, in turn, means clearing the forest to build settlements where these people can live.

The huge challenges presented by these causes of deforestation are their scale, speed and wasteful use of the forest's land and resources.

▲ Figure 6.12 Cattle grazing on deforested land

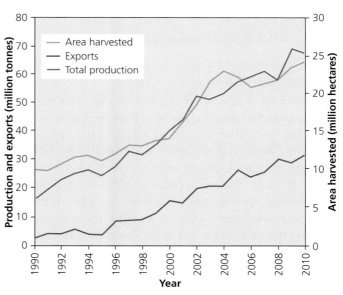
▲ Figure 6.13 Soybean production in Brazil (1990–2010)

Geographical skills

Use Figure 6.13 to calculate the percentage increase between 1990 and 2010 in a) soybean production and b) soybean exports.

→ Activities

1 Study Figure 6.11.

  (a) Describe the importance of different causes of deforestation.

  (b) Explain why logging has such a low value in Brazil.

2 Explain why the construction of roads is a serious threat to the tropical rainforest.

3 Give examples of the wasteful use of the rainforest and its land.

Case study

⭐ KEY LEARNING

➤ The global impacts of deforestation

➤ The local impacts of deforestation

The tropical rainforest in Brazil: the impacts of deforestation

What are the global impacts of deforestation?

Figure 6.14 shows the main consequences or impacts of deforestation. Two are of global significance, while the others are essentially local impacts.

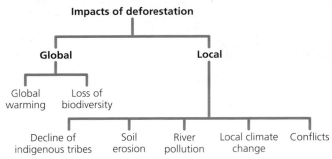

▲ Figure 6.14 Some impacts of deforestation

Global warming

The rainforest is significant at a global level. The tree canopy absorbs carbon dioxide in the atmosphere. This stops, of course, as soon as the trees are felled and more carbon dioxide remains in the air. Also, fire is often used in clearing rainforests, which means that the carbon stored in the wood returns to the atmosphere. In these ways, deforestation is a main contributor to the greenhouse effect, which is a cause of the global warming that is a threat to the survival of the human race – not just the people of the tropical rainforest.

Loss of biodiversity

Clearing tropical rainforests means that:

- the biodiversity will be reduced
- individual species will become endangered and then possibly extinct.

It has been estimated that 137 plant, animal and insect species are being lost every single day due to rainforest deforestation. That amounts to 50,000 species a year. As the rainforest species disappear, so do many possible cures for life-threatening diseases (Section 6.6). New research has shown that parts of the Amazon rainforest could lose between 30 and 45 per cent of their main species by 2030.

Local impacts of climate change

Deforestation disrupts the water cycle (Section 6.2). With the felling of trees, evapotranspiration is reduced, and so too the return of moisture to the atmosphere. The local climate becomes drier. The recycling of water is like a cooling system. Once the recycling is reduced, the local climate becomes warmer. The combination of increasing dryness and rising temperatures is not good for people or activities, such as agriculture, that follow the forest clearance.

▼ Figure 6.15 The frightening scale and devastation of deforestation

What are the local impacts of deforestation?

Soil erosion and fertility

As soon as any part of the forest cover is cleared, the thin topsoil is quickly removed by heavy rainfall. Bare slopes are particularly prone to **soil erosion**. Once the topsoil has been removed, there is little hope of anything growing again. Soil erosion also leads to the silting up of river courses (Chapter 11).

Even where the soil is protected, the soil quickly loses what little fertility it had when covered by trees. Grazing and plantations do little or nothing to keep the soil fertile. The decline in soil fertility leads to pastures and plantations being abandoned, so more areas of rainforest are cleared.

River pollution

Gold mining not only causes deforestation, but the mercury used to separate the gold from the ground is allowed to enter the rivers. Fish are poisoned, as well as people living in nearby towns. Rivers are also being polluted by soil erosion.

Decline of indigenous tribes

Not all Brazilians are benefitting from this exploitation of the rainforest resources. Most obviously, indigenous tribes have a traditional way of life that is closely geared to the resources of the natural forest. There are now only around 240 tribes left (Figure 6.16), compared with over 330 in 1900. Many indigenous people have been forced out of the rainforest by:

- the construction of roads
- logging
- the creation of ranches, plantations and reservoirs
- the opening of mines.

▲ Figure 6.16 Men of the Yawalapiti tribe fishing

Most displaced people have ended up in towns and cities. Few have adjusted to this very different environment. Addiction to drugs and alcohol has been common, and sadly many have died young. With the loss of these tribes have gone centuries of detailed knowledge of the forest, such as the medicinal value of various rainforest species (Section 6.6).

Despite all this, Brazil's tropical forest is still home to an estimated 1 million indigenous people. They still make their living through subsistence farming or hunting and gathering, or through low-impact harvesting of forest products like rubber and nuts. Today, they are less easily persuaded to move out of their traditional lands. They now know that they have the right to remain in the forest.

Conflicts

Disputes between indigenous people and loggers and other developers of the rainforest often end in open conflict. Disputes arise because people have conflicting views about the rainforest, for example, between conservationists and developers (Section 6.7).

All these causes of deforestation also apply elsewhere in the world. The mix of causes varies from country to country, but economic development and population growth seem to be the main drivers.

→ Activities

1 Describe what happens to the forest land in Brazil after it has been cleared.

2 Which of the five local physical impacts (Figure 6.14) do you think is the most serious? Give your reasons.

3 List more impacts to Figure 6.14. (Remember, impacts can be positive as well as negative.)

4 Assess which economic activity most threatens the indigenous people of the Brazilian rainforest.

→ Going further

Imagine you are either: a rancher, a plantation owner, a minister in the Brazilian government or a conservationist. Write a short statement (a) supporting your view of the rainforest and (b) identifying how your views might conflict with the others.

KEY LEARNING

- The goods the tropical rainforest supplies
- The services the tropical rainforest supplies

The value of the tropical rainforest to people and the environment

Section 6.4 showed that the tropical rainforest is a valuable provider of resources and opportunities. These fall into two different groups:

- those provided by the rainforest in its natural state
- those provided by the land once it is cleared of its forest cover.

The vast commercial value of the second group, particularly the crops and livestock, is the driving force behind much of the current deforestation. But what is the value of the tropical rainforest itself?

The resources and opportunities offered by the tropical rainforest or any other biome or ecosystem are more widely known as goods and services. In this instance, goods are things that can be obtained directly from the rainforest. Services are benefits that the rainforest can offer to both people and the environment (Figure 6.17).

Figure 6.17 Tropical rainforests: goods and services

Goods	Services
• Native food crops (fruit and nuts) • Wild meat and fish • Building materials (timber) • Energy from HEP • Water • Medicines	• Air purification (absorbing CO_2) • Water and nutrient recycling • Protection against soil erosion • Wildlife habitats • Biodiversity • Employment opportunities

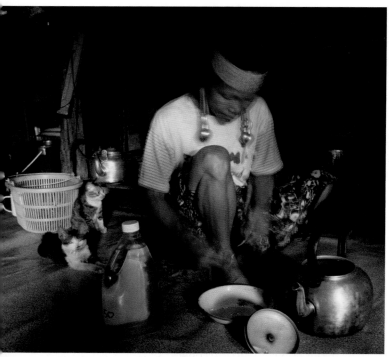

Figure 6.18 Indigenous peoples harvest medicinal plants

What goods are supplied by the tropical rainforest?

The plants of the tropical rainforest include many of the things we eat, such as cocoa, sugar and bananas. Cinnamon, vanilla and many other spices also come from the rainforest. Useful products like rubber, rope and baskets are made from rainforest plants. Some of the chemicals from rainforest leaves, flowers and seeds are used to make perfumes, soaps, polishes and chewing gum. Traditional subsistence farming is still very much about the harvesting of rainforest goods. The use of these forest products has been going on for centuries.

Finding new medicines

Today, however, we are beginning to realise that the forest has something more to offer. It is the stock of plants that **pharmaceutical companies** are finding to contain ingredients to help treat and cure diseases. Indigenous rainforest tribes have a very long tradition of using plant parts (barks, resins, roots and leaves) of various plants for this purpose (Figure 6.18).

Currently, over 120 prescription drugs sold worldwide come from plant sources. About a quarter of the drugs used today in the developed world are derived from rainforest ingredients. Less than one per cent of the tropical rainforest trees and plants have been tested by scientists to find out whether they have any medicinal value. Twenty-five per cent of the active ingredients in today's cancer-fighting drugs come from organisms found only in the rainforest (Figure 6.19).

In 1980, there were no pharmaceutical companies researching possible new drugs and cures from plants. Today, there are well over a hundred. It is in the interests of global health care to protect the tropical rainforest and its stock of medicinal plants. It is vital that these plants are not over exploited. Either the wild plants are harvested in a sustainable way or they should be deliberately cultivated as crops, perhaps on deforested land.

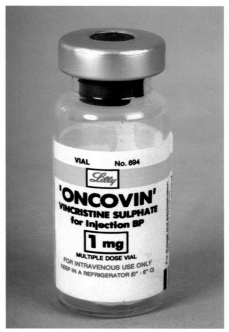

▲ Figure 6.19 Vincristine: an anti-cancer drug derived from a rainforest plant, periwinkle

What services are supplied by the tropical rainforest?

Some of the services listed in Figure 6.17, such as water and nutrient recycling (Section 6.1) and protection against soil erosion (Section 6.5), are services that benefit the environment and help to maintain the general health of the rainforest. On the other hand, the forest's biodiversity and wildlife habitats are benefits that people can enjoy, either as native settlers or as tourists. However, it is the rainforest's air purification service that is perhaps of most value to both people and the environment – not just within the tropical rainforest, but globally.

Perhaps the single most important global issue today is global warming and climate change. Global warming will only be checked by:

- greatly reducing the burning of fossil fuels and so lowering emissions of carbon dioxide
- greatly reducing the rate of deforestation to make sure that as much of the Earth as possible is covered by trees to absorb the carbon dioxide in the atmosphere.

As one of the largest carbon sinks in the world, the tropical rainforest has a critical role to play. Protecting the remaining rainforest requires doing two things:

- making sure that much of it is left untouched, so that it stays in a pristine state, for example, making large areas of rainforest into nature reserves or national parks
- allowing the resources of the rainforest, its goods and services, to be used, but only in a controlled and sustainable way (Section 6.7).

→ Activities

1 Add another example to each of the columns in Figure 6.17.

2 What is the difference between the goods and the services of an ecosystem?

3 Suggest what might be done to save the knowledge of the rainforest that indigenous tribes have.

4 List the ways in which tropical rainforests are good for human health.

5 Explain why the tropical rainforest is needed to fight global warming.

Strategies for managing tropical rainforests sustainably

Why does the tropical rainforest need to be managed sustainably?

If the goods and services of the tropical rainforest are not protected, then they will soon become lost forever. Sustainable management means using goods and services in such a way that they are still available for the benefit of people in the future. If that does not happen, then the forest's stock of renewable resources will gradually become exhausted. Further large-scale deforestation has no place in any sustainable management of the rainforest.

What actions will bring about sustainable management of the tropical rainforest? Most actions are taken at three levels: international, national and local. Some topics, such as conservation and education, crop up in all three action levels. Other actions tend to occur at only one of these levels.

What can be done at an international level?

Three different types of action illustrate what is being done at this level.

Inter-government agreements on hardwoods and endangered species

The first of these involves agreements between governments aimed at protecting the biodiversity and resources of the rainforest. They include:

- the International Tropical Timber Agreement (2006), which restricts the trade in hardwoods taken from the tropical rainforest. The very high prices paid for tropical hardwoods have encouraged a huge amount of illegal felling. This tends to take place in remote areas of the rainforest and so often goes on unnoticed by forestry officials. The 2006 Agreement restricts the trade in hardwood timber to timber that has been felled in sustainably managed forest (Section 6.8). All such timber has to be marked with a registration number (Figure 6.20).
- the CITES (Convention on International Trade in Endangered Species, 1973) treaty blocks the illegal trade in rare and endangered animals and plants. The illegal trade is still worth millions of pounds.

Debt reduction by HICs

Most of the countries with tropical rainforest are newly emerging economies or low-income countries. They may also have large debts, often resulting from overseas aid in the form of loans. Schemes known as debt-for-nature swaps are sometimes arranged. In 2010, for example, the USA signed an agreement to convert a Brazilian debt of £13.5 million into a fund to protect large areas of tropical rainforest. These swaps are all part of what is known more widely as **debt reduction**, where some high-income countries (HICs) agree to write off the debts of some poor low-income countries (LICs).

▲ Figure 6.20 Legally registered hardwood timber awaiting collection in Cameroon

Conservation and education by NGOs

The third type of international action is by **non-governmental organisations (NGOs)** such as the WWF, Fauna & Flora International, Birdlife International and the World Land Trust. All of them are charities that rely on volunteers and donations. They are not just interested in the tropical rainforests, but operate anywhere in the world where they think ecosystems are being seriously threatened. Such organisations:

■ promote the conservation message largely through education programmes in schools and colleges
■ provide training for conservation workers (another aspect of education)
■ provide practical help to make programmes more sustainable (see Section 6.8)
■ buy up threatened areas and create nature reserves.

It is vital that those involved in the exploitation and management of the rainforest should be made to understand the consequences of their activities. Important initiatives about conservation can also be undertaken at both national (see below) and local levels (Section 6.8).

What should national governments do?

In terms of conservation and education, achieving a sustainable balance between protection and development in the tropical rainforest is the prime responsibility of government. All governments have the powers to pass laws to achieve this, as for example by:

■ creating protected areas or reserves
■ stopping the abuse of the rainforest and other biomes by developers
■ making subjects, such as ecology or environmental studies, a compulsory part of the school curriculum.

There are, however, some problems. For example:

■ Few governments are willing to do anything that might slow down the rate of economic development. Citizens expect or want better living standards rather than new nature reserves.
■ Governments seem unwilling to enforce and monitor laws aimed at protecting or conserving the rainforest.
■ There is a lot of corruption in the way rainforests are treated, for instance by illegal loggers and developers paying bribes.

▲ Figure 6.21 Training conservation officers and scientists in Cambodia for Fauna & Flora International

Conservation means that natural resources such as timber can still be used, but must be used sustainably.

Protection means that the environment should be untouched and humans should not interfere, so ecosystems can find their own balance.

→ Activities

1 Describe the aims of sustainable management.
2 Design an educational leaflet aimed at people from HICs to explain why the conservation of the tropical rainforest is important.
3 Suggest how corruption threatens the tropical rainforest.
4 Explain the conflict between economic development and conservation.

→ Going further

Research one of the following NGOs and what it is doing to help the tropical rainforest:

■ Birdlife International
■ Fauna & Flora International
■ WWF (World Wildlife Fund)
■ World Land Trust.

⭐ KEY LEARNING
➤ Strategies for sustainable management

Strategies for managing the tropical rainforests sustainably

Sustainable actions start at a local level. **Sustainability** emphasises the importance of local actions, such as:

- respecting the environment and cultures of local people
- using traditional skills and knowledge
- giving people control over their land and lives
- generating income for local people
- using appropriate technology – machines and equipment that are cheap, easy to maintain and do not harm the environment.

What can be done at the local level?

In answering this question, aim to distinguish between two different situations:

- areas with logging
- areas still untouched by logging.

For the first, there are four possible actions.

Selective logging: this involves felling trees only when they are fully grown, and letting younger trees mature and continue protecting the ground from erosion. It involves a cycle lasting between 30 and 40 years.

Stopping illegal logging: given the remoteness of rainforest areas, illegal logging can easily go on unnoticed. It is still happening on a large scale. However, satellites and drones are now helping to monitor this.

Agroforestry: this involves combining crops and trees, by allowing crops to be grown in carefully controlled, cleared areas within the rainforest, and by growing rainforest trees on plantations outside the rainforest.

Replanting: a project in the Atlantic rainforest of Brazil (REGUA) has shown it is possible to recreate a forest cover almost like the original. This is done by collecting seeds from remaining patches of primary forest, growing the seeds into saplings in nurseries and then planting the saplings back in the deforested areas (Figure 6.22).

It is amazing how quickly a new forest cover develops with almost the same gene bank as the original cover. No doubt what has been learnt in this small project will help Brazil fulfil its promise to reforest 12 million hectares, agreed in a new US–Brazil climate partnership in June 2015.

For the areas untouched by logging, **ecotourism** presents a type of sustainable action. Scenery, wildlife, remoteness and culture are the main attractions (Figure 6.24). It aims to educate visitors and increase their understanding and appreciation of nature and local cultures. It is small-scale and local (controlled by local people, employing local people and using local produce). Its profits stay in the local community. It tries to minimise the consumption of non-renewable resources and the ecological impact. In this eco format, tourism becomes both a sustainable and a profitable activity.

➤ Figure 6.22 The REGUA nursery produces 70,000 saplings of 180 species a year

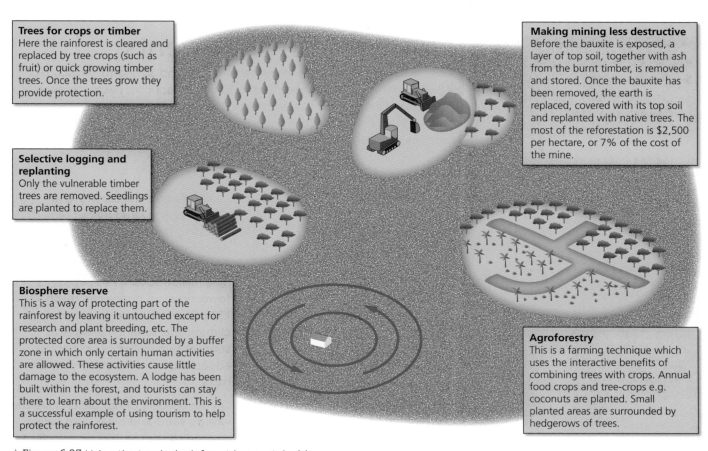

Trees for crops or timber
Here the rainforest is cleared and replaced by tree crops (such as fruit) or quick growing timber trees. Once the trees grow they provide protection.

Making mining less destructive
Before the bauxite is exposed, a layer of top soil, together with ash from the burnt timber, is removed and stored. Once the bauxite has been removed, the earth is replaced, covered with its top soil and replanted with native trees. The most of the reforestation is $2,500 per hectare, or 7% of the cost of the mine.

Selective logging and replanting
Only the vulnerable timber trees are removed. Seedlings are planted to replace them.

Biosphere reserve
This is a way of protecting part of the rainforest by leaving it untouched except for research and plant breeding, etc. The protected core area is surrounded by a buffer zone in which only certain human activities are allowed. These activities cause little damage to the ecosystem. A lodge has been built within the forest, and tourists can stay there to learn about the environment. This is a successful example of using tourism to help protect the rainforest.

Agroforestry
This is a farming technique which uses the interactive benefits of combining trees with crops. Annual food crops and tree-crops e.g. coconuts are planted. Small planted areas are surrounded by hedgerows of trees.

▲ Figure 6.23 Using the tropical rainforest in a sustainable way

For both types of area, local communities involved in projects such as replanting and ecotourism will gain a better understanding of the tropical rainforest. They will also help spread the message of sustainability to neighbouring communities. Community involvement represents yet another action falling under the broad heading of conservation and education.

Figure 6.23 shows different ways of using the tropical rainforest sustainably.

▲ Figure 6.24 An eco-lodge in Costa Rica

→ **Activities**

1 Write a campaign slogan to persuade local people to look after tropical rainforests.

2 Explain why it might be difficult to persuade locals that selective logging is a good idea.

3 Identify the resources needed for replanting in rainforests.

4 Study Figure 6.23.

a) Identify ways in which different activities help the natural environment.

b) Make a list of the goods being produced by this sustainable use of the rainforest.

5 Look at Figure 6.24. What makes this eco-lodge sustainable? Draw an annotated sketch.

7 Hot deserts

Hot desert environments

These are found in subtropical areas between 20° and 30° north and south of the Equator. The Tropic of Cancer or the Tropic of Capricorn passes through most of the world's hot desert regions. In Chapter 3 we learned that this climate is characterised by hot and dry sinking air. The extremely arid conditions occur where less than 250 millimetres of rain falls annually (Figure 7.1).

What are the physical characteristics of a desert climate?

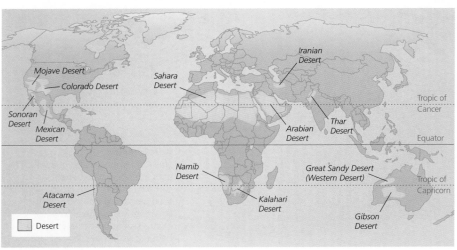

▲ Figure 7.1 The global distribution of the world's hot desert environments (semi-arid areas or drylands not shown)

Large areas of the Earth's land surface are covered by **hot deserts**, including the Australian, Thar, Arabian and Kalahari deserts. Largest of all is the Sahara, which measures nine million square kilometres. Its towering **sand dunes** can reach 150 metres. Despite its extreme climate, the Sahara manages to support two million people at its edges and in towns along caravan trails.

On the borders of hot deserts are the world's **semi-arid** areas, also called drylands or desert fringe areas (see Section 7.6). The Sahel is a long strip of semi-arid drylands that borders the south of the Sahara. It includes parts of Sudan, Chad, Burkina Faso and Niger. Water can sometimes be obtained in deserts from aquifers (see page 86), or rivers, like the Nile, which transport water from wet regions across much drier ones.

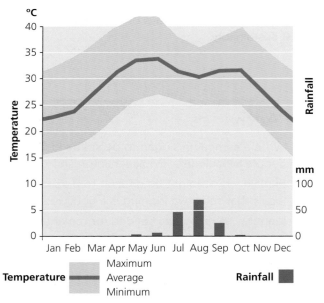

▲ Figure 7.2 The climate graph and monthly temperature data for Khartoum, Sudan

Hot desert climates

Not only is annual precipitation low in hot desert environments, it is extremely unreliable too. Parts of Chile's Atacama desert have not seen rain for 400 years; others receive just one millimetre per year! More typically, hot desert regions experience around 100–200 millimetres of rainfall in most years. Figure 7.2 shows the annual climate graph for Khartoum in Sudan. Between June and October, Khartoum records on average six days with ten millimetres or more and 19 days with one millimetre or more of rainfall. Like other hot desert areas, rainfall can be unpredictable.

Figure 7.2 also shows the extreme high temperatures Khartoum experiences. An even higher temperature of 57.8 °C was once recorded in the northern Sahara at El Azizia in Libya.

Another characteristic is the extreme range of temperatures often experienced in a single day. The cloudless skies that allow high levels of insolation in the daytime also permit rapid heat loss at night. This can bring a drastic fall in temperature, and occasionally sub-zero temperatures. Figure 7.3 shows that the **diurnal temperature range** for a desert may exceed 35 °C.

Hot desert soils

In hot climates, soil-forming processes are limited by the shortage of water and vegetation. Over time, weathering creates deep deposits of sand and loose material. There may be little organic content due to the lack of vegetation growing there. These sandy, rocky soils are typically around one metre deep, although in some places, wind action builds tall dunes where deeper soils can potentially develop. Sand dunes should not be classified as soils if there is no organic matter present there at all.

Some desert soils are potentially very fertile because important nutrients for plant growth, such as calcium, have not been leached away over time. Once irrigated, the land can become highly productive for agriculture. Large desert regions of oil-rich Saudi Arabia and the southern states of the USA have benefited greatly from irrigation.

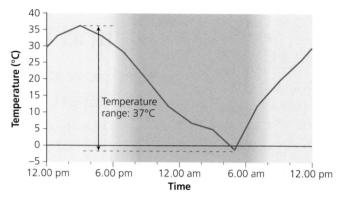

▲ Figure 7.3 Diurnal temperature range in a hot desert

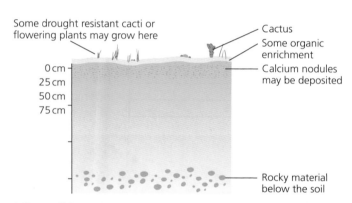

▲ Figure 7.4 A desert soil profile

→ Activities

1   This chapter reminds us that the climate of hot deserts 'is characterised by hot and dry sinking air'. Why is this the case? (Refer back to Chapter 3, pages 22–3 for the explanation.)

2   Look at Figure 7.1.

   a) Describe the global distribution of hot desert environments.

   b) Can you suggest local physical factors that might help to explain why hot desert is not spread uniformly around the world between 20° and 30° north and south of the Equator?

3   What do you think is meant by 'maximum, average and minimum' temperatures in Figure 7.2?

4   Look at Figure 7.4. Describe the main characteristics of soils in hot desert environments.

Geographical skills

Study the climate graph for Khartoum (Figure 7.2).

1   What is the total annual rainfall? This is calculated by adding all of the values for the rainfall bars together.

2   What is the annual pattern of rainfall? Are there wet or dry seasons? If so, when are they?

3   What is the annual temperature range? This is the difference in average temperature between the hottest and coldest times of the year.

4   Does the temperature show a distinctive seasonal pattern? If so, at what time of year are the hot and cooler seasons?

Hot desert ecosystems and biodiversity issues

How have plants adapted to life in hot desert environments?

Desert biodiversity is far lower than in other global ecosystems you have studied. In Chapter 6 we learned that tropical rainforests have high levels of biodiversity (page 65). In contrast, far fewer species are supported by the extreme climate of hot deserts (Figure 7.5). Environment challenges include:

- dry conditions – plants that can survive in very dry conditions are called xerophytes. They use a range of adaptations, including thick, waxy cuticles and the shedding of leaves to reduce **transpiration**, to minimise water loss
- high temperatures – some plants have the bulk of their biomass below the ground surface where temperatures are cooler
- short periods of rainfall – deserts bloom suddenly after rainfall so to complete their life cycle quickly. This is an important issue to consider when discussing biodiversity.

The relatively small numbers of plants that survive are adapted in a range of ingenious ways (Figure 7.7).

▼ Figure 7.5 Hot desert biodiversity (Grand Canyon National Park)

Group	Number of species
Mammals	56
Birds	400
Reptiles	36
Plant species	1,700
Mosses	60
Lichens	195

⌃ Figure 7.6 The prickly pear is a type of cactus that grows in the Sonoran Desert

▼ Figure 7.7 Plant adaptations that aid survival in hot desert environments

Drought-tolerant trees	Acacia trees have developed short, fat trunks that act as reservoirs for excess water. They are also fire-resistant. Their roots can penetrate 50 metres into the ground and reach out sideways to find as much water as possible.
Cacti	Cacti are 'succulents'; they store water in their tissues (Figure 7.6). The USA's saguaro cactus can grow up to fifteen metres. The cacti's spikes deter consumer species, and their small, waxy leaves minimise transpiration losses.
Flowering plants	Desert flowers have seeds that only germinate after heavy rain and can lie dormant for years in between rains. They are 'ephemerals': they can complete their lifecycles in less than a month. Plants immediately produce brightly coloured flowers to attract insects.
Lichen	Lichen appears as a flaky crust on the ground, rocks and tree trunks. Lichens do not need soil to grow and for this reason are called **pioneer species**. They can grow on a bare rock surface in high temperatures. Lichen survive by breaking down the rock chemically using their own organic acids. This helps them to extract nutrients they need.

How have animals adapted to life in hot desert environments?

In the hottest desert regions, few animals have adapted to survive besides tough scorpions and small reptiles. In areas with a greater supply of water, the level of biodiversity rises as tough grasses, shrubs, cacti and hardy trees begin to form the basis of a larger food web that will include mammals (foxes, coyotes) and raptors (buzzards, hawks). Because deserts are found in most continents, different species have evolved to fill the niches in African, Asian and American deserts.

A simplified food web for the Mojave Desert in the USA is shown in Figure 7.8. This is the hottest and driest hot desert region in North America. The coyote and hawk are this environment's top carnivores (Figure 7.9). In desert regions around the world, consumer species have adapted to their environment:

- Kangaroo rats do not need to drink water; they get it from food. They live in burrows during the day to avoid extreme heat. They do not perspire and have highly efficient kidneys that produce very little urine.
- Desert foxes have thick fur on the soles of their feet, protecting them from the hot ground. The light-coloured fur on their bodies reflects sunlight and keeps them cool.

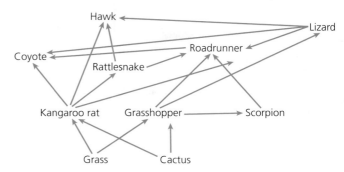

▲ Figure 7.8 A simplified food web for part of the Western Desert

▲ Figure 7.9 Coyotes in the Western Desert

What interdependence exists in hot desert environments?

The biotic and abiotic components of an ecosystem are interdependent. Living or 'biotic' creatures play an important role in maintaining a healthy environment (the 'abiotic' parts) and vice versa. Abiotic components of hot deserts include the soil, underlying rocks and water supplies.

The interdependency between different biotic and abiotic parts of hot deserts is shown by:

- links between different parts of the food web (animals eating plants that have gained nutrients from soils and water, for example)
- the role that vegetation roots play in stabilising sandy soils in semi-arid areas at the edges of deserts. The plants stop the soil from being blown away by the wind. From page 92, you will learn how the destruction of desert vegetation sometimes triggers a process called **desertification**.
- Increasingly unsustainable human use of deserts threatens the interdependence between the physical environment and people (pages 92–3).
- The fragility of hot deserts is an important issue affecting biodiversity.

> → **Activities**
>
> 1 How have hot desert animals adapted to survive in an extreme environment? In your answer, make reference both to their appearance and to their behaviour.
>
> 2 Study Figure 7.8. Using examples from the hot desert food web, explain what the difference is between primary consumers and secondary consumers.
>
> 3 Study Figure 7.5.
>   a) Calculate the total biodiversity of the hot desert environment.
>   b) Estimate the percentage of all species that are mammals.
>   c) Can you explain why only a relatively small number of mammal species are supported by the hot desert environment? You can refer back to Chapter 5 pages 58–59 to support your answer.
>
> 4 Why might a study of desert biodiversity made immediately after rainfall come to a different conclusion to a study made during a long dry period?

⊘ KEY LEARNING

➤ Migration to the USA's Western Desert
➤ Economic development opportunities of the Western Desert

Development opportunities in the Western Desert

The USA's Western Desert region is actually made up of three different hot deserts. These are the Mojave Desert, part of the Sonoran Desert and part of the Chihuahuan Desert. In total, the Western Desert covers 200,000 square kilometres.

How has migration brought different groups of people to the USA's Western Desert?

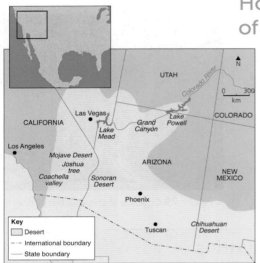

▲ Figure 7.10 The location of the Western Desert, USA

The Western Desert's indigenous people are made up of many different cultural groups. Arizona is home to the Navajo people, and the Havasupai people live by the Grand Canyon, which is a spiritually significant place for some indigenous peoples. The Sonoran Desert's Cocopah Tribe, also known as the River People, settled by the Colorado River centuries ago. People of European descent first began to migrate into this region in the 1800s.

The Western Desert includes parts of several states: California, Nevada, Utah, Arizona and New Mexico. Population is low and nearly half live in the large cities of Phoenix and Tucson (both in Arizona), and Las Vegas and Henderson (both in Nevada). The majority of people in these cities are of European descent.

What are the Western Desert's development opportunities?

The Las Vegas region is home to 2 million people, and Phoenix has 4.5 million residents. Urban residents can always find work in retailing and service industries. In the less populated, more inhospitable areas of the Western Desert, people earn their living from farming, mineral extraction, energy and tourism industries.

Farming

High temperatures and sunlight are generally favourable for agriculture, provided water can be found for **irrigation**. Two important sources of irrigation water are:

■ aquifers: large stores of water lie beneath some hot desert regions. Sometimes, a layer of permeable rock lies on top of impermeable rock. Rainwater and groundwater seep into the permeable layer and become trapped. This water can be brought to the surface by digging a well. Despite being part of the Sonoran desert, aquifer-based farming in California's Coachella Valley produces lush crops of vegetables, lemons, peppers and grapes (and in turn, a wine industry).

■ canals: most canals are used for large-scale industrialised agriculture. Farmers are allocated 80 per cent of Colorado water, even though they make up just 10 per cent of the economy.

▲ Figure 7.11 Irrigated agriculture in the Coachella Valley

Mineral extraction

The Western Desert states are rich in minerals, including copper, uranium, lead, zinc and coal. Not all these opportunities have been exploited, due to possible conflicts with other land uses, like tourism and farming. A plan for uranium mining near the Grand Canyon, Arizona was recently halted due to a campaign by the Havasupai people. As uranium is used in nuclear power plants, they were concerned about the risk to wildlife and endangered species, and the contamination of water supplies.

Copper mining has taken place for centuries in the Sonoran Desert near Ajo, Arizona. The lack of water discouraged large-scale mining and settlement until underground water was found in an ancient lava flow north of Ajo. Today, opencast mining is carried out on a large scale (Figure 7.12).

▲ Figure 7.12 The Ajo copper mine, Arizona

Energy

The strong insolation in desert regions provides a fantastic opportunity for solar power. The entire Western Desert region is predicted to benefit from the construction of new solar power plants. The Sonoran Solar Project in Arizona is a new solar power plant project that will ultimately produce energy for 100,000 homes and requires 360 workers to help build it.

Hydroelectric power (HEP) plants also supply Western Desert communities with some of their electricity. These are powered by water leaving Lake Mead. At the peak of its construction in the mid-1930s, the Hoover Dam employed 5,000 people.

Fossil fuels bring opportunities to the Western Desert too. People have been drilling for oil in Arizona since 1905. Today, there are 25 active oil production sites, all of which are on land owned by the Navajo people. Since 1998, the Navajo Nation Oil and Gas Company (NNOGC) has exploited this economic opportunity for the benefit of local Navajo people. More than 100 employees work to produce oil worth US$50 million.

Tourism

As US society has grown to have more money and leisure time, tourism has become the Western Desert's most important source of income:

- The national parks offer visitors a chance to experience a **wilderness area**. Important areas include the Grand Canyon and California's Joshua Tree National Park (named after the dominant plant type).
- The heritage and culture of Native Americans are celebrated at the Colorado Museum in Parker, Arizona.
- The entire economy of Las Vegas is built around entertainment, attracting 37 million visitors per year.
- Two major lakes have been created as part of water management projects: Lake Mead and Lake Powell. Combined, they attract two million visitors a year and offer sailing, power boating, water-skiing and fishing (see Chapter 7.3).

→ Activities

1 Describe how the Western Desert is distributed across different parts of the USA.

2 Look at Figure 7.11. Explain how aquifers create important opportunities for people living in hot deserts.

3 Explain why mining opportunities in the Western Desert have not always been exploited.

4 Describe how energy opportunities have benefited different groups of people in the Western Desert.

5 Write a short advert about the Western Desert for a travel magazine. What different kinds of people and interest groups would your advert be targeted at?

6 'The environmental impacts of all human activities in hot deserts outweigh the economic benefits they bring.' Do you agree with this statement? Refer back to the opportunities you wrote about in Activities 2, 3 and 4.

⭐ KEY LEARNING

➤ The uneven development of the Western Desert

➤ Why accessibility is a challenge in the Western Desert

➤ How people have adapted to the climate

Development challenges in the Western Desert

What explains the uneven development of the Western Desert?

Adapting to the hot desert environment of the Western Desert is a challenge, for both traditional Native American communities and more recent settlers. Figure 7.13 shows population distribution in the Western Desert. There is no settled population in some areas, mainly due to very high temperatures. In the Mojave Desert's Death Valley, temperatures approach 50 °C in July. This is reaching the survival limit for plants. The absence of people in places like Death Valley reflects the low **carrying capacity** of the land (Figure 7.14).

The first Native Americans settled in areas where temperatures were at least tolerable, both for themselves and for their crops or animals. In all three desert areas (Mojave, Sonoran and Chihuahuan), indigenous communities formed near sources of water, either rivers or aquifers. Some groups developed an economy based on subsistence farming based on maize and corn, or alfalfa grown for livestock. Others hunted wild animals which thrive in areas with greater water supply and plant growth, such as rabbits.

Why is accessibility a challenge in the Western Desert?

The low population density of less than one person per square kilometre means that parts of the Western Desert lack surfaced roads. Accessibility is thus severely limited in areas of Nevada north of Las Vegas. Tourists and explorers must find their own way. Even where roads and rough tracks have been provided, the extreme temperatures make this a dangerous place if your car breaks down. In 2015, an elderly tourist died of dehydration in the Los Coyotes Reservation near the edge of the Mojave Desert. He had become lost after attempting to drive off-road. However, accessibility is less of a challenge than it used to be.

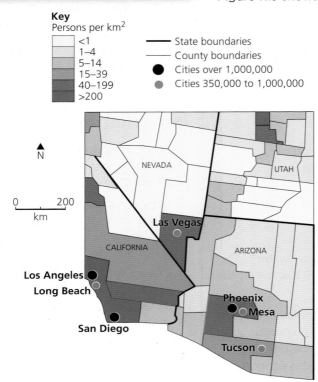

Key
Persons per km²

<1	
1–4	
5–14	
15–39	
40–199	
>200	

⬛ State boundaries
— County boundaries
⚫ Cities over 1,000,000
🔵 Cities 350,000 to 1,000,000

0   200
km

▲ Figure 7.13 Population distribution in the Western Desert

▲ Figure 7.14 Some parts of the Western Desert are uninhabitable, like Death Valley

- By the late 1800s, railroad developers moved in. Their choice of sites for stations influenced the growth of future key settlements. For instance, developers determined that the water-rich Las Vegas Valley would be a perfect location for a train station. Soon after, the first saloon bars, shops and hotels were built.
- Better roads were laid in the 1900s. Soon people were driving through the desert in buses or their own cars (Figure 7.15).
- Major cities can now be reached directly by air. Las Vegas airport receives over 40 million people annually.

How have people adapted to the climate?

Traditional Native American housing was adapted to the extreme climate of hot deserts. Cocopah people lived in earth houses, made with a wooden frame packed with clay and thatched with grass. The thick earth walls kept the house cool in the daytime heat, and warm in the cold of night: perfect for life in a hot desert!

Nineteenth- and early twentieth-century migrants quickly adapted to the climate too. Before the arrival of air conditioning and improved water supplies, their houses:

- had flat roofs to help collect rainwater
- were small, to reduce sunlight and keep temperatures low inside
- had whitewashed walls to reflect sunlight and keep buildings cool, a tradition which continues today (Figure 7.16).

Outdoors, they wore wide-brimmed cowboy hats to prevent sunburn.

Recent water shortage concerns have led people to change their behaviour accordingly. Some old sports pitches have been replaced with fake grass. More people have adopted drought-resistant 'desert landscaping' in their gardens.

▲ Figure 7.15 Route 70 cuts through the Western Desert in Utah. Since 1926, Route 66 has connected Chicago to California via the Western Desert.

▲ Figure 7.16 A whitewashed building in Taos, New Mexico

→ Activities

1   Look at Figure 7.13. Describe how population is distributed in the Western Desert. Refer to data in Figure 7.13 to support your answer.

2   a) State what is meant by the carrying capacity of land.

    b) Suggest reasons why the carrying capacity of the land may vary from place to place in the Western Desert.

3   Look at Figure 7.15. Write a brief article for a tourist website, advising motorists on the risks they face crossing the desert and steps they should take to look after themselves.

4   Look at Figure 7.16. Explain ways in which this urban landscape shows signs of adaptation to the hot desert environment where it is located.

⭐ KEY LEARNING

➤ The costs and benefits of irrigating the Western Desert
➤ Future water supply and population growth issues in the Western Desert

The Western Desert's water crisis

Until now, cities in the Western Desert have prospered thanks to massive water transfers (see page 348). Vast volumes of water have been transferred from the Colorado River, but there are limits to what can be achieved. Further population growth may not be possible.

What are the costs and benefits of using the Colorado River's water to irrigate the Western Desert?

Twentieth-century migrants could see plenty of opportunities in the Western Desert's sunny skies. Farming and tourism would flourish if they could tackle the issue of water shortages. The solution was close at hand: the Colorado River. This massive, 2,300 kilometre continental river brings meltwater from the Rockies and Wind River Mountains across the USA and down to Mexico.

On a small scale, for centuries, the Cocopah people had long been drawing Colorado water through canals to irrigate their fields. But while the snowmelt brings huge volumes of water in summer, the Colorado has a very low flow between September and April. In the most extreme years of the early 1900s, the Colorado's discharge was 13 times higher in mid-summer compared with winter.

In 1935, work began on the Hoover Dam, which stores the equivalent of two years' river flow in Lake Mead. The Glen Canyon Dam followed in 1963 (Figure 7.18). Together, the two dams and their reservoirs smooth out the Colorado's flow through the year and remove its flood peaks, and bring additional benefits.

Reservoir water is piped along aqueducts, including the US$4 billion Central Arizona Project (Figure 7.20). Where required, it feeds the homes, farms and golf courses of the Western Desert. This water transfer has brought many benefits, but there are also costs, particularly to the Colorado River's ecosystem.

▲ Figure 7.18 Lake Powell and the Glen Canyon Dam

▼ Figure 7.17 Economic benefits and ecological costs of taking water to the Western Desert

Benefits	Costs
• Colorado's giant reservoirs bring water to cities throughout the Western Desert area, including Phoenix, Tucson, Albuquerque, San Diego, Las Vegas (see Figure 7.19) and Los Angeles. • The Colorado's aqueducts bring life-giving water to farms growing fruits and vegetables in places like Coachella. In total, more than 1.4 million acres of irrigated land throughout the Colorado River Basin produce about 15 per cent of the USA's crops and 13 per cent of its livestock. The total agricultural benefits are calculated to be US $1.5 billion per year.	• Silts and sands get trapped behind both dams. This makes the water that leaves the dam colder (silt heats up in sunlight, warming water around it). As a result, the river ecosystem has changed and many species have been lost. • Sandbanks along the sides of the river in its lower course have been starved of sediment and are smaller. Plants and animals that live on the sandbanks – such as the Kanab snail and the willow flycatcher – have also declined. The sandbanks that were once used for fishing and rafting have disappeared.

▲ Figure 7.19 Las Vegas

What are the issues around future water supply and population growth in the Western Desert?

Already, 30 million people in the southwest USA depend on water from the Colorado. Phoenix takes the maximum share of water it is allowed, but is predicted to double its population by 2050 (Figure 7.21). Between 2000 and 2010, several states experienced rapid population growth, more than double the national average. These were Nevada (35 per cent), Arizona (25 per cent) and Utah (24 per cent).

Yet while the cities of the Western Desert continue to grow, there is a physical limit to how much water can be taken from the Colorado. There is a political limit too, because of an international agreement which states that water must also be allowed to flow into Mexico (where the Colorado ends).

The region's water security is further threatened by climate change. Scientists have suggested that reduced rainfall could occur in places where water is already naturally scarce for part or all of the year. In 2014, Lake Mead reached a record low level. The Western Desert region is projected to warm faster than the whole world in coming decades. By 2100, average annual temperatures in many areas could be five degrees higher than they were in the 1970s. The combination of a changing climate and the region's rapid population growth means that even greater water scarcity is expected in the future.

▲ Figure 7.20 Part of the Central Arizona Project

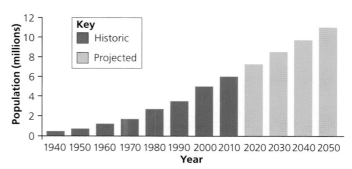

▲ Figure 7.21 Population growth in Arizona

➔ Activities

1 State three economic ways in which the Western Desert benefits from the Colorado River's water.

2 Describe three problems created by the management of the Colorado River.

3 a) State what is meant by 'water security'.

   b) Describe the population growth of Arizona shown in Figure 7.21.

   c) Suggest what the implications of Figure 7.21 are for the future water security of people living in the Western Desert region.

4 Make an assessment of whether or not new development such as golf courses should be permitted near cities like Las Vegas and Phoenix. Try to make arguments both for and against the case before arriving at a final judgement. Refer to ideas taken from both physical and human geography in your answer.

Desert fringes and desertification

What are the characteristics of desert fringe areas?

Desert fringe areas have many names, including semi-deserts, semi-arid areas and drylands. In some places, rain falls in a fairly predictable pattern, making settled agriculture possible.

At the borders of hot deserts, desert fringe areas support greater biodiversity and larger plants. Grasses grow in higher rainfall desert fringe areas, such as the American Prairies, and there are more drought-resistant trees, such as the baobab tree in Africa's Sahel region, and Australia's eucalyptus tree.

Despite their higher rainfall, desert fringes are classified, alongside hot deserts, as **fragile environments**. Catastrophic ecological and environmental consequences can be caused by climate change or poor land management. As a result, desert fringes are at constant risk of desertification (Figure 7.22).

What is the link between desertification and natural climate change?

Desertification is a major problem in many parts of the world, as Figure 7.23 shows. Around one billion people, or 15 per cent of the world's population, either experience or are threatened by habitable desert fringe areas turning into hot desert areas. This includes desert fringes in Australia, China, the USA and large parts of Africa.

The Sahel (see page 82) is where the human risks created by desertification are greatest. This is because 50 million poor and vulnerable people live there. There is very little money and technology available in Sahel countries to help people adapt to the challenge.

Climate data for the Sahel suggest that desertification may have a physical cause: a long-term reduction in rainfall has taken place (see Figure 7.24). The most likely explanation for this is a natural rainfall cycle. The African continent has a long history of rainfall fluctuations of varying lengths, including cycles lasting decades.

▼ Figure 7.22 A sandstorm approaches the desert fringe in Niger

The decline in rainfall may be occurring anyway, independent of human-induced global warming. We do not know whether global warming caused by humans will create even greater rainfall deficiencies in the Sahel or other desert fringes. While many scientists are certain that global temperatures will rise this century, they are less confident when it comes to predicting how rainfall patterns will change. Some think there is a possibility of rain returning to the Sahel as a result of global warming, as the heating of oceans adds more water vapour to the atmosphere. This could, in turn, bring more rainfall that leads to the 'greening' of the Sahel. Equally, it may be that the climate becomes even drier, leading to the spread of sand dune systems across valuable crop land.

The climatic system is complex and there remains much uncertainty over what will happen in future.

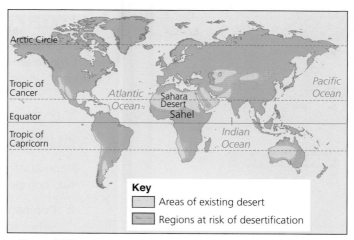

▲ Figure 7.23 World regions at risk of desertification

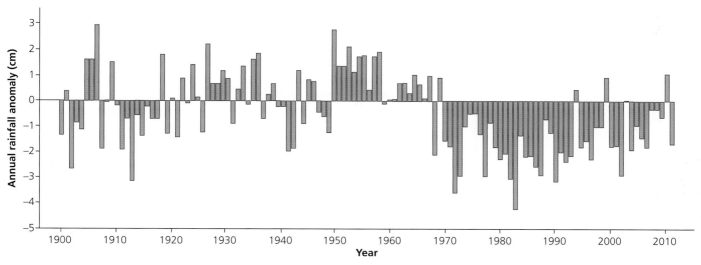

▲ Figure 7.24 Annual rainfall in Sahel countries, 1900–2011. Each bar shows how much the year's rainfall was above or below the long-term average

→ Activities

1 Draw a table to show differences between desert and semi-desert areas. You can include physical characteristics (climate, vegetation and water supply) and human characteristics (population and economic activities).

2 Look at Figure 7.23. Describe the global distribution of areas at risk of desertification. Your answer could refer to lines of latitude, names of continents and proximity (nearness) to desert areas.

3 Look at Figure 7.24. What does it suggest could be happening to the climate of the Sahel region?

4 Why might desertification pose a greater challenge for people in some affected areas than it may in others? Think about the numbers of people who could be affected and also which countries they live in.

➤ The role of population growth and human factors in desertification

➤ Desertification in Darfur

Human causes of desertification

In addition to natural pressures, desertification occurs when fragile land in desert fringe areas is overexploited by humans. Physical and human causes of desertification are often interlinked with one another (Figure 7.25).

What role do population growth and human factors play in desertification?

Population growth in desert fringe areas, as in other places, has two components:

- Population growth remains very high in the poorest parts of the Sahel, for reasons linked with poverty, such as lack of education. There were just 30 million people living in the Sahel in 1950. Today, the figure is closer to half a billion. By 2050, it is expected to reach one billion. Most of this growth is caused by the high number of children being born, and people living longer than they did in the past (Figure 7.26).

- Migration brings even greater population pressure. Drought and desertification in one region will displace people to another fragile environment. Desertification occurs there too, as the number of people increases, so the problem gets spread from place to place. In addition to 'climate change refugees', millions of people have been forced to move into desert fringe areas by armed conflicts in the Sahel region.

Figure 7.27 shows the human causes of desertification that follow on from population growth.

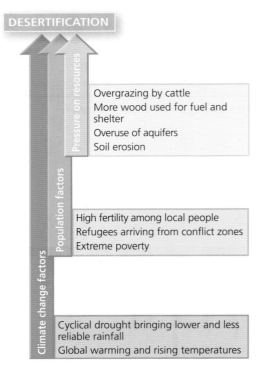

▲ Figure 7.25 The causes of desertification

Why has desertification become a problem in Darfur?

You may have heard of Sudan's Darfur region, where 250,000 people have been killed and around three million made homeless by conflict since 2003. Previously, the environment was already under pressure:

- One year out of every five brings drought, crop failure and livestock loss to Sudan.

- The Sahel's ability to produce food has not kept pace with its growing population (Figure 7.26).

The environment has played a major role in creating conflict. In 2003, nomadic cattle herders and settled farmers began to fight over water supplies and land. Herders were deliberately prevented from reaching water sources by farmers, which led to **overgrazing**. Once the vegetation was gone, their cattle died. In revenge, some herders chased farmers from their villages, and cut down their crops and trees.

Country	Population (millions)	Fertility rate (children per woman)	Pop. growth (% per year)	Pop. density (people per sq km)
Burkina Faso	17.9	5.9	3.1	65
Chad	13.3	6.6	3.3	10
Eritrea	6.5	4.7	2.6	56
Gambia	1.9	5.6	3.1	169
Guinea-Bissau	1.7	5.0	2.5	48
Mali	15.9	6.1	2.9	13
Mauritania	4.0	4.1	2.6	4
Niger	18.2	7.6	3.9	14
Senegal	13.9	5.3	3.2	71
Sudan	38.8	5.2	2.5	21

▲ Figure 7.26 Population data for Sahel nations, 2013

Millions of people fled their land and homes. They were housed in refugee camps, with help from the UN. But refugee camps create new environmental stress wherever they are located and cause desertification to spread (Figure 7.29). When people finally return home, further desertification is expected. One charity estimates that each returning family will need 30–40 trees to rebuild their houses and fences. The desert fringes of Sudan do not have enough trees left to support this.

▼ Figure 7.27 Population and land use trends in Darfur prior to conflict

	1973	2003
Population	1.3 million	6.5 million
Type of land use	Percentage in 1970s	Percentage in 2000s
Bush and shrub	23.8	17.5
Rain-fed agriculture	22.7	34.4
Wooded grassland	11.8	7.1
Closed forest (natural)	10.7	7.9
Grazing/pastures	9.0	6.8

▼ Figure 7.28 Human causes of desertification in populated areas

Overcultivation	Overcropping land can exhaust soil's fertility. Part of the problem is due to small-scale subsistence agriculture. As health has improved over time, more of the children born into farming families are surviving infancy. More crops are planted and some aquifers have been drained dry. Commercial farming makes this situation worse. European companies are using large areas of fragile land in Ghana to grow water-hungry cash crops such as jatropha (which is used to make vegetable oil).
Overgrazing	If too many goats and cattle are grazed for too long on one site, all the vegetation is eaten and may be unable to regrow. Nomadic groups used to wander freely, following the rain wherever it fell. They would leave land before all the vegetation was gone, giving it a chance to recover. Now they cannot, due to new political boundaries, or because large companies have bought up the land rights in a region. Civil war and political instability also force herders to stay too long in places.
Soil erosion	Overcultivation and overgrazing both result in soil erosion. If vegetation has been eaten by cattle or killed by drought, the exposed topsoil becomes baked hard by sunlight. When it finally arrives, intense rain washes over the soil rather than soaking into the ground. As it flows, it carries the topsoil away. Once the soil has eroded, it becomes impossible for the vegetation to grow back.

▲ Figure 7.29 Refugee camps greatly increase the population density of some fragile desert fringes

Geographical skills

Describing data sets

Figure 7.26 is a large and complex table. Imagine a question that asks you to 'Describe the variations shown in the table'. You could:

- For each column, say if it varies a lot or a little.
- Identify the maximum and minimum value in each case (you might even want to subtract the two to find the range of data is).
- If the maximum value is unusually high, point this out (see Sudan).
- Finally, look for patterns horizontally as well as vertically. Is there a country which is highest- or lowest-scoring in most categories?

→ **Activities**

1 How are the problems of overgrazing and soil erosion linked?

2 Explain how a combination of physical and human factors leads to desertification taking place in some parts of the world.

3 Look at Figure 7.27.

a) Describe how the population and environment in Darfur changed between 1973 and 2003.

b) Explain how the population change you have described may have led to two environmental changes.

Tackling desertification

How can better land management help combat desertification?

The majority of the 50 million people who live in the Sahel region suffer from poverty. Niger is losing 250,000 hectares of farmland every year through desertification. Millet crops have failed and sand dunes are advancing. Women in some villages now walk as far as 25 kilometres a day to fetch water for their families. But a range of land management measures can help to preserve soil quality and water supplies, such as:

- tree-planting schemes to bind and protect the soil
- planting grass on slopes to help stabilise the topsoil
- building small rock dams to trap rainwater in gullies
- collecting rainwater on roofs by designing a flat roof with a surrounding lip
- building terraces (flattened sections with a retaining wall) on farmed slopes.

One successful strategy, introduced in the Sahel countries of Mali and Burkina Faso, is the construction of low stone walls called bunds (Figure 7.30). The stones are planted in lines parallel to the slope gradient. They help to prevent soil erosion, and slow down the flow of rainwater over the baked ground. When water pools behind a bund instead of running fast over the land, it has time to soak into the ground.

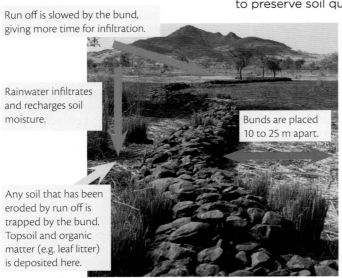

Run off is slowed by the bund, giving more time for infiltration.

Rainwater infiltrates and recharges soil moisture.

Bunds are placed 10 to 25 m apart.

Any soil that has been eroded by run off is trapped by the bund. Topsoil and organic matter (e.g. leaf litter) is deposited here.

▲ Figure 7.30 How bunds work

How can planting trees help to stop desertification?

Tree roots help to stabilise soil, while their decomposing leaf litter adds valuable nutrients (see Section 5.1). This makes the planting of trees a practical way to tackle desertification. The African Union's proposed 'Green Wall' is a plan to plant a wall of trees across the entire Sahel region, running from the Atlantic Ocean in the west, all the way to the Indian Ocean in the east (Figure 7.31). It will be decades before the Green Wall reaches maturity, but it offers hope for sustainable development among communities. In addition to making the physical environment more secure, the project will also generate work for desperately poor communities in all the Sahel countries. Finally, it will help bring about political co-operation in the region. This might reduce conflict and the number of refugee camps, which, unfortunately, contribute to desertification. However, climate change projections suggest increased aridity may threaten the survival of the trees in the long term.

How can appropriate technology be used to change the way people cook?

Removing trees for firewood is one of the most damaging human activities in desert fringe areas like the Sahel. For millions of years, people have been using wood as a cooking fuel. Yet population growth in the Sahel has meant that wood is a vanishing resource. When trees are cut down in large numbers, the resulting soil erosion effectively prohibits any future regrowth. This is unsustainable land management, but people in rural areas have no access to gas or electricity infrastructure.

Recently, however, an alternative way of cooking has begun to be adopted, using appropriate technology called 'efficient stoves'. One example is the Toyola stove in Ghana, and another is the Upesi stove in Kenya. The stove designs are being distributed to rural desert fringe communities by charities like Practical Action and the Global Alliance for Clean Cookstoves. The key to their success is that the stoves can be made locally using more available materials like clay, and much smaller amounts of wood and charcoal.

Some stove designs also incorporate a thermocouple, which generates sufficient electricity from the heat to charge a mobile phone – which growing numbers of Sahel farmers own. In turn, mobile phone access is helping farmers to gain access to weather forecasts, which can help them prepare for drought or rain.

Another important energy development for desert fringe areas is the move towards solar power (see page 52). As well as providing energy for cooking and other needs, earnings from solar power could provide Sahel nations with money to tackle desertification even more effectively. Hot deserts and their fringes places may eventually be seen as the world's most resource-rich places.

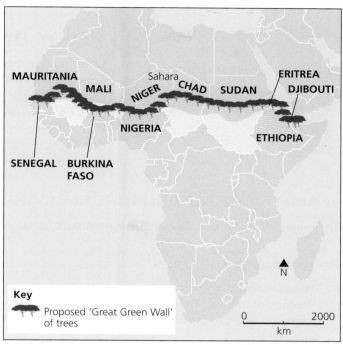

△ Figure 7.31 The proposed Great Green Wall, with a total distance of 7,775 km and a total area of 11.6 million hectares.

△ Figure 7.32 This Ghanaian woman is cooking on a stove that uses far less fuel than a traditional fire

→ Activities

1 Study Figure 7.30. (a) Describe the characteristics of a bund. (b) Explain how bunds can help tackle the problem of desertification. It would be helpful to think carefully about what information you use to answer the two parts of this question. The command words 'describe' and 'explain' have different meanings and require you to use your knowledge of bunds selectively.

2 Study Figure 7.31. What kind of problems could be encountered by the African Union as it tries to carry out the Great Green Wall project (think of physical and human challenges)? Can you suggest any possible solutions for the problems that may develop?

3 a) Explain what is meant by appropriate technology.
  b) To what extent can the use of appropriate technology help to tackle the problem of vegetation removal in desert fringe areas?

4 Some people have criticised the steps being taken to tackle the problem of desertification as 'too little, too late'. Discuss why the process of desertification, once it has begun, can become an irreversible problem.

8 Cold environments

Polar and tundra environments

Covering one quarter of the Earth's land surface, the world's cold environments are high-latitude world regions where cold, sinking air generates freezing winds and sunlight is thin. At the highest latitudes, the Sun does not even rise for several months of the year. Few people want to live in such extreme conditions. Alert, on the northeast coast of Ellesmere Island, Canada, is the world's darkest populated settlement. Located at 82 degrees north, its tiny handful of residents receive no sunlight for 50 days annually.

Where are polar and tundra environments located?

There are two types of cold climate environment: **polar** environments and **tundra** environments.

Polar environments

■ These are found in inland areas, far from the warming influence of the sea. They include Greenland, Northern Canada, Northern Russia (Siberia) and Antarctica.

■ The average monthly temperature is always below freezing. This allows snow and ice to accumulate overtime.

■ Most polar regions are partly or completely covered with ice caps (see Figure 8.1).

Tundra environments

■ These are found south of the ice caps in the northern hemisphere.

■ They occupy around one-fifth of the Earth's land surface, including enormous areas of Russia and Canada. These places lack permanent ice cover, but experience very cold weather for most of the year.

■ Most of the ground is permanently frozen. The treeless tundra ecosystem is composed of low-lying shrubs and mosses.

What are the characteristics of polar and tundra climates?

Polar climates

Figure 8.2 shows the annual climate graph for Antarctica's McMurdo research station. You can also see the **temperature range** for Antarctica's summer months. Daytime temperatures at the surface of Antarctic glaciers very occasionally reach 10 °C. This is rare, however, and the average temperature usually remains below freezing in all polar climates.

Precipitation in a polar climate falls mostly as snow. Overall, there is very little precipitation, the same as in hot deserts. This is because cold air cannot hold much water vapour.

▲ Figure 8.1 Ice caps in a polar environment

	Oct	Nov	Dec	Jan	Feb	Mar
Record high °C	0.6	2.8	10.5	8.3	2.2	−2.2
Average high °C	−15.5	−6.7	−0.8	−0.2	−6.3	−14
Daily average °C	−18.9	−9.7	−3.4	−2.9	−9.5	−18.2
Average low °C	−23.4	−12.7	−6.0	−5.5	−11.6	−21.1
Record low °C	−40.0	−26.1	−14.4	−15.0	−25.0	−43.3

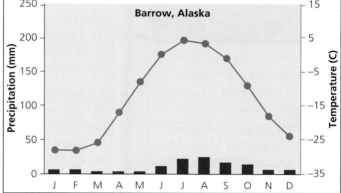

	Apr	May	Jun	Jul	Aug	Sep
Record high °C	6	8	22	26	24	17
Average high °C	−13	−3	5	8	7	2
Daily average °C	−16.8	−6.1	2	4.9	3.9	0.1
Average low °C	−21	−8	−1	2	1	−2
Record low °C	−41	−28	−16	−6	−7	1−17

▲ Figure 8.2 The climate graph and monthly summer temperature data for McMurdo, Antarctica (a polar climate)

▲ Figure 8.3 The climate graph and monthly temperature data for Barrow, Alaska (a tundra climate)

Tundra climates

Figure 8.3 shows the climate graph for the northern hemisphere town of Barrow, Alaska. The **thermal growing season** lasts just six to ten weeks. This is a short, cool summer, although the Sun does shine 24 hours a day! In December, in contrast, there is complete darkness. In some years, temperatures have fallen below −40 °C. The tundra climate is very harsh in winter. In the coldest tundra environments, the average monthly temperature nudges above freezing for just one month of the year. Precipitation falls as snow in winter and as rain during the brief summer. The total annual amount is low, due to the cold air temperature.

Figure 8.3 shows that daytime temperatures in Barrow occasionally rise above 20 °C in the summer months. However, the monthly average temperature is always much lower, due to colder temperatures at night.

Permafrost and polar ground

In cold climates, most of the ground is permanently frozen. This condition is called **permafrost**. Around one-quarter of the Earth's surface is affected by continuous or sporadic permafrost, including tundra, polar and some mountain regions. In tundra regions, ice in the uppermost **active layer** of the soil thaws for one or two months of the year during the brief summer, but there is still ice below the active layer. The ice acts as an impermeable barrier to the downward movement of melted water in the soil layer above. This results in waterlogged conditions.

→ Activities

1 Explain why the following are found in cold environments (refer to Chapter 3, pages 22–23 to help you):
   a) cold sinking air   c) thin sunlight.
   b) freezing winds

2 Use the information on these pages to locate the polar and tundra environments on a blank map of the world.

Geographical skills

Describing a climate graph

Study Figure 8.3.

1 What is the total annual rainfall?

2 What is the annual pattern of rainfall? Are there wet or dry seasons? If so, when are they?

3 What is the annual temperature range?

⭐ KEY LEARNING
➤ Cold environment animals and their adaptations
➤ Tundra plants and their adaptations

Interdependence of people, plants and animals

- Tundra birds and small mammals use moss to line their nests for warmth against the icy wind.
- Traditionally, Inupiat and Yup'ik people of the Arctic Circle depended on animal skin and feathers for their clothing (see page 102).
- Historically, indigenous people in coastal areas have depended on marine species (including fish, sharks and whales) for food and other uses (see page 102).

Cold climate ecosystems and biodiversity

How have animals adapted to survive in cold environments?

Many different animal species live in cold climate regions. The polar bear is perhaps the most well-known. Much of the polar bear's time is spent hunting for seals along the northern edges of the Arctic Ocean, in places where there is very little plant life (Figure 8.4). When seals are scarce, the polar bears roam inland into the tundra, in the very far north of Russia and Canada. In contrast, brown bears graze and hunt at the southern edges of the tundra, close to the forest boundary (Figure 8.6). To escape freezing conditions and food scarcity in winter, they hibernate, insulated by fat from the cold.

Between the northern and southern edges of the tundra lie vast expanses of land, where the Arctic fox and tundra wolf are the environment's top carnivores (Figure 8.5). A simplified food web for part of this region is shown in Figure 8.7. The consumer species are adapted to their environment in ways which aid survival:

- Snowshoe rabbits have white fur. This means they cannot easily be seen against the winter snow.
- Caribou and musk ox have two layers of fur to help them survive the bitter cold. They also have large hooves to help them travel over soggy ground and break through ice to find drinking water during the winter months.

Some species, like the musk ox, are permanent tundra residents. Many birds and mammals are not, however. They use the tundra as their summer home and migrate elsewhere in winter. Biodiversity is relatively low overall (Figure 8.8).

▲ Figure 8.4 The northern edge of the tundra, where polar environments take over

▲ Figure 8.5 An Alaskan tundra wolf

▲ Figure 8.6 Looking south from the edge of the tundra towards the treeline

How are plants adapted to tundra environments?

At the base of the tundra food web are many producer species (Figure 8.7). These include low-lying shrubs, mosses and lichens. Tundra plants are adapted in a range of ingenious ways to help them overcome environment challenges (Figure 8.9). In turn, the biotic (plants) and abiotic (soils) components of ecosystems have become interdependent. Low-lying vegetation helps to protect the soil from wind erosion, for instance.

▼ Figure 8.7 A simplified tundra food web

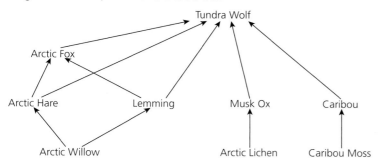

▼ Figure 8.8 Tundra biodiversity

Group	Number of species
Mammals	75
Birds	240
Insects	3,300
Flowering plants and shrubs	1,700
Mosses	600
Lichens	2,000

▼ Figure 8.9 Tundra plant adaptations

Permafrost	The permafrost chills the ground and is a barrier to root growth. Only plants with shallow root systems can survive, including mosses, lichens, some flowering plants and low-growing shrubs. Trees, which rely on deep roots for stability in the wind, cannot survive here (the word 'tundra' means 'treeless' in Finnish).
Poor drainage	In the flat, low-lying tundra regions, summer melting of the active layer leaves many areas waterlogged. This favours hardy organisms like mosses that can tolerate both extremely dry and wet conditions. This allows them to cope with the seasonal change from very dry to wet conditions, as the active layer melts.
Low insolation	Due to the high latitude, light is weak even with long summer days. Snow covers plants for many months of the year. They have therefore adapted in ways that maximise photosynthesis during the short growing season. For instance, most tundra shrubs are perennials. As a result, photosynthesis can begin immediately because the plants do not need time to regrow their leaves. The snow buttercup and Arctic poppy have adapted to the short growing season by producing flowers very quickly, while the snow is still melting. They have cup-shaped flowers that face up towards the Sun so the Sun's weak rays are directed towards the centre.
Strong wind	High air pressure over the North Pole generates strong, cold winds that blast tiny particles of ice southwards over the tundra. Plants have adapted in ways that keep them warm and minimise **transpiration** loss to the wind. They grow close together and near ground level (few species reach 40 centimetres in height). This allows plants to trap pockets of warmer air. Their leaves are small and fringed with tiny hairs to capture heat. The seeds of some plants have woolly covers. The Arctic willow is a widespread dwarf shrub which grows no more than six centimetres in height to avoid the cold wind.

→ Activities

1. a) Draw up a list of five tundra animals.
   b) Explain how each is adapted to survive in an extreme cold environment. In your answer, make reference both to their appearance and to their behaviour.
2. Study Figure 8.7. Using examples from the tundra food web, explain the difference between primary consumers and secondary consumers.
3. a) Calculate the total biodiversity of the tundra environment using Figure 8.8.
   b) Estimate the percentage of all species that are mammals.
   c) Can you explain why only a relatively small number of mammal species are supported by the tundra environment?
4. Compare and contrast the dominant vegetation of the tundra with that of the tropical rainforest (Section 6.1). How does the climate of each region help explain the similarities and differences you have identified?

⊕ KEY LEARNING

➤ Migration to Alaska
➤ The economic development opportunities of Alaska's environment

Alaska's development opportunities

How has migration brought different groups of people to Alaska?

Covering nearly two million square kilometres, the US state of Alaska borders Canada and the Arctic Ocean (see Figure 8.10). Like many of the Earth's coldest regions, Alaska has been settled for thousands of years, despite challenging environmental conditions.

- Alaska's indigenous peoples include the Inupiat and Yup'ik tribes. These Native Americans are part of an even larger ethnic group called the Inuit.

- During the last Ice Age, the Inuits' ancestors spread widely throughout the Arctic Circle into Canada, Alaska, Russia and Scandinavia. Then, the USA and Russia were joined by a land bridge.

- Since the 1800s, around 100,000 Inuit have been joined by more recent permanent settlers, mainly European, bringing the population to almost 750,000.

- Alaska is also home to economic migrants travelling north temporarily to work for oil and mining companies.

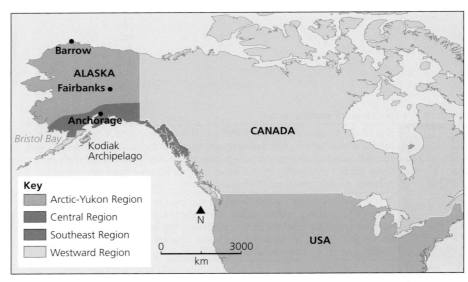

▲ Figure 8.10 The location of Alaska, the 49th state of the USA

What are Alaska's development opportunities?

Given its great size, Alaska is one of the most sparsely populated places in the world. This reflects the lack of economic opportunities in this extreme environment. Nearly half of the residents live in the city of Anchorage. Some urban residents work in retailing or education, health and government services. Others, along with much of Alaska's rural population, earn a living from fishing, mineral extraction, tourism or energy industries.

The fishing industry

The 3,000 rivers, three million lakes and 10,686 kilometres of Alaskan coastline provide many economic opportunities linked with fishing. There are two main sectors of the industry:

- **Commercial fishing**. Since the 1870s, the sector has grown to employ one in ten Alaskans. Some of the biggest salmon, crab, and whitefish fisheries in the world are in Alaska. They provide 78,500 jobs and

add US$6 billion to the state economy annually. Some jobs are only seasonal, however.

- **Subsistence fishing**. Native American communities remain dependent on fish for several uses. Fish provide food, oil (for fuel), and bones (used to help make clothing and tools).

Alaska's fisheries are widely viewed as a successful example of sustainable management.

▼ Figure 8.11 Fishing opportunities in different Alaskan regions

Arctic-Yukon region	Subsistence fishing is very important here. There are many Inuit fishing villages that rely on salmon and herring for their survival.
Central region	The world's most important commercial sockeye salmon fisheries are found in Bristol Bay and Copper River. Inuit communities depend on shrimp and scallop fishing in Prince William Sound.
Southeast region	Plenty of commercial fishing takes place here. Important species include salmon, herring and red king crab.
Westward region	This region includes all Pacific Ocean waters extending south from the Kodiak Archipelago, in addition to Bering Sea waters. The largest crab and Pacific cod fisheries in the state are here.

▲ Figure 8.12 The Pebble Mine generated opposition from the Bristol Bay fishing industries and communities

Mineral extraction

In the late 1800s, Alaska was known as 'the gold rush state'. Today, one-fifth of the state's mining wealth still comes from gold (although silver, zinc and lead mining are also very important).

Large gold mines must be managed carefully to minimise environmental impacts. Humans and ecosystems can be harmed by the toxic chemicals used to process gold ore (such as mercury, cyanide and nitric acid).

Mining development has sometimes been halted due to environmental campaigns. In 2013, the Pebble Mine gold project was closed down. It would have been North America's largest open-pit operation. Native American communities ran an effective 'No Dirty Gold' campaign. Fifty businesses supported the campaign by saying they would not buy Pebble Mine gold (see Figure 8.12). Anglo American, one of the world's biggest mining companies, walked away from a half a billion dollar investment due to the scale of opposition.

Tourism

Tourism attracts between one and two million summer visitors each year, making tourism one of Alaska's biggest employers, although some work is seasonal and poorly paid. Some tourists enjoy fishing, while others merely view the wildlife, with popular activities including whale watching and kayaking. Approximately 60 per cent of summer visitors are cruise ship passengers.

Hiking, skiing, rock climbing and sightseeing by helicopter are also available. The state has numerous national parks, preserves, refuges and monuments. There are historical sites for those interested in the Inupiat and Yup'ik heritage. These tourists arrive mostly by air.

Energy

Energy production is another big employer, especially the oil industry (see pages 106–107).

■ More than 50 hydroelectric power (HEP) plants supply Alaskan communities with one-fifth of their electricity. Previously glaciated U-shaped valleys in Alaska are a perfect site for HEP generation.

■ Geothermal energy is also being harnessed in tectonically active parts of the state. Alaska's coastline is part of the Pacific 'Ring of Fire'. A tourist resort at Chena Hot Springs near Fairbanks is now powered entirely by geothermal power.

→ Activities

1 Compare the four main economic activities in Alaska by carrying out an environmental impact assessment. Consider impacts on water, ecosystems and the landscape. Give a score between 0 (no impact) to 5 (major impact) for each one. Then add them together.

2 Explain how different groups of people benefit from economic activities in Alaska.

3 Write a short advert about Alaska for a travel magazine. What different kinds of people and interest groups would your advert be targeted at?

4 How does the theory of plate tectonics help us to understand why Alaska can generate its own geothermal power? What risks may accompany this opportunity?

⭐ KEY LEARNING

➤ The uneven development of Alaska
➤ Why accessibility is a challenge in Alaska
➤ Protecting buildings and infrastructure

Alaska's development challenges

What explains the uneven development of Alaska?

Adapting to the cold tundra environment of Alaska is a challenge, for both traditional Inuit communities and more recent settlers. There is virtually no settled population in the northern interior (see Figure 8.13), due to perilously low temperatures and months without sunlight.

The absence of people reflects the low carrying capacity of the land there: permafrost and the short thermal growing season rule out crop production. Instead, the first indigenous people settled along Alaska's coastal margins developed a subsistence economy based on fishing and seal hunting.

Life was still very challenging for the first settlers. Traditionally, Inupiat and Yup'ik people coped with the cold by making coats from caribou skin and sealskin boots. Goose down was used as a lining. Over time, however, they have increasingly adopted the use of modern textiles like Gore-tex.

Why is accessibility a challenge in Alaska?

The low population density of less than one person per square kilometre means that most of Alaska lacks surfaced roads. Hunters, miners and explorers must make their own way across the tundra. Even where roads and rough tracks have been provided, physical processes make their use difficult and dangerous.

- Snow and ice make some roads and tracks unusable for months of the year.
- A process called **solifluction** takes place in summer. On slopes, the soil's active layer starts to flow downhill. The thawed soil slides easily over the impermeable frozen layer below. Large amounts of soils and mud can collect at the base of slopes, covering highways that run along valley floors, cutting places off for months.
- Permafrost underlies most of Alaska (Figure 8.14). The seasonal melting of the active layer means that off-road travel cannot take place during summer.
- Over time, the seasonal melting and re-freezing of the active layer results in great expanses of uneven ground surface called **thermokarst** (Figure 8.15) making travel impossible in some places.

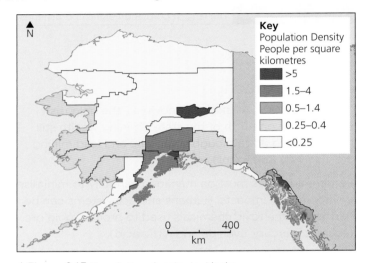

Key
Population Density
People per square kilometres

- >5
- 1.5–4
- 0.5–1.4
- 0.25–0.4
- <0.25

0   400
km

▲ Figure 8.13 Population density in Alaska

- Frost heave – where pebbles and stones slowly rise upwards to the ground – can make tracks dangerous.

What can be done to protect buildings and infrastructure?

Indigenous people and newcomers alike use high-pitched steep roofs for their homes so snow can slide off. Triple-glazed windows help to keep the cold at bay.

The active layer melting causes the most serious challenge for building. The heat that buildings and settlements create – known as the 'urban heat island' effect – can make this worse. Many buildings erected by early European settlers in the 1880s soon became unusable (Figure 8.16). Escaping heat from the underside of properties led to the melting of the frozen icy ground beneath. As ice loses volume when it turns to water, the land under many homes subsided.

Millions of kilometres of permafrost have been damaged due to insufficient attention being paid to the sensitivity of soil conditions in Alaska. Today, new buildings are always raised on piles to prevent melting. These piles can lift a structure several metres above the surface and are sunk deep into the land, well below the lower limit of the active layer. Increasingly, telescopic piles are used. These expand or contract if there is any residual ground movement. The same principles are applied to the vital **infrastructure** that connects places together.

■ Roads are now built on gravel pads one to two metres deep that stop heat transfer from taking place.

■ Utilities such as water, sewerage and gas cannot be buried underground or they would freeze too. Instead, they are carried by utility corridors or 'utilidors'.

■ Airport runways are painted white to reflect sunlight and stop them from warming up too much on sunny days.

▲ Figure 8.16 House destroyed by permafrost melting, North America

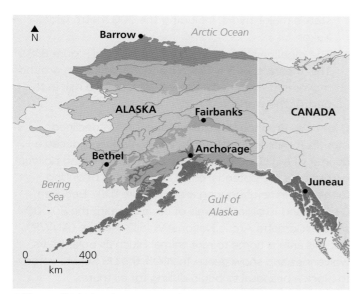

▲ Figure 8.14 The distribution of permafrost in Alaska

Key	
	Underlain by thick permafrost
	Generally underlain by continuous permafrost
	Generally underlain by discontinuous permafrost
	Generally underlain by isolated masses of permafrost
	Generally free of permafrost

▲ Figure 8.15 A thermokarst land surface in Alaska

→ **Activities**

1 Describe the population distribution of Alaska. Refer to data in Figure 8.13 to support your answer.

2 State what is meant by the carrying capacity of land. Why does it vary from place to place in Alaska?

3 Explain how solifluction, permafrost melting and frost heave all create development challenges in Alaska.

4 Look at Figure 8.14. Suggest reasons why some parts of Alaska remain free of permafrost.

5 Suggest why climate change could mean that measures taken to protect roads and buildings in Alaska may not prove to be sufficient in coming years.

⊘ KEY LEARNING

➤ The costs and benefits of Alaska's onshore oil fields

➤ The issues surrounding future development of offshore oil fields

The challenges and opportunities of Alaskan oil

Alaska has vast reserves of oil and gas, both on land and also offshore, in the Beaufort sea and the Arctic ocean. The industry provides 100,000 jobs: in other words, it employs one in seven Alaskans. Oil and gas contribute one-third of the state's annual earnings of around $US40 billion.

What are the costs and benefits of Alaska's onshore oil fields?

In 1968, vast onshore oilfields were discovered near Alaska's north coast. Oil production began at Prudhoe Bay on the North Slope in 1977. In the early days, almost two million barrels a day were produced. The 800 kilometre trans-Alaskan oil pipeline was built to transport the oil to the southern coast port of Valdez (see Figure 8.17). This was made necessary by the challenge of ice in the northern seas, which meant that oil tankers could not be used.

The pipeline took five years and US$8 billion to build. This was seen as a price worth paying in the 1970s. Rising oil prices and political problems in the Middle East had left the USA desperate to improve its own energy security.

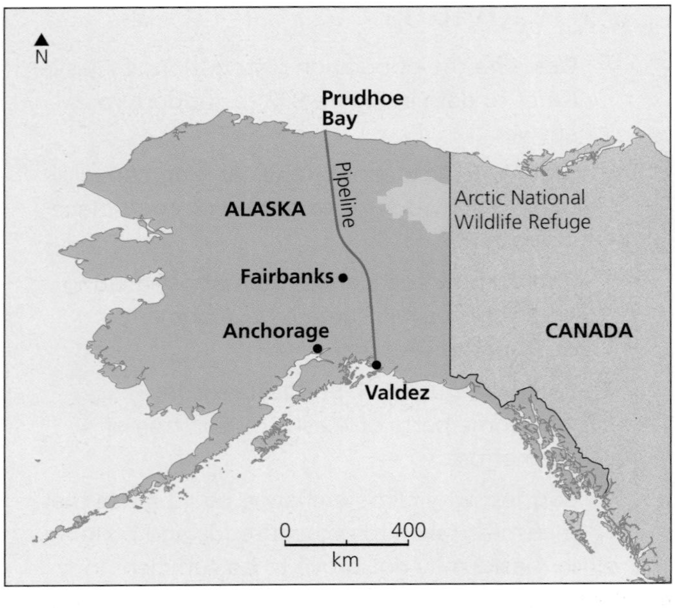

Engineers were careful to modify the trans-Alaskan oil pipeline in ways that helped overcome Alaska's environmental challenges:

■ The pipeline was raised off the ground on stilts (they are eleven metres deep and cost US$3,000 each to build back in the 1970s).

■ Pipeline suspension bridges were used to cross the state's major rivers, including the 700 metre wide Yukon.

■ The pipeline zigzags in some places (see Figure 8.18). This means that it is flexible and can adjust to ground movement from earthquakes (which are a risk in this part of the world).

Over time, Prudhoe Bay's oil production has declined though. It peaked in the 1980s, when Alaska produced a quarter of all US oil. Today, there are fierce ongoing political battles over the potential costs and benefits of drilling for new oil in neighbouring areas (Figure 8.19). Between 6 billion and 16 billion barrels of oil lie beneath the 80,000 square kilometre Arctic National Wildlife Refuge (ANWR). But this area is home to rare animals such as polar bears, wolverines and snow geese. In 2005, the US Senate voted to block a proposal to begin drilling for oil there.

▲ Figure 8.17 The trans-Alaskan oil pipeline and Arctic National Wildlife Refuge

▲ Figure 8.18 Pipeline bridge over Yukon

▼ Figure 8.19 Costs and benefits of oil production in Alaska

Benefits	Costs
Many working Alaskans rely on the oil and gas industries for their income. More than 90 per cent of taxes raised by the Alaskan state come from this sector, so it pays for education, health, policing and important community services.	Migrant workers take the majority of jobs created by the oil industry, spend little locally and often only have short-term contracts. In Prudhoe Bay, locals take just 400 of the 2,000 available jobs.
In some places along its route, the trans-Alaskan pipeline passes underground so that it does not disturb the migration routes of the tundra caribou. The pipeline is thickly insulated to protect it from freezing and stop the permafrost from melting.	In 1989, an oil tanker, Exxon Valdez, ran aground on the southern Alaskan coast. Only 15 per cent of the 1.2 million spilled barrels was ever recovered. Around 5,000 sea otters and many seals and eagles were killed. A broken pipeline spilled 1 million litres of oil in the fragile North Slope region in 2006.

What are the issues surrounding future development of offshore oil fields?

In addition to the dispute over the ANWR, Alaska's offshore waters are also a source of controversy. There are believed to be 30 billion barrels of recoverable oil, and many trillions of cubic metres of gas beneath the Beaufort Sea and Arctic Ocean (Figure 8.20).

No drilling is allowed off the coast of Kaktovik and Barrow. These waters are home to bowhead whales, and Inupiat residents do not want the whales to be disturbed. Inupiat people are allowed to conduct carefully controlled whale hunts. Sharing whale meat among families is an enduring tribal tradition. There is also a drilling ban in Bristol Bay, on the south coast of Alaska, to protect the area's sockeye salmon fishery.

Most of the Beaufort and Chukchi seas remain open to exploration. Big companies like Shell, Chevron and Statoil began exploratory undersea drilling there in 2012. For the Inupiat, supporting offshore drilling is a tough decision. Like many in Alaska, they are dependent on the oil industry for jobs.

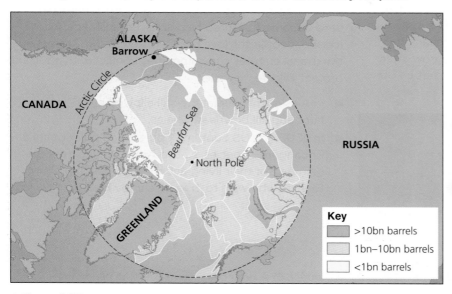

Key
>10bn barrels
1bn–10bn barrels
<1bn barrels

▲ Figure 8.20 Oil and gas fields under the Arctic Ocean

→ Activities

1 Why is it potentially difficult and expensive to make use of Alaska's oil and gas reserves?

2 Suggest why wilderness areas like Alaska face an increased threat from economic development.

3 a) Explain how the cold environment of Alaska creates physical risks for the trans-Alaskan oil pipeline.
   b) How have these risks been dealt with?

4 Look at Figure 8.20.
   a) Identify countries with offshore oil and gas fields near their coastline.
   b) List the names of the oil companies that have begun to explore the Arctic region.
   c) Suggest why these offshore oil and gas deposits could become a source of political conflict between the USA (Alaska) and other countries.

5 Make an assessment of whether or not the exploitation of offshore oil should be allowed near the Alaskan coastline.

→ Going further

Other cold environment countries have fossil fuel supplies that you can investigate. Reserves of 'tar sand' in Canada dwarf the oil supplies of Saudi Arabia. Commercial use of tar sands has devastated the local environment in parts of northern Canada: www.theguardian.com/environment/oil-sands

Wilderness environments

Why is there a need to protect wilderness environments?

Wilderness areas are unspoilt and remote regions of the world. They include truly isolated regions, such as Antarctica, and populated areas of the tundra that retain some wilderness characteristics. Increasingly, the world's wilderness areas are under pressure. Extreme climate and inaccessibility used to keep mass tourism and economic development at bay. Now, modern transport gives easy access to previously inaccessible areas. In our 'shrinking world', wilderness areas have been opened up to tourism and businesses. Travel companies are keen to market new 'exotic' locations, while energy and mining companies hope to discover new **natural resources**.

There are several justifications for preserving wilderness areas and protecting them from development:

- There is a planet-wide need to maintain the 'gene pool' of wild organisms, to make sure that genetic diversity is maintained over time.
- Scientists need to have access to undisturbed animal and plant communities for their studies.
- There should always be some places on Earth that are left in their natural state, so we can understand how much developed places have changed.

Wilderness areas perform vital **ecosystem services** that the whole world relies on. The white snow and ice cover in polar regions reflects sunlight and helps to regulate Earth's temperatures. The permafrost keeps enormous volumes of methane, a potent greenhouse gas, locked in ice (if released, it would contribute significantly to global warming).

The tundra is a fragile environment too, and thus in need of special protection. Extreme climates give rise to physical environments that are extremely sensitive to change. The slow growth of tundra plants means that it takes many years for the ecosystem to regrow after damage. It can take 50 years for tyre tracks left by off-road vehicles to disappear.

Why are wilderness cultures under threat?

It is not just the physical environment of wilderness areas that is threatened by economic development. 'Cultural erosion' can occur too, due to tourism and in-migration. Outside influences may cause a culture to lose unique characteristics, such as its language. In the past, 20 native languages were spoken in Alaska. European languages such as English have been adopted by the youngest generations of tribes, however. Native names have sometimes been replaced by English names like Ed and Steve. In the 1970s, American schooling insisted on classroom use of English. Now, some languages such as Eyak have lost their last speaker, while others are on the verge of dying out.

Other cold countries have their own highly distinctive cultures too. Life in a challenging environment has evolved over time so that practices born in hardship have become treasured traditions. For instance, in Iceland, people still love to eat 'rotten shark'. Protecting traditional cultures can sometimes conflict with efforts to protect wildlife, however.

- In Alaska, the Inupiat are still allowed by US law to hunt and kill bowhead whales.
- In 2008, the US Supreme Court made polar bears an Endangered Species. Native American Inuit people gain part of their living acting as guides for rich tourists who wish to 'bag a bear' by hunting.

◄ Figure 8.21 A polar bear cub relaxes with her mother

How is the Antarctic wilderness managed?

Antarctica remained unexplored until just over a century ago, although large-scale slaughter of seals and whales in the Southern Ocean dates back to the 1700s. Public awareness of the Antarctic wilderness first began to develop when the Shackleton expedition of 1914–17 brought back moving film images of spectacular glaciers and wildlife.

Concerned that Antarctica could be spoiled by unchecked commercial exploitation, several leading nations signed the Antarctic Treaty in 1961. It has become one of the most successful international agreements of all time. The more recent 1998 Protocol on Environmental Protection to the Antarctic Treaty is one of the toughest sets of rules for any environment in the world. Under the agreement, no new activities are allowed in Antarctica until their potential impacts on the environment have been properly assessed and minimised. Tourist boat operators taking visitors there have to follow incredibly strict guidelines.

Even so, there has still been a tremendous growth in tourism over the last two decades. Between 2000 and 2010, the number of tourists setting foot on Antarctica tripled (see Figure 8.22). Just 100 years after the very first explorers set foot there, Western tour operators now charge up to £10,000 per person for an Antarctic excursion. Some tourists merely gaze at the coast from icebreaking ships, while more active individuals engage

▲ Figure 8.23 A tourist photographs emperor penguins in Antarctica

in outdoor pursuits such as mountaineering or wildlife photography (Figure 8.23). For US$50,000 it is even possible to take a guided ski trek to the South Pole.

→ Activities

1 Write a list of possible characteristics that 'wilderness' areas have. Can you identify a place in the UK that could be described as a 'wilderness' area? Explain your answer.

2 Give three reasons why the world's wilderness areas face an increased threat from economic development.

3 Explain why ecotourism could be a valuable model of economic development for wilderness regions in cold environments and elsewhere in the world.

4 Look at Figure 8.22.
   a) Describe the trend in visitor numbers over time in Antarctica.
   b) Suggest reasons for the trend you have described.

5 To what extent do you believe the management of tourism in Antarctica can be regarded as a success? In your answer you should provide arguments and evidence that both support and reject the suggestion, before reaching a final viewpoint.

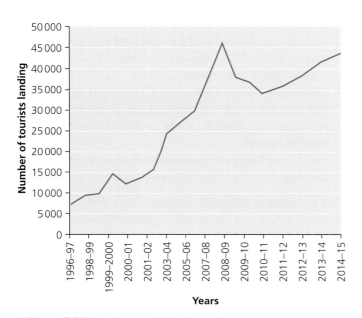

▲ Figure 8.22 Tourist growth in Antarctica

KEY LEARNING

➤ How technology can help development of wilderness communities

➤ Why global technology companies are relocating to cold environments

Balancing development and conservation using technology

Economic development in cold environments has often had disastrous impacts for their ecosystems. Hunting, whaling and fur trapping have pushed some land and marine species close to extinction. Pollution from oil and mining industries has added further stress. Indigenous tribes struggle to maintain their traditions, and out-migration of the young in search of education and employment threatens the sustainability of some communities. Yet some technological developments have helped make life easier (see Figure 8.24) and have encouraged the young to stay, particularly since the arrival of information and communications technology (ICT). Digital technology offers exciting new development opportunities which may minimise environmental damage and cultural erosion.

Then	Now
Animal skins and hides were used as clothing. People had to hunt animals and make clothes.	Modern materials such polyester are used. People buy clothes online.
Traditional Inuit games were played outside, depending on the season.	Today's younger Inuit generation often play computer games and watch TV.
People travelled by walking, kayaking or using dogsleds (some still do).	The Inuit now use modern transport (see Figure 8.25).

▲ Figure 8.24 Technology has changed Inuit lifestyles

▲ Figure 8.25 Inuit people have adopted modern technology including the snowmobile

In what ways can ICT help the development of wilderness communities?

Scandinavian countries were amongst the first to see the potential of ICT for community survival. By the late 1980s, isolated villages in Finland, Sweden and Norway had been provided with shared computer and internet facilities by their governments. For these isolated communities inside the **Arctic Circle**, the internet became a survival lifeline.

In recent years, Inuit communities in Alaska and Canada have also discovered the value of ICT for their sustainable development.

Things first began to change in the 1990s when wireless radio and satellite links were provided for all the main populated areas, although coverage remains very uneven. As a result, many isolated parts of Alaska have been able to 'leapfrog' forwards in terms of their access to technology. People can now receive a mobile phone call in places where landlines are still absent. The tiny Inuit village of Little Diomede is home to 120 people. On an island in the Bering Strait, with Alaska to the east and the Russia to the west, mobile phone service is available despite the fact that the mail only arrives once a week.

■ In some places, two-way video conferencing is transforming how communities can access education and health services. In Canada, local government has collaborated with the company Cisco to allow remote Inuit schools to be taught in real time by teachers in other schools. They also collaborate with students of the same age throughout Canada.

■ The University of Alaska offers a range of degrees and courses which may be completed entirely online by students in isolated areas, provided they have internet access.

■ The Alaska Native Knowledge Network is an online database that collects and preserves Inuit culture.

More progress is around the corner. Soon, even the most remote Alaskan and Canadian villages will have access to high-speed broadband networks. As the Arctic ice thins, it is becoming easier to lay fibre optic cable. In 2015, work began threading a 15,600 kilometre fibre optic cable along the Arctic coastline of North America. Data will run from Tokyo to London in 154 milliseconds, bringing globalisation to the world remotest places.

Why are global technology companies relocating to cold environments?

Some of the world's largest internet companies have relocated their data centres to cold environments. These offices have a low environmental impact and also offer employment to people in some of the world's most remote places. The global relocation of Facebook and Google to northern Scandinavia provides a good example of how development and conservation can be balanced.

Facebook users upload 150 million new photos each day, while 1,000 status updates are made every second. All this information must be stored and backed up in the company's data centres. As a result, data centres are very expensive to run. Energy is used to power the computers and also to cool them down with fans or air conditioning.

Consequently, more companies are placing their data centres in Arctic areas because the climate cools down the machinery. Google has a data centre in Finland. Facebook has a giant data centre in Luleå, Sweden – a long way from its desert headquarters in California! The site covers 30,000 m² (the size of ten football pitches) and cost around US$750 million to build. Luleå is at the same latitude as Fairbanks, Alaska, USA and is part of Sweden's coldest region. Physically, the site of Luleå offers Facebook:

- **a cold climate** – located on the edge of the Arctic Circle, Luleå has short, mild summers and long, cold and snowy winters. Winter temperatures are well below freezing. For eight months of the year, the high power computer equipment will cool itself at no cost
- **low HEP costs** – Luleå lies near hydroelectric power stations at the mouth of the River Lule. This provides cheap electricity for lighting (the heat from the computers is used to warm the Facebook staff's offices)
- **flat land** – Luleå is built on a flat, glacially eroded valley floor.

Facebook is not the only technology company in Luleå. The town is now home to 2,000 employees working for several large technology industries that have clustered there.

▲ Figure 8.26 The Facebook data centre in Luleå

→ **Activities**

1. Where else in the world could ICT be used to help isolated communities? Think of other communities you have learnt about so far on this course.

2. Suggest how the arrival of ICT could threaten Inuit culture as well as helping communities to develop.

3. Explain how and why global companies like Facebook and Google are helping to create economic opportunities for people living in the world's cold environments.

❂ KEY LEARNING

➤ Protective actions at different scales

➤ The balance between conservation and development in the Arctic

➤ The Arctic and climate change

▲ Figure 8.27 A bowhead whale

Can conservation and development be balanced in the Arctic?

Figure 8.28 shows how international organisations, national governments and NGOs have varying views about how the Arctic should be managed.

Managing cold environments

What actions at different scales can protect cold environments?

Managing the balance between development and conservation is rarely easy to achieve, particularly in cold environments. A range of management actions have been taken at global, national and local scales.

- **International agreements** and treaties can influence what happens to cold environments and their ecosystems. For instance, the number of bowhead whales in Arctic waters has been growing at three per cent per year since the 1970s (see Figure 8.27). They were heavily exploited by whaling in the past. But population size has recovered since a global ban on commercial whale hunting was introduced by the International Whaling Convention in 1986. The Antarctic Treaty is another international success story (see page 109). Countries in the Arctic Circle have created a similar organisation called the Arctic Council, which wants to deliver sustainable development throughout the entire Arctic region.

- **National governments** sometimes struggle to manage their own regions because they are expected to support the conflicting interests of many different groups, from indigenous people to big businesses. Nowhere is this clearer than in Alaska. The state is running short of money due to low world oil prices. Some of its politicians want to increase oil production to increase Alaska's income. But, while in office, US President Barack Obama wanted to maintain what he called 'the integrity' of the Alaskan wilderness. He then banned oil exploration from taking place in 12 million acres of the Arctic National Wildlife Refuge (ANWR).

- **Non-governmental organisations (NGOs)** support the interests of groups of people who may not otherwise be heard. The Inuit Circumpolar Council (ICC) is a non-profit organisation that represents indigenous people from Nunavut and other northern regions. Throughout the Arctic Circle, Inuit NGOs have frequently taken action to save the environment protect themselves. The campaign of native Alaskans against Pebble Mine is one example (see page 103). In 2014, the UK's Greenpeace sent campaigners to Russia's Arctic Ocean to protest against oil exploration. The Russian government responded by arresting and imprisoning them.

▼ Figure 8.28 Three Arctic management approaches

The Arctic Council (an international organisation)	National governments	Greenpeace (an NGO)
This represents eight countries and the indigenous people of the Arctic. It was established in 1996 to promote co-operation. Sustainable development and environmental protection are priorities, but it is important that any protection measures do not harm the economies of indigenous people. In the future, the Arctic Council could become an organisation with legal powers, potentially setting fishing or hunting quotas.	National governments are currently entirely responsible for managing their own ecosystems. Some have given areas protection as National Parks (see Figure 8.29). Just over ten per cent of all Arctic land now has some level of special protection. Each country has its own laws controlling pollution from mining and oil industries. International agreements and laws are not always followed though. Norway continues to hunt whales, for instance.	Greenpeace has called for a 'global sanctuary' to be established in the Arctic, stating that: 'The best way is to make its resources off limits. That's why we're campaigning for a global sanctuary and a ban on oil drilling and industrial fishing.' But would this approach limit indigenous people's freedom to use Arctic resources for their own economic development?

Can the Arctic be saved from climate change?

Polar and tundra regions face an uncertain future. Climate change may already be causing permanent harm. One estimate shows that the tundra climate zone and ecosystem has shrunk in size by about 20 per cent since 1980. For complex reasons, global warming is felt most in polar regions. According to scientists at NASA, temperatures in Newtok, Alaska, have risen by 4°C since the 1960s, and by as much as 10°C in winter months.

Traditional life for the Yup'ik population is now under threat from global warming. The permafrost in the Alaskan village of Newtok is melting, causing buildings to subside, tilt and sink. The ice pack on the Bering Sea has thinned by 50 per cent. It is no longer possible for them to fish because dog teams and sledges cannot move safely across this surface. During building construction, Yup'ik workers must now push building piles four metres into the ground to guarantee stability. In the past, three metres had been sufficient. Over the next 100 years, the temperature in the Arctic is predicted to rise by another 4–7°C. The local wildlife will change as a result, possibly causing the extinction of some threatened species.

→ Activities

1 State two international agreements that help protect cold environments. Provide evidence to show the effectiveness of each agreement.

2 Suggest why the management of cold environments might be an important political issue at election time for some countries.

3 Look at Figure 8.28. Compare the different management approaches used by the Arctic Council, national governments and Greenpeace. Explain what you think their strengths and weaknesses are.

4 Look at Figure 8.29. Describe the distribution of protected areas. Think carefully about the language you will use to help you write your description. It may help you to briefly re-sketch the diagram and write 'North Pole' on it. You may also want to draw the Arctic Circle and label the continents.

5 Explain why all three management approaches in Figure 8.28 may fail if climate change predictions come true.

→ Going further

The intergovernmental Panel on Climate Change (IPCC) has published reports which include predictions about how cold environments may be affected by climate change by 2100. Two reports you can examine are:

1 Findings for Alaska – www.climatechange.alaska. gov/docs/GovCCmtg_walsh.pdf

2 Polar regions (Arctic region and Antarctic) – www.ipcc.ch/pdf/assessment-report/ar4/wg2/ ar4-wg2-chapter15.pdf

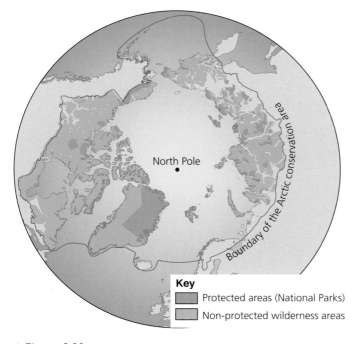

Key

Protected areas (National Parks)

Non-protected wilderness areas

▲ Figure 8.29 Protected areas already make up ten per cent of Arctic wilderness areas

Unit 1 Section B

1 Study the data in Figure 1. It shows population changes for three animal species in an area of deciduous forest in Scotland.

Which species shows the greatest proportional change between 2014 and 2016?

Species	2010	2012	2014	2016
Fox (carnivore)	55	55	40	50
Rabbit (small herbivore)	4,000	2,500	2,300	3,000
Deer (large herbivore)	80	80	100	95

▲ Figure 1 Population changes for three animal species

[1 mark]

2 Describe **one** natural reason why the deer population in Figure 1 rose between 2012 and 2014.

[2 marks]

The word 'natural' appears in the question, meaning that human management cannot be suggested as a possible answer.

3 Which **one** of the following describes the mean monthly temperature range for a deciduous forest? Select **one** letter only.

    A 18 °C in July and 5 °C in January

    B 28 °C in July and 10 °C in January

    C 18 °C in July and –5 °C in January

    D 32 °C in July and 14 °C in January

[1 mark]

4 Which **one** of the following statements best describes the soils of a tropical rainforest? Select only **one** answer.

    A Deep with well-developed soil horizons

    B Infertile and rapidly recycles nutrients

    C Alkaline, thin and fertile

    D Moderately deep and slowly recycles nutrients

[1 mark]

5 Describe and explain the distribution of deforestation in the Amazon basin shown in Figure 6.10.

'Describe and explain' means you should set out the characteristics and the purposes or reasons for that characteristic.

[6 marks]

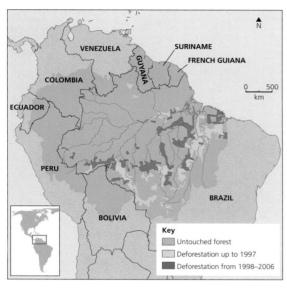

▲ Figure 6.10 Deforestation in the Amazon basin

6 Study Figure 6.2. What is the difference in the total rainfall between the wettest and driest months?

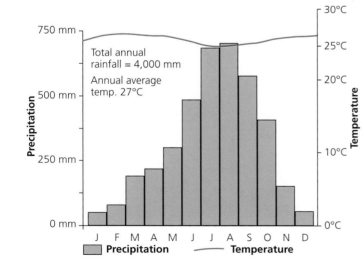

▲ Figure 6.2 The climate at Douala (Cameroon)

[1 mark]

7 Outline **one** social impact of deforestation of the Amazon rainforest.

[2 marks]

8 Suggest **one** way of making use of the tropical rainforest more sustainable.

[2 marks]

9 For a hot desert environment **or** a cold environment that you have studied, to what extent has human activity threatened that environment?

[9 marks]

9 The physical diversity of the UK

⭐ KEY LEARNING

➤ The location of UK's uplands

➤ The location of UK's lowlands

If a line is drawn from the River Tees to the River Exe, the area to the northwest of the line may be considered as upland UK, and the area to the southeast, lowland UK.

▲ Figure 9.1 Relief map of the UK

Where are the UK's uplands?

Mountains of Scotland: the North West Highlands are separated from the Grampians to the south by Glen More. This lowland area contains several lochs (lakes), including Loch Ness. The mountains form a plateau which has been dissected by steep-sided glaciated valleys. The west coast has fjords. To the south of the Grampians lies the 80 kilometre wide rift valley of the Central Lowlands. The Southern Uplands to the south is another plateau at around 600 metres, where the source of the River Clyde is found.

Mountains of Northern Ireland: the Antrim Plateau is located in northeast Ireland and is made of basalt. Along the coast, this rock has weathered to form the hexagonal blocks of the Giant's Causeway. Just south of Belfast lies the Mourne Mountains.

Cumbrian Mountains: located between the Solway Firth to the north and Morecambe Bay to the south, this area is commonly known as the Lake District. Rivers and glaciated **ribbon lakes** radiate from the centre of the 900-metre dome. Shap Fell ridge links the Lake District to the Pennines.

The Pennines: this 600 metre high plateau runs from the River Tyne in the north to the northern edge of the Midlands in the south. The Pennines form a west–east watershed, with most rivers flowing in an easterly direction. The Rivers Tyne and Aire have forged gaps through the Pennines, while the Trent flows to the south.

Cambrian Mountains of Wales: this dissected plateau has rivers which radiate from its centre. The Rivers Severn and Wye, have their source here. They then flow in a large loop to the estuary in the Bristol Channel.

Moorlands of the South West Peninsula: the granite moors of Dartmoor and Bodmin Moor rise to over 600 metres. In places, bare rock pillars called tors rise above the poorly drained moors. The rivers Exe and Tamar start close to the north coast of the peninsula and flow southward. To the north of the peninsula are Exmoor and the Quantocks.

Where are the UK's lowlands?

Scarps and vales: central and southern England has low-lying clay vales alternating with chalk ridges, with their distinctive steep scarp slopes and gentle dip slopes. Examples of chalk ridges include the North and South Downs and the Berkshire Downs. Limestone hills include the Cotswolds, the Chilterns and the Mendips. The clay vales stretch across Wiltshire and Oxfordshire. The River Thames has its source in the Cotswolds and flows southeast through the low-lying London Basin into the North Sea.

Fens and East Anglia: the most extensive areas of lowland are found in East Anglia, especially in the Fens south of the Wash. Here, much of the land is below sea level, where the Rivers Nene and the Great Ouse are found.

In northeast England the extensive river system of the Derwent, Ouse, Wharfe, and Trent drain into the Humber Estuary.

The following three chapters focus on the coastal, river and glacial **landscapes** of the UK. Landscapes are areas characterised by the result of action and interaction of natural and human factors. These chapters explore the various features of the UK, including physical landforms and how humans have used the land.

→ Activities

1 State the compass direction from:
   - the North York Moors to Dartmoor
   - Exmoor to the Cumbrian Mountains
   - the Mendip Hills to the Lincoln Wolds.

2 Use the scale on the map to measure, in a straight line, the distance from the mouth of the River Tees to the mouth of the River Exe.

3 Give the latitude and longitude for the mouth of the River Tweed.

4 Working from north to south, name the upland areas that longitude 4 degrees west goes through.

10 Coastal landscapes

⭐ KEY LEARNING

➤ The cause of waves
➤ Why some waves are stronger than others
➤ Constructive and destructive waves

▲ Figure 10.1 A stormy sea

Waves

What causes waves?

Waves are caused by the transfer of energy from the wind to the sea due to friction of the wind on the water's surface. This chapter shows how waves shape the coastline through **erosion** and **deposition**. Examples are mostly drawn from the Hampshire/Dorset coast.

Wave terminology

Crest: the top of a wave

Trough: the base of a wave

Wave height: the vertical distance from trough to crest

Wave length: the horizontal distance between two successive crests

Wave frequency: the number of waves breaking per minute

In deep water, water molecules within a wave move in a circular movement. It is only in shallow water that the water itself is moving forward (see Figure 10.2).

Why are some waves stronger than others?

The amount of energy in a wave depends on:

■ the speed of the wind: strong winds result in stronger waves because more energy is transferred into waves
■ how long the wind has been blowing: the longer the wind has been blowing, the more energy is transferred and the stronger the waves
■ the fetch (this is the maximum distance of open sea that a wind can blow over): the longer the fetch, the greater is the possibility of large waves if the storms are widespread.

▼ Figure 10.2 Waves in shallow water

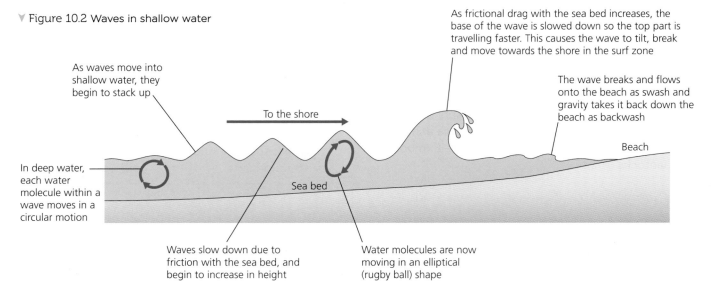

As waves move into shallow water, they begin to stack up

To the shore

As frictional drag with the sea bed increases, the base of the wave is slowed down so the top part is travelling faster. This causes the wave to tilt, break and move towards the shore in the surf zone

The wave breaks and flows onto the beach as swash and gravity takes it back down the beach as backwash

Beach

In deep water, each water molecule within a wave moves in a circular motion

Sea bed

Waves slow down due to friction with the sea bed, and begin to increase in height

Water molecules are now moving in an elliptical (rugby ball) shape

What are constructive and destructive waves?

Characteristics of constructive waves

Constructive waves are found in sheltered **bays** and spits where they build up sandy **beaches**. They are more common in summer than in winter. The swash spills forward over the beach, covering a large area. The wave loses energy as it is in friction with the sand, and some of the water soaks into the sand. Its swash is therefore relatively stronger than its backwash. This means that constructive waves build up beaches rather than destroy them.

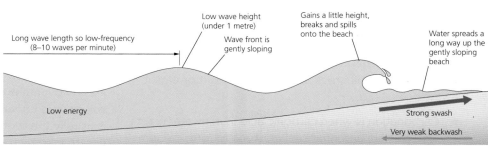

The strong swash transports sand and shingle up the beach to construct it

▲ Figure 10.3 Characteristics of a constructive wave

Characteristics of destructive waves

Destructive waves are found in more exposed bays, where they build up pebble beaches. They are more common in winter than in summer. Although a destructive wave's swash is much stronger than that of a constructive wave, the relative strength of a destructive wave's swash is much weaker than its backwash. This means that destructive waves can comb beach material back into the sea and lower beaches in winter. The force generated by a breaking destructive wave is also sufficient to erode a **headland**.

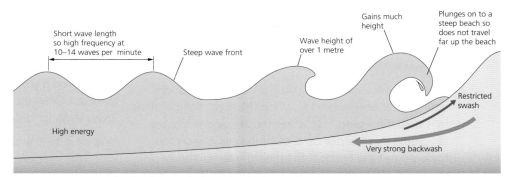

The relatively stronger backwash drags beach material out to sea

▲ Figure 10.4 Characteristics of destructive waves

→ Activities

1 State three differences between a constructive wave and a destructive wave.

2 Explain what makes a wave break on a beach.

3 Study Figure 10.1.

 a) Using the yacht and sailors as a size guide, estimate the height and wave length of the waves.

 b) Would you expect the wave frequency to be high or low? Explain your answer.

 c) Describe the impact that these waves will have as they break on the shore.

Fieldwork: Get out there

1 Devise a method for testing the frequency of waves.

2 Devise a method for determining how far up the beach the swash flows.

3 How would you ensure you were safe on the beach while collecting this data?

Weathering and mass movement

How does weathering weaken a cliff face?

Weathering is the breaking down of rock in situ (where it is). It is caused by day-to-day changes in the atmosphere, such as extremes of temperature and **precipitation**.

Chemical weathering

Chemical weathering is caused by a chemical reaction when rainwater hits rock and decomposes it or eats it away.

- Carbonation is when carbonic acid in rainwater reacts with calcium carbonate in limestone to form calcium bicarbonate. This is soluble, so limestone is carried away in solution.
- Hydrolysis is when acidic rainwater breaks down the rock, causing it to rot.
- Oxidation is when rocks are broken down by oxygen and water.

Mechanical (physical) weathering

Mechanical weathering results in rocks being disintegrated rather than decomposed. It is usually associated with extremes of temperature:

- **Freeze-thaw weathering** (frost shattering) happens when water enters cracks. When the night temperature falls below freezing, this water freezes and increases in volume by nine per cent, putting pressure on the rock around the crack. If the daytime temperature rises above freezing, the ice will thaw and relieve the pressure. Constant repetition of this diurnal (daily) freezing and thawing cycle causes angular rock fragments to break away and collect as scree at the base of the cliff.
- Salt weathering is when salt spray from the sea gets into a crack in a rock. It may evaporate and crystallise, putting pressure on the surrounding rock and weakening the structure.

How does mass movement happen?

Mass movement is the downslope movement of rock, soil or mud under the influence of gravity. Heavy rainfall is usually the trigger, but the scale of movement is determined by the extent of weathering on the slope.

Sliding

Landslide is the generic term for the downhill movement of a large amount of rock, soil and mud. Slides occur on steep cliffs previously weakened by weathering. Heavy rain infiltrates the soil and percolates down into the rock. The now heavier, saturated mass falls away along a distinct slip plane, which is a line of weakness such as a fault or bedding plane. A slide happens quickly. It starts by tearing away the vegetation on the top edge of the cliff. Once the slide has begun, its descent is aided by lubrication from the wet rocks below. The cliffs in the Jurassic Coast near Durdle Door in Dorset have suffered from large landslides in recent years.

A rock slide is where a large amount of rock slides down a cliff. This happens along a fairly straight slip plane, where rock falls as a block which maintains contact with the cliff. The leading edge of the slide collects as a pile of rocks on the beach or in the sea.

Mud slides are usually wet, rapid and tend to occur where slopes are steep (over ten degrees). Monmouth Beach at Lyme Regis, Dorset is prone to mudslides. They usually occur where vegetation cover is sparse so cannot hold the soil in place. They happen after a period of heavy rain. At the base of a mudslide, the saturated soil spreads out to make a lobe (see Figure 10.5).

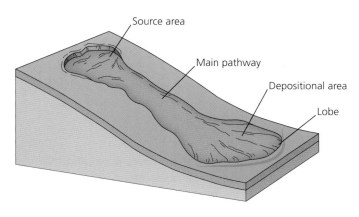

Figure 10.5 Mud slides

Source area

Main pathway

Depositional area

Lobe

Rock falls

Bare, well-jointed rocks are prone to freeze-thaw weathering, which results in falling rocks losing contact with the cliff face. At the bottom of the cliff they fan out to form a scree slope. Rock falls are common on vertical cliffs such as at Burton Bradstock, Dorset, where 400 tonnes of rock fell from the 49 metre vertical cliff in July 2012 (see page 128).

Slumping

While a slide takes a fairly straight path down a cliff, a **slump** has a concave slip plane so material is rotated backwards into the cliff face as it slips. At Barton on Sea, Hampshire, the cliff is slumping at up to 30 centimetres a day (Figures 10.6 and 10.7).

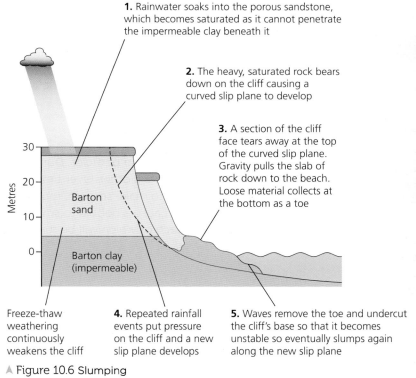

1. Rainwater soaks into the porous sandstone, which becomes saturated as it cannot penetrate the impermeable clay beneath it

2. The heavy, saturated rock bears down on the cliff causing a curved slip plane to develop

3. A section of the cliff face tears away at the top of the curved slip plane. Gravity pulls the slab of rock down to the beach. Loose material collects at the bottom as a toe

Barton sand

Barton clay (impermeable)

Freeze-thaw weathering continuously weakens the cliff

4. Repeated rainfall events put pressure on the cliff and a new slip plane develops

5. Waves remove the toe and undercut the cliff's base so that it becomes unstable so eventually slumps again along the new slip plane

Figure 10.6 Slumping

Figure 10.7 Rotational slumping at Barton on Sea, Hampshire

→ Activities

1 Draw a series of three or four labelled diagrams to show how freeze-thaw weathering widens cracks in a rock.

2 a) Define the terms weathering and mass movement.

  b) Explain the link between weathering and mass movement.

3 With the help of diagrams, explain the difference between a rock slide and a rock fall.

4 Draw a sketch of the cliffs at Barton on Sea (Figure 10.7). Annotate it to describe and explain what is happening to the cliff.

✪ KEY LEARNING

➤ The processes of coastal erosion

➤ How material is transported by waves

➤ Causes of deposition

Marine processes

What are the processes of coastal erosion?

Marine erosion is the removal of material by waves. Waves erode by hydraulic action, abrasion and attrition.

Hydraulic action

Hydraulic action is the relentless force of destructive waves pounding the base of cliffs. This causes repeated changes in air pressure as water is forced in and out of joints, faults and bedding planes. The forward surge of water compresses air in these cracks, and, as the wave retreats, there is an explosive effect as pressure is suddenly released. This onslaught is aided further by the weakening effect of weathering. Material breaks off cliffs, sometimes in huge chunks.

Abrasion (corrasion)

Destructive waves have enough energy to hurl sand and shingle at a cliff. The resulting scratching and scraping of the rock surface is called **abrasion**. This is concentrated between the high and low watermarks and is particularly effective in high-energy storm conditions.

Attrition

Attrition is the grinding down of load particles. During transport, pebbles collide with each other. Over time, this wears away jagged edges to make smooth, rounded pebbles. Some collisions may cause a pebble to smash into several smaller pieces, each of which will be further smoothed and rounded.

▲ Figure 10.8 Seven Sisters cliffs in East Sussex, UK. Note the rounded pebbles due to attrition

The rate of erosion will be higher where:

■ the coastline is exposed to a large fetch, such as the Needles on the Isle of Wight (Hampshire), which have an 8,000 kilometre fetch across the Atlantic Ocean

■ strong winds blow for a long time and create destructive waves. These conditions are common in winter

■ an area has no beach to act as a buffer between the sea and the cliffs

■ a headland juts out into the sea. Waves converge on a headland (wave refraction) and gain height and erosive energy

■ there are soft rocks. The average annual rate of erosion of the unconsolidated, soft, boulder clay rocks of the North Norfolk coast is five metres a year. Contrast this with the hard granite rocks of South West England which erode at 0.001 metres a year

■ a rock has many joints.

How is material transported by waves?

Load is transported material. Most marine load originates from river deposits, from eroded headlands and from the seabed. Load varies from fine silt to large rocks.

Transport onto the beach

Load is transported by waves. The larger and heavier the load particle, the greater is the velocity needed to transport it. The lightest load is carried in suspension or saltated, while heavier load is moved onto a beach by traction. These processes are outlined in Section 11, Figure 11.3.

Transport along the beach parallel to the shore

Load is transported along the shore by **longshore** (littoral) **drift**. The direction is determined by the **prevailing wind**. Along the Dorset coast, the prevailing southwest wind causes a drift in an easterly direction, making the swash surge up the beach at an oblique angle. In response to gravity, the backwash goes back down the beach at right angles to the shore. Suspended load is therefore carried easterly in a zig-zag manner.

What conditions cause deposition?

Deposition is when waves drop and leave behind the load they were transporting. The deposited load is called **sediment**. Deposition results in more sediment staying on the beach than is taken away by the backwash. This will happen:

- in low energy, sheltered bays, where constructive waves are dominant
- if there is a large source of sediment **updrift**, such as a rapidly eroding headland
- where there are large expanses of flat beach so the swash spreads out over a large area. This weakens the wave so that its backwash is not strong enough to transport the sand back out to sea
- when, on an outgoing tide, tidal material is trapped behind a **spit** (see page 136)
- where engineered structures like groynes trap sediment on the updrift side (see page 138).

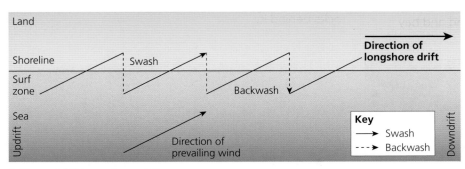

▲ Figure 10.9 Longshore drift

→ Activities

1 Draw an annotated diagram to explain how hydraulic action causes erosion.

2 Draw a series of diagrams to explain the effect of attrition on pebbles.

3 Create a fully annotated diagram to show longshore drift where the prevailing wind is from the east.

4 Suggest why local citizens in a seaside resort may be concerned about longshore drift in their area.

Fieldwork: Get out there!

Using two markers, such as rulers, a tape measure, a floating object and a digital watch, devise a method for measuring the speed of longshore drift.

Example

KEY LEARNING

➤ How landforms are affected by hardness of rock

➤ How rock structure affects landforms

➤ The location of major landforms of the Dorset coast

Geology and rock structure on the Dorset coast

Throughout this chapter, examples are used from the Dorset coast where differing geology, or rock types, and rock structures have led to a variety of coastal landforms.

How do hard and soft rocks affect landforms?

As hard rocks are less easily eroded than soft rocks, they project into the sea as headlands (see page 126) and form high cliffs. In the Isle of Purbeck, the hard Portland limestone forms steep cliffs while the softer Bagshot Beds, Wealden beds and sandstone form low-lying bays. Hard chalk rocks have produced arches and stacks (see page 130).

How does rock structure affect landforms?

Rock structure includes:

■ how rocks are aligned in relation to the coast (concordant and discordant coasts)

■ how rocks dip down to the sea as a result of folding.

Concordant and discordant coasts

Along the east coast of the Isle of Purbeck, the alternating layers of hard and soft rock run at right angles to the shore giving rise to headland and bay

formation (see page 126). This is typical of a discordant coastline. In contrast, the southern coast is fairly smooth in shape where the rock is uniform. This is typical of a concordant coastline with alternating layers of hard and soft rock running parallel to the coast. This can be seen in particular in the Kimmeridge clays and Lower Purbeck (limestone) rocks. However, at Lulworth, a cove has formed where waves have broken through a weakness in the hard Portland limestone and scooped out the softer rocks behind it.

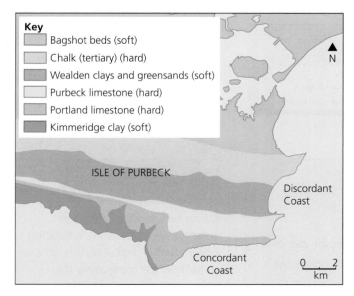

▲ Figure 10.10 Concordant and discordant coast, Isle of Purbeck, Dorset

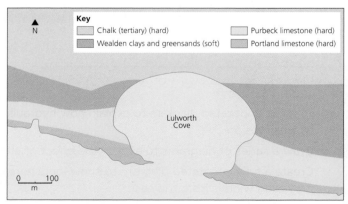

▲ Figure 10.11 Geology map of Lulworth Cove, Dorset, a concordant coastline

Rock's angle of dip

Sedimentary rocks are formed on the sea bed and are raised by mountain building processes, which folded the rocks over millions of years. In places along the Dorset coast an up-folded area called an anticline can be seen in a headland. This can be seen on the eastern edge of Stair Hole near Lulworth Cove. From left to right, there is a slight anticline, then a syncline (or downfold) moving into a clear anticline nearest the sea. The rising angle of dip at the coast results in a steep cliff profile.

▲ Figure 10.12 Folding at Stair Hole

What are the major landforms at the Dorset coast?

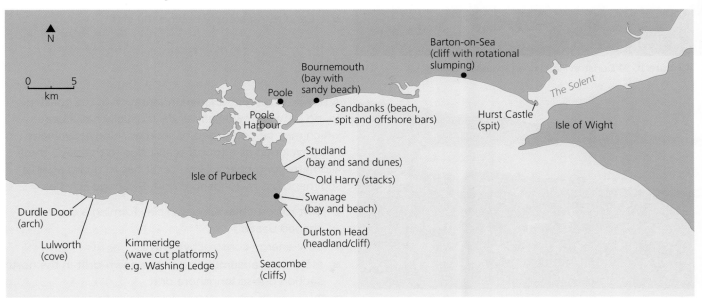

▲ Figure 10.13 The location of major landforms on the Dorset/Hampshire coast

→ Activities

1 Define the terms geology and rock structure.
2 Outline the difference between a concordant coast and a discordant coast.
3 Use Figure 10.13 to compile a table for coastal erosion landforms and coastal deposition landforms along the Dorset coast.

Geographical skills

Draw a geological sketch of Figure 10.12. Annotate it to show how this cove was formed.

⭐ KEY LEARNING

➤ The characteristics of headlands
➤ The characteristics of bays
➤ How headlands and bays are formed

Headlands and bays

The geology of a coast affects how the coastline is shaped by marine processes (pages 118–19). A discordant coastline is where bands of different rock types meet at near right angles to the coast, which results in the formation of **headlands and bays**.

What characterises a headland?

A headland is a **cliff** that juts out into the sea so it is surrounded by water on three sides. Headlands are composed of hard rock such granite, chalk or limestone, which are difficult to erode.

Characteristics of Durlston Head

- Portland limestone
- Near vertical cliff face
- High-energy area affected by destructive waves
- Hard rock jutting out into the sea.
- Caves forming in its sides
- Stacks and stumps
- Land rising steeply behind the cliff

▲ Figure 10.14 Durlston Head near Swanage, Dorset

What are the characteristics of a bay?

A bay is a crescent-shaped indentation in the coastline found between two headlands. It usually has a beach, which may be composed of sand or shingle. Swanage has developed as a tourist resort due to the bay's three kilometre-long beach. However, it has to guard against sediment loss by longshore drift. Swanage Bay has:

- Soft rock such as sand and clay forming a crescent-shaped beach
- Low-energy constructive waves
- More sand accumulating in the down-drift in the north section due to longshore drift
- Two headlands marking the edges of the bay

▲ Figure 10.15 Swanage Bay, Dorset

How are headlands and bays formed?

Initial development

A discordant coastline has rocks of different hardness aligned at right angles to the coast. Differential erosion means soft rocks are eroded at a faster rate than hard rocks. Figure 10.16 explains the formation of headlands and bays in part of Dorset's coast. Notice how the soft clays and sands are eroded more easily than the harder chalk and limestone. This has resulted in the formation of bays at Studland and Swanage, and headlands.

Later stages of headland and bay development

Once a headland and bay pattern has emerged, the processes operating on the coastline are reversed. The now sheltered bays become low energy environments in which deposition occurs, and the exposed headlands become targets for erosion. This is explained by wave refraction (Figure 10.17).

Before and during erosion

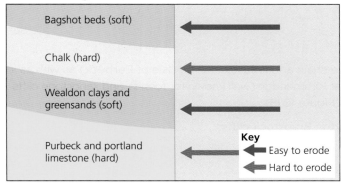

The section of Dorset's coast today

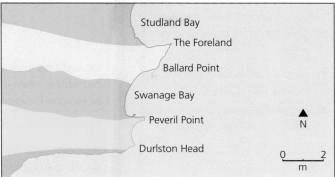

▲ Figure 10.16 Differential erosion of a discordant coastline

Stage 4: On either side of the headland, the faster moving parts of the wave now refract (bend) towards the headland to keep the line of the crest intact. This concentrates more wave energy on the headland. As a result, the rate of erosion is increased.

Meanwhile, low-energy waves are building up beaches in the adjoining bays. This is helped by the creation of diverging longshore drift currents at the headland, where constructive waves carry eroded material towards the bays.

Stage 3: The part of the wave crest that is on course to break at the headland is the first to experience shallow water and to experience frictional drag. This slows down the lower part of the wave and makes it higher and steeper.

Stage 2: As the wave crest approaches a discordant coastline, it begins to take on the shape of the coastline.

Stage 1: The wave crest is in deep water where it does not reflect the shape of the land.

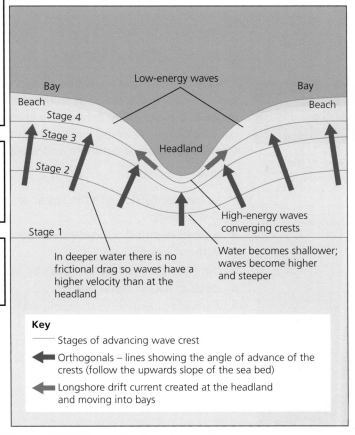

▲ Figure 10.17 Wave refraction at a headland

→ **Activities**

1 Draw a sketch of Figure 10.14 and label it to show the characteristics of a headland.

2 Study Figure 10.15 (the camera is facing south).

   a) Draw a sketch of Figure 10.15 to show the characteristics of Swanage Bay.

   b) Draw an arrow to show the direction of longshore drift.

3 a) Explain how geology influences the formation of headlands and bays.

   b) Explain why erosive wave energy is now concentrated on headlands rather than bays.

Fieldwork: Get out there!

1 Suggest how you might test the hypothesis that headlands have destructive waves and bays have constructive waves?

2 How would you present this evidence?

⭐ KEY LEARNING

➤ The characteristics of cliffs

➤ The characteristics of wave cut platforms

➤ How cliffs and wave cut platforms are formed

Cliffs and wave cut platforms

What are the characteristics of cliffs?

Dorset has impressive chalk and limestone cliffs (Isle of Purbeck) as well as crumbling shale cliffs at Kimmeridge (well known for fossils), and sandstone cliffs at Burton Bradstock (Figure 10.18).

Characteristics of Burton Bradstock cliffs:

- height: 45 metres
- hard Bridport sandstone
- horizontal bedding
- layers of harder rock jutting out
- near-vertical bare rock cliff face
- wave-cut notch at the base
- fallen rocks at the base in some places.

▼ Figure 10.18 Cliffs near Burton Bradstock, Dorset

The rate at which a cliff recedes depends on the hardness of the rocks. Contrast the 0.001 metre annual rate of erosion of the hard granite cliffs at Land's End (Cornwall) with the five metre annual rate of erosion of the soft boulder clay at Happisburgh (North Norfolk). Soft rocks are prone to slumping, but hard rocks create high cliffs and **wave cut platforms**.

What are the characteristics of a wave cut platform?

A wave cut platform is an area of bedrock visible at the base of some cliffs. It slopes to the sea at a very gentle angle, and is generally only visible at low tide. The wave cut platform of Washing Ledge at Kimmeridge Bay, Dorset is formed of dolomite and shale.

A wave cut platform:

- slopes gently down to the sea at an angle of 3–4 degrees
- an overall pitted appearance of bare rock interspersed with rock pools
- bare rock, smoothed in places by attrition
- deep cracks in some places
- barnacles clinging to the rock, with seaweed thrown there by rough seas
- pools that form in larger depressions. Shingle, shells and crabs collect there
- covered at high tide and exposed at low tide.

◄ Figure 10.19 A wave cut platform at Kimmeridge, Dorset

How are cliffs and wave cut platforms formed?

Cliffs are formed by coastal erosion. Weathering initially weakens the rock, and mass movement transports material to the beach.

Stages in the formation of a cliff and wave cut platform

Stage 1: The land slopes down to the sea. Freeze-thaw weathering weakens the rock.

Stage 2: Marine erosion is concentrated between the high and low watermark. Wave pounding by hydraulic action, and abrasion by shingle hurled at the land, erode the base.

Stage 3: Continued erosion causes rock to break away and collect at the base of the cliff. Destructive waves remove this material. The cliff now has a notch at its base. The section of cliff above the notch is now unsupported and becomes more precarious as erosion continues.

Stage 4: Eventually, the notch is enlarged to the point where the overhanging cliff can no longer defy gravity. It breaks away and falls onto the beach. The resulting steep drop is called a cliff.

Aided by freeze-thaw weathering this process of erosion, notch formation and cliff collapse continues. This causes the cliff to retreat back towards the land and, if the land is still sloping, to increase in height.

Stage 5: As the cliff retreats, the former base of the cliff is left as a wave cut platform. In places, it will be continuously smoothed by shingle grinding over it. Elsewhere, it may become encrusted with barnacles or covered with algae. Joints will become enlarged by abrasion and hydraulic action.

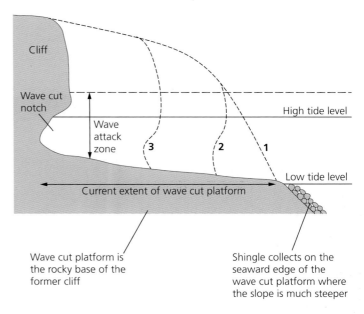

▲ Figure 10.20 Cliff retreat and wave cut platform formation

→ Activities

1. Draw a sketch of the cliff at Burton Bradstock (Figure 10.18).
   a) Label the sketch to show its characteristics.
   b) Indicate where weathering processes will operate on the cliff.
   c) Indicate where marine processes will operate on the cliff.

2. Suggest why the cliffs at Burton Bradstock are almost vertical.

3. Draw a series of five annotated diagrams to explain how an area of land along the coast may be eroded into a cliff and a wave cut platform.

4. a) Study Figures 10.19, 10.20 and the bullet points on page 128. Draw an annotated sketch of a wave-cut platform to show its characteristics.
   b) Describe the processes that continue to shape the wave cut platform.

➤ How caves, arches and stacks are formed
➤ The characteristics of caves, arches and stacks

Caves, arches and stacks

Section 10.5 showed how headlands project into the sea and how they are eroded by destructive waves associated with wave refraction. So, despite the hardness of rock, headlands are constantly being re-shaped by the waves. Landforms located at headlands include **caves**, sea **arches**, **stacks** and **stumps**.

How are caves, arches and stacks formed?

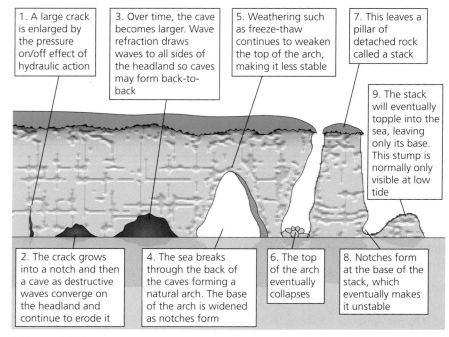

1. A large crack is enlarged by the pressure on/off effect of hydraulic action

3. Over time, the cave becomes larger. Wave refraction draws waves to all sides of the headland so caves may form back-to-back

5. Weathering such as freeze-thaw continues to weaken the top of the arch, making it less stable

7. This leaves a pillar of detached rock called a stack

9. The stack will eventually topple into the sea, leaving only its base. This stump is normally only visible at low tide

2. The crack grows into a notch and then a cave as destructive waves converge on the headland and continue to erode it

4. The sea breaks through the back of the caves forming a natural arch. The base of the arch is widened as notches form

6. The top of the arch eventually collapses

8. Notches form at the base of the stack, which eventually makes it unstable

▲ Figure 10.21 Formation of caves, arches and stacks

What characterises a cave?

Caves at a headland may be several metres high at their entrance and taper back a long way (Figure 10.21). The Dorset coast has caves in the limestone cliffs at Durlston Head and also in the chalk stack at Ballard Down and also in some of the chalk stacks at the Foreland (Figure 10.24).

A blow hole (gloup) may form in the roof of the cave towards the back.

Pressure from waves may push water up the blow hole so that it emerges on the cliff above

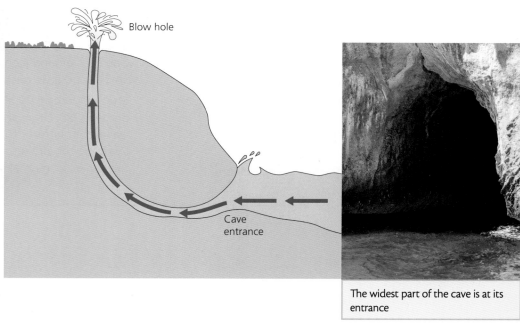

Blow hole

Cave entrance

The widest part of the cave is at its entrance

▲ Figure 10.22 Characteristics of a cave

What characterises a sea arch?

- Unsupported top of the arch
- Wave-cut notches at the base of the arch, so wide at the base
- Water going right through the gap
- The arch is an extension of the headland

What characterises stacks?

- Detached blocks or pillars of rock located off a headland
- Some may be pinnacle shaped like the needles of the Isle of Wight and some stacks off Ballard Down
- Often several metres high
- Hard rock
- Wave-cut notches at the base

▲ Figure 10.23 Characteristics of Durdle Door, Dorset – a sea arch

What characterises stumps?

Some headlands also have stumps, which are the bases of collapsed stacks. They can only be seen at low tide. Old Harry's Wife is a stump that lies just beyond Old Harry. The stump is submerged.

▲ Figure 10.24 Old Harry Rocks – sea stacks at the Foreland, off Ballard Down, Dorset

→ Activities

1 a) Describe a cave.
  b) Draw a series of three to four diagrams to explain how a cave is formed.
  c) Explain why caves may form on either side of a headland.

2 a) Imagine you were to take a photo of the Old Harry Rocks in 200 years' time. Sketch what you think this headland would look like.
  b) Explain the changes you have made to the original photo.

3 a) Draw a diagram to show the characteristics of a stump.
  b) Explain how a stump is formed.

4 Draw a sketch of Durdle Door arch (Figure 10.23). Label the arch to show its characteristics.

5 Draw a sketch of Figure 10.24. Annotate with the following labels: a broad stack composed of hard chalk, a new small arch, the steep pillar-shaped stack of Old Harry, rock debris from the previous collapse of an arch, the headland.

KEY LEARNING

- How beaches are formed
- The characteristics of beaches
- Beach profiles
- Why beach profiles change

Beaches

A beach is a landform of coastal deposition that lies between the high and low-tide levels. Most beaches are formed of sand, sand and shingle or pebbles, as well as mud and silt. A beach that forms in a bay is crescent- shaped (see Figure 10.25), but its shape is distorted by longshore drift, so the beach is narrower updrift than downdrift.

How are beaches formed?

Section 10.3 showed how waves transport materials from the sea to the shore. Two distinct types of beaches can be formed.

Sandy beach

In sheltered bays, low-energy constructive waves transport material onto the shore (see Section 10.1). The swash is stronger than the backwash, so sediment is slowly but constantly moved up the beach. Once the tide has gone out, there is more material on the beach than before. An example is Sandbanks Beach, Poole, Dorset (Figure 10.25 A).

Pebble beach

Exposed beaches such as West Bay, part of Chesil beach in Dorset (Figure 10.25 B), sometimes have a large fetch. The plunging nature of destructive waves (see Section 10.1), along with their stronger backwash, means that pebbles are not moved far up the beach, which makes the beach profile steep. A storm beach may form when there is wild, stormy weather and waves hurl boulders and large pebbles to the back of a beach.

What characterises beach?

Figure 10.25 Contrasting Dorset beaches

	A: Sandy beach	B: Pebble beach
Characteristics	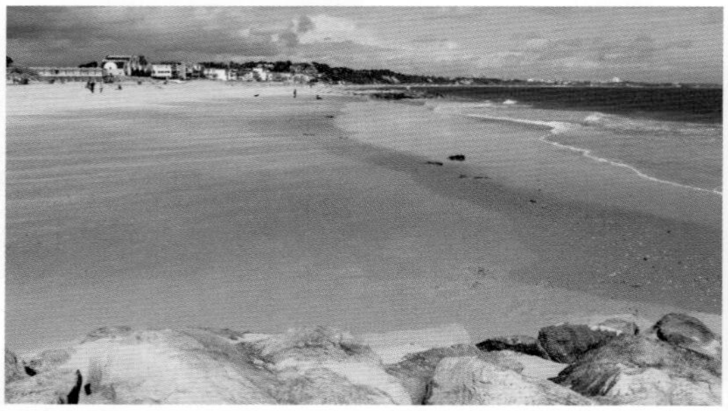	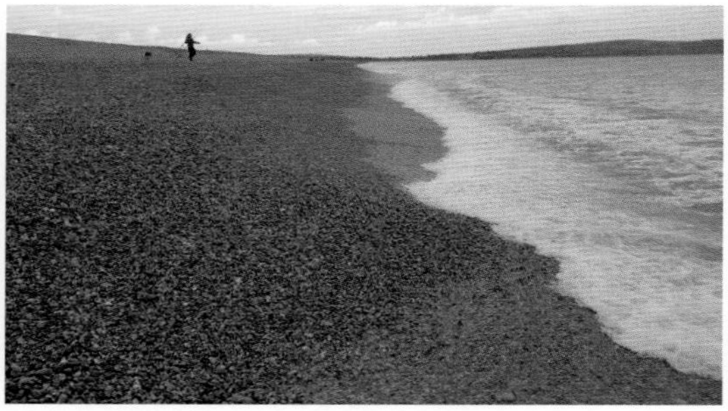
Gradient	Generally shallow, almost flat	Generally steep
Dominant waves	Constructive	Destructive
Distance stretches inland	A long way	Not far
Back of beach	Sand dunes (sometimes)	Storm beach with large pebbles
Other characteristics	At low tide, small water-filled depressions called runnels form. These are separated by small sandy ridges running parallel to the shore. The wet sand may have a rippled appearance	Pebbles increase in size towards the back of the beach

What is a beach profile?

A beach profile shows the gradient from the back of the beach to the sea. A sandy beach generally has a gentle, fairly flat profile, whereas a pebble beach has a steep, stepped profile.

Why do beach profiles change?

A berm is a terrace on a beach that has formed in the backshore, above the water level at high tide. On broad beaches there may be three or more subparallel berms, each formed under different wave conditions. Berms are formed in calm weather when constructive waves transport material onto the beach. While an existing berm is moved up the beach by storms and spring tides, a new berm may develop and change the beach profile.

- In winter, berms, and sometimes the sand dunes at the back of the beach, are eroded by destructive waves which drag beach deposits offshore to create an **offshore bar**. This lowers the height of a beach.
- In late spring and summer, so long as longshore drift is not depleting the beach of sand, constructive waves will rebuild the beach. The offshore bar is worked by the waves to rebuild the berms, and dunes are replenished by saltation by the wind (see Section 10.9).
- Destructive waves often result in winter profiles that are narrower and steeper.

▲ Figure 10.26 A sandy beach at low tide

Labels on figure: Water-filled trough called a runnel | Sand dunes | Berm | Dry backshore | Ridge | Ripples in the sand made by waves | Wide sandy beach shows a large tidal range

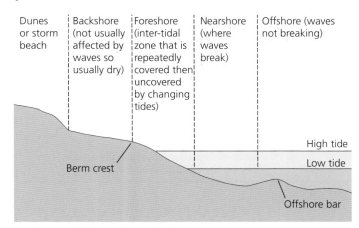

Labels: Dunes or storm beach | Backshore (not usually affected by waves so usually dry) | Foreshore (inter-tidal zone that is repeatedly covered then uncovered by changing tides) | Nearshore (where waves break) | Offshore (waves not breaking) | Berm crest | High tide | Low tide | Offshore bar

▲ Figure 10.27 Beach profile in summer

Fieldwork: Get out there!

Research how to measure the gradient (profile) of a beach using a clinometer.

→ Activities

1   Draw a labelled sketch of photo B in Figure 10.25 to show the characteristics of a pebble beach.

2   Explain why pebble beaches are steep.

3   The following descriptions are mixed up. Can you correct them?

Nearshore	The inter-tidal zone repeatedly covered, then uncovered by changing tides
Backshore	The breaker zone where waves break
Foreshore	Fairly far out to sea where the waves do not break
Offshore	An area that is not usually affected by waves, so the sand is usually dry.

4   a)   Define the term beach profile.
    b)   Draw the shape of a beach profile in summer (Figure 10.27).
    c)   On the same diagram, use a different colour to draw how you would expect this profile to look in winter.
    d)   Explain the change you have made to the original profile.

5   a)   Explain how a sandy beach is formed.
    b)   Study Figure 10.26. Describe the changing nature of this sandy beach if you were to walk from the sand dunes towards the sea.

➤ How sand dunes are formed
➤ The characteristics of sand dunes
➤ Dune succession

Sand dunes

Sand dunes (large heaps of sand) form on the dry backshore of a sandy beach.

How are sand dunes formed?

For a sand dune to form, it needs:

- a large flat beach
- a large supply of sand
- a large tidal range, so there is time for the sand to dry
- an onshore wind to move sand to the back of the beach
- an obstacle such as drift wood for the dune to form against.

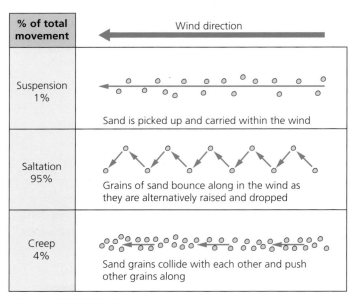

% of total movement	Wind direction →
Suspension 1%	Sand is picked up and carried within the wind
Saltation 95%	Grains of sand bounce along in the wind as they are alternatively raised and dropped
Creep 4%	Sand grains collide with each other and push other grains along

▲ Figure 10.28 Wind transport of sand

Wind moves sand in three ways (Figure 10.28):

- When there are obstacles, such as driftwood, the heaviest grains of sand will settle against the obstacle to form a small ridge. Lighter grains may be transported and will settle on the other side of the obstacle.
- Eventually, the area facing the wind begins to reach a crest. This is because the pile of sand becomes so steep that it becomes unstable and begins to collapse under its own weight.
- When this happens, the lighter grains of sand fall down the other side on the lee (slip) face. Sand stops slipping once a stable angle has been reached at 30–34 degrees.
- The repeated cycle of wind blowing up the windward side and slipping down the leeward side causes a sand dune to migrate inland over time.
- A sand dune itself becomes an obstacle, so more dunes may form in front of it. The height of dunes depends on the strength of the wind. Stronger winds create higher dunes.

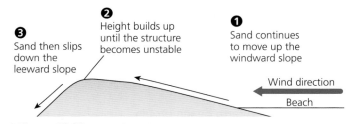

❸ Sand then slips down the leeward slope

❷ Height builds up until the structure becomes unstable

❶ Sand continues to move up the windward slope

Wind direction

Beach

▲ Figure 10.29 Sand dune formation

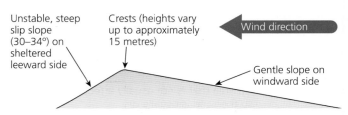

▲ Figure 10.30 A sand dune near Studland, Dorset

What characterises sand dunes?

Unstable, steep slip slope (30–34°) on sheltered leeward side

Crests (heights vary up to approximately 15 metres)

Wind direction

Gentle slope on windward side

▲ Figure 10.31 Characteristics of a sand dune

How do dunes change inland?

Several lines of dunes may run parallel to the shore (Figure 10.32). The change in vegetation with increased distance inland is known as a dune succession:

- Dunes grow taller. Embryo dunes are only a few metres high whereas mature dunes may be up to fifteen metres high.
- Size increases inland as long-rooted marram grass and other vegetation bind the sand together, thereby preventing further migration. Marram grass grows quickly and aids sand accretion. Its long roots bind the sand and help build up the height of the dunes.
- Dunes closest to the beach have a yellow, sandy colour and not much vegetation. Dunes further back look grey and less sand-like.

- Inland, the dunes become increasingly colonised by vegetation.
- Each line of dunes is separated by a trough called a slack. Slacks are formed by the ongoing removal of sediment from the leeward base of one line of dunes and up the windward side of the next dune line. Sometimes, slacks are eroded so much that they reach down as far as the **water table**, resulting in the formation of salty ponds.
- Occasionally, a dune may develop a huge depression called a blowout, when strong winds remove sand from an area that has lost its protective vegetation cover.

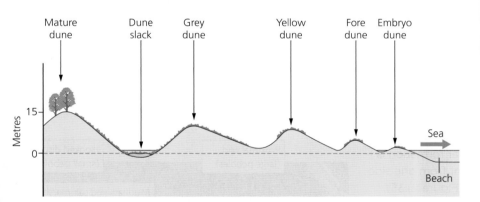

▲ Figure 10.32 How sand dunes change inland

Fieldwork: Get out there!

1 Describe how you might use a quadrat (a square frame) and a tape measure to carry out systematic sampling of vegetation over lines of sand dunes.

2 How would you expect the density of vegetation to change inland?

3 Suggest a reason for the change.

→ Activities

1 The following definitions have been mixed up. Can you match the correct term and definition?

Embryo dune	The slope that faces away from the wind
Marram grass	The upper horizontal limit of wet sand
Saltation	Where there is a trough or low point in a line of dunes
Crest	A newly formed sand dune closest to the sea
Water table	The slope that faces the wind
Dune slack	A plant found in sand dunes that has long, binding roots
Leeward slope	How sand is bounced along by the wind
Windward slope	The top of a sand dune

2 Explain why a large tidal range aids the formation of sand dunes.

3 Which form of wind transport is dominant in sand dune formation? Can you suggest why?

4 Draw an annotated diagram to show how sand initially builds up behind an obstacle on the beach.

5 Explain why sand dunes are steeper on one side.

Spits and bars

If unchecked, longshore drift (see page 123) can deplete updrift beaches of their sand and create new landforms downdrift. Two landforms associated with longshore drift are spits and bars.

How is a spit formed?

A spit is a sand or shingle beach that is joined to the land but projects downdrift into the sea. Spits form where the coastline suddenly changes shape or at the mouth of an estuary.

A spit is an unstable landform. It will continue to grow until the water becomes too deep or until material is removed faster than it is deposited. Hurst Castle spit in Hampshire is growing in length, but losing shingle from its main ridge. Sandbanks Spit in Poole, Dorset, is not growing, due to tidal currents and dredging.

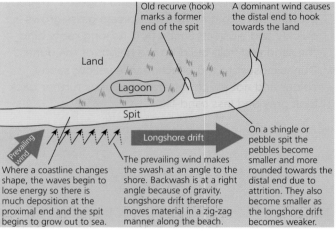

▲ Figure 10.33 How longshore drift forms a spit

How are bars formed?

A bay bar (or barrier beach bar) is a ridge of sand or shingle that stretches from one side of a bay to the other, forming a lagoon behind it. Barrier beach bar formation is due to longshore drift transporting sediment from one side of a bay to the other (see Figure 10.33 and Figure 10.35).

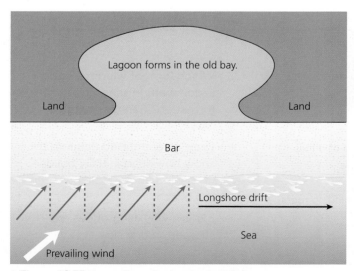

▲ Figure 10.35 Formation of a barrier beach bar

What are the characteristics of a spit?

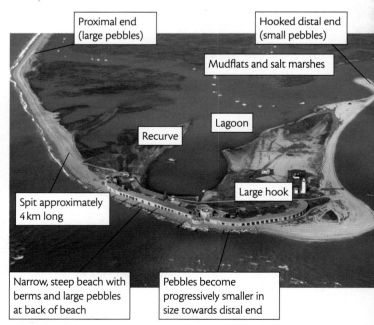

▲ Figure 10.34 Characteristics of Hurst Castle spit

A submerged offshore bar is a raised area of the seabed that lies a little offshore. Submerged bars form in shallow waters where there is a lot of sediment on a beach. These bars are formed by the transport of sediment off and then back onto a beach. In stormy weather, destructive waves drag beach material out to sea to form an offshore bar. When calm conditions are resumed, constructive waves steadily transport bar sediment towards the shore, thus moving the material back onto the beach.

Barrier islands are visible offshore bars that form parallel to the coast, often in chains. Formation varies. They may be:

- offshore bars formed by waves churning up the sand to make a vertically forming submarine bar
- due to waves breaching a spit, e.g. Scolt Head, North Norfolk
- coastal submergence along a discordant coast, which leaves higher land as islands.

Bars can also form out to sea where tidal currents result in a build-up of sediment. Rising sea levels due to ice melting have driven some offshore bars onshore.

What are the characteristics of bars?

Characteristics of a bay bar

One of the UK's best-known bay bars is Slapton Sands in Devon. Unlike a submerged bar, at least some part of a barrier beach bar is visible at all times.

Characteristics of a submerged bar

A submerged bar is often completely detached from the land and parallel to it. It can be several metres long, but is usually no more than ten metres wide. You can see submerged bars off Poole's beaches in Dorset.

Characteristics of an offshore barrier island

Scolt Head barrier island in north Norfolk is completely detached from the shore and aligned parallel to it. It has sand dunes, salt marshes and mudflats.

▲ Figure 10.36 Slapton Sands bay beach bar

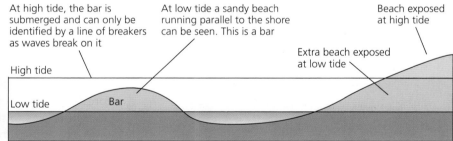

At high tide, the bar is submerged and can only be identified by a line of breakers as waves break on it

At low tide a sandy beach running parallel to the shore can be seen. This is a bar

Beach exposed at high tide

Extra beach exposed at low tide

High tide

Low tide

Bar

▲ Figure 10.37 Characteristics of a submerged bar

→ Activities

1. Explain the role of longshore drift in the formation of a spit.

2. Draw a sketch of the bar at Slapton Sands (see Figure 10.36). Annotate the sketch to show the characteristics of the bar.

3. Use annotated diagrams to explain why a submerged bar is usually only a temporary feature.

4. State two differences between a submerged bar and a barrier beach bar.

5. a) Use an atlas or local OS map to locate the nearest spit to your home. Use the internet to obtain a photo of that spit.
   b) Label the photo to show the spit's characteristics.
   c) Add arrows to show the direction of longshore drift and the cross wind.

KEY LEARNING:

➤ How sea walls, rock armour, gabions and groynes reduce erosion

Coastal management: hard engineering

Exposed coastal areas take a battering from erosive, destructive waves. If the coastline contains high-value buildings, then the local authority tries to protect the coastline from erosion.

Hard engineering is when expensive artificial structures are used for protection. They are effective, but do not blend in well with the natural environment. Hard engineering includes **sea walls**, **groynes**, **rock armour** and **gabions**.

▲ Figure 10.38 Sea wall at Highcliffe, Hampshire

What are sea walls?

A sea wall provides a barrier between waves and the land. It is placed along the back of a beach. Recurved sea walls (Figure 10.38) are more expensive than flat sea walls, but are more effective in reflecting waves and reducing overtopping. Steps are often added to the base to give extra stability.

The recurved face rotates the wave backwards so that some of its energy is reflected back out to sea. This impedes the next wave and reduces its energy, reducing its erosive power.

What are groynes?

Groynes are wooden or stone structures built in the foreshore; they look like fences or walls. They are built at right angles to the beach and are spaced at regular intervals, approximately 50 metres apart. Traditionally, groynes were made of hardwood timber, but stone groynes are now more popular.

Groynes trap sediment transported by longshore drift. This builds up the beach on the updrift side of a groyne (Figure 10.39). A larger beach provides a more effective buffer as it absorbs the waves' energy, and reduces the impact of waves on the sea wall. Groynes are particularly effective when used in conjunction with **beach nourishment** (see page 142).

▲ Figure 10.39 A wooden groyne traps sediment

What is rock armour?

Rock armour (rip rap) is made up of thousands of tonnes of huge boulders of hard rock like granite, to act as a barrier between the sea and the land. Boulders are generally big enough not to be moved by storm waves. Their downward slope arrangement to the sea deflects the waves' energy. As water enters gaps between boulders, pressure is released and this reduces the waves' energy, so there is little scouring of the base. This form of hard engineering is therefore highly effective.

▲ Figure 10.40 Rock armour at the back of a beach

What are gabions?

Gabions are steel wire-mesh cages filled with pebbles or rocks. They are placed at the back of a sandy beach to create a low, wall-like structure (see Figure 10.41). Water enters the cages and this absorbs and dissipates some of the waves' energy, thus reducing the rate of erosion. Gabions may also be placed in front of a cliff, where they may be covered with vegetation. This gives stability to a cliff and reduces the risk of landslides.

▲ Figure 10.41 Gabions near Hengistbury Head, Hampshire

→ Activities

1. a) In pairs, draw an imaginary map of a 500 metre stretch of coastline. Add features such as roads, houses, some named key buildings and a beach. Indicate the direction of longshore drift.
   b) Put forward an argument for protecting this coastline.
   c) One person then writes a proposal for building groynes, while the other person writes a proposal for a sea wall. (Include social, economic and environmental aspects.) Present the proposals to another pair of students and gain their opinions on the strength of the proposals. Agree on a solution.

2. Study Figure 10.39. Assuming the camera was pointing south, in which direction is longshore drift travelling? Explain your answer.

3. Draw an annotated sketch of Figure 10.39 to show how a groyne reduces coastal erosion.

4. Draw annotated diagrams to show how rock armour and gabions reduce the rate of coastal erosion.

Fieldwork: Get out there!

Devise a way of measuring the effectiveness of a groyne in trapping sediment.

Suggest how the data collected could be presented.

Justify this choice of presentation.

● KEY LEARNING

➤ The benefits and costs of hard engineering

Benefits and costs of hard engineering

What are the benefits and costs of sea walls, groynes, rock armour and gabions?

▼ Figure 10.42 Benefits and costs of hard engineering

	Benefits	Costs
Sea walls	• **Social:** A sea wall gives people a sense of security. It often has a promenade on top of it, which doubles up as cycle route outside peak walking periods. Steps at the base of a wall act as seating areas for beach users. • **Economic:** If well maintained, sea walls can last for many years. • **Environmental:** Sea walls do not impede the movement of sediment downdrift, so they do not disadvantage other areas.	• **Social:** They restrict people's access to the beach and if waves break over the sea wall (overtopping), coastal flooding may occur. • **Economic:** At about £5,000 per linear metre, sea walls are expensive to build. Repairs are also expensive. Reflected waves scour the beach in front of a sea wall and this undermines its foundations. If damage is not repaired quickly, the result may be devastating. In Dawlish, Devon (February 2014), the sea wall carrying the main south coast railway collapsed causing £35 million repairs to the wall and track (see Figure 10.43). To reduce scouring, rock armour (page 139) and beach nourishment (page 142) may be needed. This adds to the cost. • **Environmental:** From the beach, a wall of concrete is ugly to look at. Sea walls can also destroy habitats.
Groynes	• **Social:** Rock groynes at Sandbanks, Poole, have concrete crests for people to walk along to reach a viewing or fishing point. Groynes also act as windbreaks. • **Economic:** At £5,000 each, groynes are relatively cheap and, if well maintained, can last up to 40 years. A larger beach, with more space for activities, attracts more tourists, which boosts the local economy.	• **Social:** Groynes are barriers, which impede walking along a beach. They are also dangerous, as they have deep water on one side and shallow water on the other. This is a particular hazard to children who find it hard to resist climbing on them. Groynes may also be a danger to wind surfers, who may collide with them. • **Economic:** By trapping sediment, groynes restrict the supply of sediment down-drift. For example, the new groynes at Poole restrict sediment movement towards Bournemouth. The problem is merely passed on to incur more cost. Groynes are ineffective in stormy conditions and need regular maintaining so they do not rot. • **Environmental:** Groynes may be considered unattractive, especially degraded ones.
Rock armour	• **Economic:** It is relatively cheap. Rock armour costs £1000–3000 a metre, compared to £5,000 a metre for a sea wall. • The structure is quick to build and easy to maintain. It can be built in weeks rather than the months it takes to make a sea wall. If well maintained, rock armour lasts a long time. • It is versatile, as it can be placed in front of a sea wall to lengthen its lifespan or used to stabilise slopes on sand dunes.	• **Social:** Rock armour makes access to the beach difficult, as people have to clamber over it or make long detours. People may have accidents when clambering over it as rocks may be unstable and, if rocks are regularly covered by the tide, they may collect slippery seaweed, which accentuates the hazard. • **Economic:** Highly resistant rocks from Norway and Sweden are often used in preference to rocks from local quarries. This may cause resentment and it inflates the cost considerably. Also, heavy storm waves will move rocks and so the armour needs regular maintaining. • **Environmental:** Rock armour is ugly and it often covers vast areas of a beach. Driftwood and litter become trapped in the structure and imported rocks do not blend in with the local geology.
Gabions	• **Economic:** At £110 a metre, they are relatively cheap and easy to construct. Gabions are often constructed on site using local pebbles. This makes them much cheaper than sea walls, rock armour or groynes. It also makes them ideal as a quick-fix solution. For the cost, they are good value for money, as they may last 20–25 years. • **Environmental:** They blend in better than other hard engineering methods, especially when sand is blown into them or when they are covered by vegetation.	• **Economic:** The use of gabions is restricted to sandy beaches, as shingle hurled at them would quickly degrade them. Gabions are easily destroyed, so regular maintenance is needed. Repair of embedded, vegetation-covered gabions can be expensive. The gabions built at Thorpeness, Suffolk in 1976 had their covering of topsoil and vegetation washed away by storms in 2010. It cost £30,000 to repair them. • **Social:** In a damaged state, gabions are dangerous. People may trip over them or cut themselves on the broken steel wire mesh. • **Environmental:** Damaged gabions are unsightly and sea birds may damage their feet in them.

◄ Figure 10.43 Overtopping of the seawall at Dawlish, Devon

➤ Figure 10.44 Collapsed gabions are unsightly and dangerous

➔ Activities

1 Suggest what damage is being done by overtopping of the seawall at Dawlish (Figure 10.43).

2 Suggest why the sea wall is not effective.

3 a) Imagine you live close to the beach in Figure 10.44. Write a letter to the council about the state of the gabions. Include economic, social and environmental concerns.

   b) Write a report, supported by a labelled diagram, to put a case for one named alternative type of hard engineering for this coastline.

4 a) Select either a UK beach that you have visited where hard engineering has been used, or the beach at Lyme Regis, Dorset. Use your own knowledge and/or the internet to create a sketch map showing the hard engineering structures on the beach.

   b) Write a report describing and commenting on the level of coastal protection that has been given by these structures.

   c) Explain why coastal protection was needed.

KEY LEARNING

➤ Beach nourishment, beach reprofiling and sand dune regeneration
➤ The benefits and costs of soft engineering

Coastal management: soft engineering

Soft engineering works more in sympathy with nature than hard engineering. It is generally less expensive, but often less effective.

What is beach nourishment?

Beach nourishment is a broad term for the replacement of lost sediment. A nourished beach means fewer waves reach the back of a beach. As more wave energy is absorbed and dissipated by the beach, the rate of erosion is reduced. The following techniques show how beaches are nourished.

Beach recharge

This is where sediment is taken from a bay and placed on a beach that is losing sand. This happens every summer at Pevensey (East Sussex), where longshore drift removes 20,000 cubic metres of beach sediment a year. A dredger collects shingle from the seabed and, on the high tide, comes in twice daily to pump out the sand. At Sandbanks in Poole (Dorset), recharge takes place every ten years. Bulldozers are often used to spread out the sand.

▲ Figure 10.45 Beach recharge at Poole

Beach recycling

This is the removal of sand from a down-drift area, which is building up sand and returning it up-drift. At some beaches, e.g. Seaford, East Suffolk, trucks move around 100,000 cubic metres of shingle twice every year.

What is beach reprofiling?

Beach reprofiling is the artificial re-shaping of a beach using existing beach material. In winter, a beach is lowered by destructive waves (see page 119). After winter storms, bulldozers move shingle back up the beach (see Figure 10.46). Like beach nourishment, reprofiling ensures that the beach is large enough to be an effective buffer between land and sea.

▲ Figure 10.46 Reprofiling

What is sand dune regeneration?

Sand **dune regeneration** is the artificial creation of new sand dunes or the restoration of existing dunes. Sand dunes act as a physical barrier between the sea and the land. They absorb wave energy and water. In this way, they protect the land from the sea.

Sand dunes near Studland, Dorset are often regenerated.

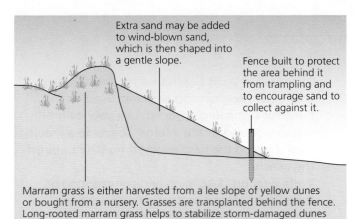

Extra sand may be added to wind-blown sand, which is then shaped into a gentle slope.

Fence built to protect the area behind it from trampling and to encourage sand to collect against it.

Marram grass is either harvested from a lee slope of yellow dunes or bought from a nursery. Grasses are transplanted behind the fence. Long-rooted marram grass helps to stabilize storm-damaged dunes when planted on their eroded windward faces.

▲ Figure 10.47 Sand dune regeneration

What are the benefits and costs of these soft engineering strategies?

▼ Figure 10.48 Benefits and costs of soft engineering

	Benefits	Costs
Beach nourishment	**Social**: A wider beach means more room for beach users. People living along the seafront are more protected from coastal flooding.	**Social**: During re-nourishment, access to the beach is restricted for several weeks. Beach recycling may cause resentment from residents living close to the donor area.
	Economic: At Sandbanks, the wider, nourished beach protects very expensive properties. The buffer of a widened beach reduces sea wall maintenance costs. A broader beach may also attract more tourists.	**Economic**: Although cheaper than hard engineering options, this has high overheads as it costs around £300,000 to hire a dredger. The 137,000 m³ of nourishment at Sandbanks in 2014 cost £1.95 million.
	Environmental: A nourished beach is natural and blends in with the environment.	
Beach reprofiling	**Social**: At Pevensey, the residential area behind the beach is now protected so residents feel safe.	**Social**: Bulldozers restrict access to Pevensey's beach, especially in winter.
	Economic: If the shingle ridge at Pevensey is breached, the estimated repair cost would be about £125 million whereas the combined cost for nourishment and frequent reprofiling is £30 million over 25 years.	**Economic**: Major reprofiling costs can be expensive. Further west along the coast at Selsey (West Sussex), £200,000 a year was paid to realign the beach prior to the Medmerry Scheme (see Section 10.15).
	Environmental: In preventing a breach, the Pevensey Levels has been protected and the beach still looks reasonably natural.	**Environmental**: A steep, high crested beach may look unnatural and uninviting to tourists.
Sand dune regeneration	**Social**: Sand dunes protect land uses behind them. Once established, they are popular as picnic and walking areas.	**Social**: While becoming established, regenerated sand dunes are fenced off and signs tell people to keep out. This may deter tourists.
	Economic: Small planting projects often use volunteer labour and local grass for transplants so costs are minimal.	**Economic**: ● Dune regeneration has to be checked twice a year and have fertilisers applied. ● Expensive systems have to be put in place to protect planted areas from trampling. Studland beach receives up to 25,000 people a day in summer. To reduce the risk of damage, boardwalks have been built through the dunes, fire warnings and fire beaters put in place, and all tourist facilities are contained around a solitary car park.
	Environmental: At Studland sand dune, regeneration has helped maintain a habitat for rare Dartford warblers, nightjars and chiffchaffs. Six species of reptiles, including adders and sand lizards, inhabit the dunes, along with damselflies and dragonflies, which hover in the slacks.	**Environmental**: It must also be remembered that sand dunes are a dynamic environment. Once regenerated, there is no guarantee that they will be stable. The grass may soon be damaged by storms, and even in favourable conditions it will take two to three years before grasses becomes established and begin to spread. If the coast is highly exposed, then additional, less attractive engineering such as rock armour is needed.

➜ Activities

1 Outline the difference between beach nourishment and beach reprofiling.

2 Devise an instruction manual for creating a new sand dune. Each stage needs a cartoon-type diagram.

   a) State two reasons why regenerated sand dunes are fenced off.

   b) Explain why there may be public opposition to fencing off the dunes.

3 In groups or pairs, present a case for and against continuing to defend the North Norfolk Coast around Happisburgh. You can find more information at the Happisburgh village website, www.happisburgh.org.uk

Coastal realignment

Creating an engineered new position of a coastline is called coastal realignment. In the context of managing coastal flooding in the UK, this involves moving the boundary inland.

What is managed retreat?

Managed retreat is when a decision is made to no longer follow a 'hold the line' strategy for managing coastal flooding and erosion. People are moved out, buildings are demolished then a breach is made in the existing sea defences so the sea can inundate the land and create new intertidal habitats. If the area is low-lying then, prior to the breach, an inland embankment is built to protect the area.

At Wallasea Island in the Thames Estuary (Figure 10.49), the intention was to create 60 hectares of mudflats, 25 hectares of salt marsh, 20 hectares of saline lagoons and 10 hectares of transitional grasses. The photo was taken six years after the breach was made.

With a prospect of rising sea levels associated with global warming, coastal realignment offers a sustainable long-term solution. It is better to be pro-active and manage a planned inundation of water than to keep reacting to breaches. Defences may still be installed further inland where they can operate more effectively.

▲ Figure 10.49 Managed realignment at Wallasea Island, Thames Estuary

What are the benefits and costs of managed retreat?

Benefits

- **Social:** It may help take the pressure off areas further along the coast and reduce their risk of flooding.
- **Economic**: It is often cheaper in the long term to use managed realignment than to continue to maintain hard engineering defences.
- **Environmental**: Managed realignment is designed to conserve or enhance the natural environment. It creates new intertidal habitats that compensate for those lost through coastal squeeze. At Wallasea, 38 species of bird have been recorded on the new mud flats, some in considerable numbers. These include ringed plover, dunlin, Brent geese and shellduck.

Costs

- **Social**: Relocation of people to new homes causes disruption and distress. If the long-term plan for the realignment of 40 square kilometres of the North Norfolk coast goes ahead, this will destroy six villages. Hundreds of people will have to be re-housed and whole communities will be split up. This could be a reality in 20–50 years' time. People feel 'let down' by managed retreat and feel the battle against the sea should continue.
- **Economic**: Short-term costs may be high, as relocation costs have to be paid. The recent Medmerry realignment scheme, in West Sussex cost £28 million, when it only cost £0.2 million a year to realign the shingle beach (Section 10.15).
- **Environmental:** Large areas of agricultural land are lost. Habitats of coastal birds such as bitterns, cranes and marsh harriers would be affected, so bird numbers would initially decline. It may take a long time to reach their previous numbers.

→ Activities

1 Define the term coastal realignment.
2 Draw a sketch of Figure 10.49. Label the position of the new coastline, the position of the old coastline, a lagoon, salt marshes, mudflats, and fields growing oilseed rape (indicated by yellow patches).
3 In pairs, conduct a 'for and against' argument for managed retreat.

⊘ KEY LEARNING

➤ Why the scheme was needed
➤ What the strategy was
➤ Positive effects
➤ Conflicts

Coastal realignment in Medmerry

Medmerry in West Sussex is the largest managed coastal realignment scheme in Europe.

Why was the scheme needed?

The **Environment Agency (EA)** considered the area to the west of Selsey (West Sussex) to be the area of South East England most at risk of flooding due to climate change. A shingle ridge was the only protection from the sea, and from the 1990s beach reprofiling (page 142) took place every winter, at an annual cost of £200,000. This was becoming unsustainable. If breached then 348 properties in Selsey, a water treatment plant and the main road between Chichester and Selsey would be flooded, along with many holiday homes and rental cottages. The last breach, in 2008, caused £5 million of damage.

What strategy was used?

Following public consultation, work to realign the coast began in 2011 and was completed in 2014. Managed retreat was achieved by the following:

■ Building a new embankment, up to two kilometres inland from the shore, using clay from within the area. This embankment enclosed the future intertidal area and protected the properties behind it.

■ Behind the embankment, a channel was built along its whole length to collect draining water. Four outfall structures were built into the embankment to take the water into the intertidal area.

■ Rock armour was then placed on the seaward edges of the embankment, where it linked up with the remaining ridge. This used 60,000 tonnes of hard rock from Norway.

■ Once the embankment and rock armour were in place, a 110 metre breach was made in the shingle bank to allow the sea to flood the land to create a new intertidal area.

▲ Figure 10.50 Map of the south coast

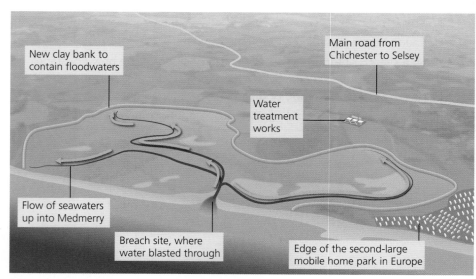

▲ Figure 10.51 Medmerry realignment

What happened as a result?

Positive effects

Social:

- Selsey now has a 1 in 1,000 chance of coastal flooding which provides the best level of protection in the UK. Disruption from a possible breach during the 2013 winter storms was avoided.
- A maintenance access track behind the embankment provides a cycle route and footpath. Today, there are ten kilometres of footpaths, seven kilometres of new bike paths and five kilometres of new bridleways in an area that previously only had two small footpaths.

Economic:

- Tourism, a main contributor to the local economy, is expected to increase. Two new car parks and four viewing points give easy access. An increase in visitor numbers to the nearby holiday village is expected as the area gains a reputation with birdwatchers and people interested in nature.
- The newly flooded area is expected to become an important fishing nursery that will boost the local fishing industry in Selsey. The salt-marsh vegetation will also be used for extensive cattle farming, to produce expensive salt-marsh beef.

Environmental:

- By carrying out a detailed environmental assessment prior to flooding, designers were able to take measures to protect existing species, such as water voles, crested newts and badgers.
- 300 hectares of new intertidal habitats are forming seaward of the embankment. Mudflats, salt marshes and transitional grasses have already attracted large numbers of ducks and lapwings. The area is turning into a huge nature reserve managed by the RSPB.

Controversy and conflicts

Social:

- Some local residents still feel that the EA should not have given up the land so easily and insist they should have looked into other options, such as offshore reefs or continued beach realignment.
- Some opponents of the scheme came from outside the area; they resented such an expenditure in a sparsely populated area. Would the money not have been better spent draining the Somerset Levels, for example? The need to compensate for coastal squeeze would have been a strong consideration at Medmerry.

Economic:

- At £28 million, the scheme was very expensive. Can this be justified if it only cost £0.2 million a year to maintain the shingle wall? However, with rising sea levels, continued reprofiling was not a viable long-term option.
- For the realignment scheme to take place, three farms growing oilseed rape and winter wheat had to be abandoned. Losing good agricultural land was regarded by some people as being wasteful, and this raises questions regarding the priority given by the EU for protecting buildings over agricultural land. Is this a short-sighted approach?

Environmental:

- Despite planning, habitats of existing species such as badgers would have been disturbed.

→ Activities

1 In groups of four, engage in a role-play of a meeting of four local residents of Medmerry when they first heard of the scheme. Possible roles: farmer, local fisherman, RSPB member, holiday homes manager. As a community, decide whether you are for or against the scheme.

2 Explain why the newly created environment may increase tourism in the area.

3 Place tracing paper over Figure 10.51. Draw a frame; trace the new coastline, the embankment, the waterworks, the B2145 road and outline of the area of holiday homes. Annotate your map to describe and explain the stages of the realignment scheme.

11 River landscapes

★ KEY LEARNING

➤ How a river erodes
➤ How a river transports its load
➤ How and why a river deposits its load

Fluvial processes

A river carries excess water on land, mainly from precipitation, to the sea. The journey a river makes to the sea is called its **long profile** (see Section 11.2). Along this journey, **fluvial** (river) **processes** of erosion, transport and deposition occur in the channel. These processes help shape the river's channel and its valley. The channel is the groove through which a river flows, and consists of its banks and bed.

How does a river erode?

Fluvial erosion is the process by which a river wears away the land. The ability of a river to erode depends on its velocity. Erosion takes place in four ways: hydraulic action, abrasion, solution and attrition.

Hydraulic action

This is when the sheer force of fast-flowing water hits the river banks and river bed and forces water into cracks. This compresses air in the cracks. Repeated changes in air pressure weaken the channel. **Hydraulic action** is responsible for **vertical erosion** in the upper course of a river. In the lower course, it contributes to **lateral erosion** of the banks, especially when fast-flowing water hits the outside bend of a **meander**. Lateral erosion is partly responsible for the migration of meanders across the **flood plain** (see page 158).

Abrasion

This is also called corrasion. Small boulders and stones may scratch and scrape their way down a river during transport, thereby wearing down the river banks and bed.

Stones which have fallen into the channel quite recently will be angular and have sharp, jagged edges. These are particularly effective tools of **abrasion**. Ongoing abrasion is responsible for both vertical erosion and lateral erosion of the channel.

Solution

This is also called corrosion. **Solution** refers to the dissolving of rocks such as chalk and limestone. Rivers travelling over these rocks will erode them in this way.

Attrition

Attrition affects a river's load. When stones first enter a river, they will be jagged and angular. As they are transported downstream, stones collide with each other and also with the river banks and bed. This gradually knocks off the stones' jagged edges so they become smooth and more rounded. Some collisions may cause a stone to smash into several smaller stones. These re-sized stones will be further smoothed and rounded on their journey to the sea. Figure 11.2 shows how erosion affects both the bed and banks of a river channel, causing vertical and lateral erosion.

■ Vertical erosion is the deepening of the river bed, mostly by hydraulic action. It is most evident in the upper course of a river. Here, what little energy the river has left over after overcoming friction is used to deepen its channel.

■ Lateral erosion is 'sideways' erosion. It wears away the banks of the river. This is most evident in the lower course of a river.

▲ Figure 11.1 Rounded boulders on a riverbed

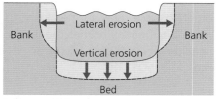

▲ Figure 11.2 Vertical and lateral erosion

How does a river transport its load?

Fluvial transport is the process by which a river carries its load. Load differs in size, from large, angular boulders in the upper course, to fine, suspended silt in the lower course. Load mostly comes from material that has weathered and tumbled down the hillside, though some also comes from eroded river banks.

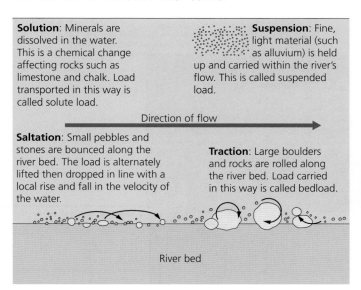

Solution: Minerals are dissolved in the water. This is a chemical change affecting rocks such as limestone and chalk. Load transported in this way is called solute load.

Suspension: Fine, light material (such as alluvium) is held up and carried within the river's flow. This is called suspended load.

Direction of flow

Saltation: Small pebbles and stones are bounced along the river bed. The load is alternately lifted then dropped in line with a local rise and fall in the velocity of the water.

Traction: Large boulders and rocks are rolled along the river bed. Load carried in this way is called bedload.

River bed

▲ Figure 11.3 River transport processes

Why do rivers deposit sediment?

Deposition is the process by which a river drops its load. Material deposited by a river is called sediment. The bigger the load particle, the greater the velocity needed to keep it moving. When velocity falls, large boulders are therefore the first to be deposited. The finest particles are deposited last. This explains why mountain streams have boulders along their bed, while close to the river's mouth there is only fine silt. Along its course, a river will deposit its load wherever the velocity falls. This could be at the base of a **waterfall**, on the inside bend of a meander, or where the river enters a sea or lake. It will also deposit more of its load in a period of drought when the **discharge** is low. The Hjulstrom Curve (Figure 11.4) shows the different critical velocities at which erosion, transport and deposition occur.

- The blue section shows that large 10 mm diameter particles are transported between approximate velocities of 85 and 110 cm/sec. When velocity falls below the critical velocity of 85 cm/sec (and enters the green section) these 10 mm diameter particles will be deposited.

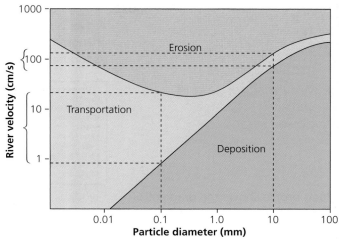

▲ Figure 11.4 The Hjulstrom Curve

→ Activities

1. Design a postcard-sized diagram to represent hydraulic action, abrasion and solution.

2. Explain how fluvial processes cause vertical and lateral erosion.

3. a) Draw a series of three diagrams to describe how a rock fragment that has fallen into a river will change as it moves downstream. Use appropriate adjectives, such as angular, jagged, rounded, rough and smooth.
   b) Explain the changes you have shown. Include the following terms: attrition, transported, abrasion, traction, collide, banks and bed.

4. a) Define the term 'load'.
   b) State two origins of load found in a river.

5. Study Figure 11.4.
   a) At what range of velocities will a 1.0 mm particle be transported?
   b) Below what velocity will a 1.0 mm particle be deposited?

6. State the general relationship between particle size and the velocity at which a particle is deposited.

- A 0.1 mm diameter particle will be transported at approximate velocities of between 0.9 and 35 cm/sec. When velocity falls below 0.9, that sized particle will be deposited.

⊗ KEY LEARNING

➤ A river's long profile
➤ How and why the long profile changes
➤ Why discharge and velocity increase downstream

The long profile of a river

A river is nature's way of removing excess water from the land. As it does so, it changes the landscape and creates landforms. The area drained by a river and its tributaries is called a river basin or a drainage basin.

What is a long profile?

A long profile shows the gradient of a river as it journeys from source to mouth. The source of a river is where it starts, and the mouth is where it reaches the sea. The River Severn travels 354 kilometres from its source in the Plynlimon Hills in the Cambrian Mountains to an **estuary** mouth in the Bristol Channel (Figure 11.5).

A river tries to achieve a smooth curve in order to reach its base level at the sea. This is called a graded long profile. Figure 11.7 shows the long profile of the River Severn. Notice how the gradient falls steeply at first, then becomes concave and then almost flat.

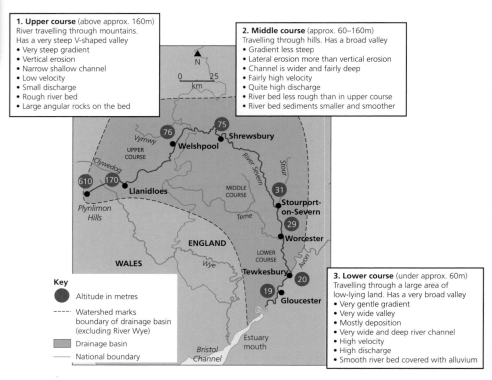

1. Upper course (above approx. 160m)
River travelling through mountains. Has a very steep V-shaped valley
• Very steep gradient
• Vertical erosion
• Narrow shallow channel
• Low velocity
• Small discharge
• Rough river bed
• Large angular rocks on the bed

2. Middle course (approx. 60–160m)
Travelling through hills. Has a broad valley
• Gradient less steep
• Lateral erosion more than vertical erosion
• Channel is wider and fairly deep
• Fairly high velocity
• Quite high discharge
• River bed less rough than in upper course
• River bed sediments smaller and smoother

3. Lower course (under approx. 60m)
Travelling through a large area of low-lying land. Has a very broad valley
• Very gentle gradient
• Very wide valley
• Mostly deposition
• Very wide and deep river channel
• High velocity
• High discharge
• Smooth river bed covered with alluvium

Key
● Altitude in metres
----- Watershed marks boundary of drainage basin (excluding River Wye)
▨ Drainage basin
— National boundary

▲ Figure 11.5 Drainage basin of the River Severn

How and why does the long profile change?

Figure 11.7 shows how the long profile can be split into three courses based on gradient.

▼ Figure 11.6 Processes operating in the three courses

	Upper course	Middle course	Lower course
Erosion	Mostly vertical erosion by hydraulic action	Less vertical erosion, more lateral erosion. Much attrition and abrasion, some solution	Very little erosion, only lateral erosion
Transport	Mostly traction. Large boulders moved	Mostly suspension, increased traction. Load becomes smaller and less angular	Mostly suspension and solution. Very small particles of load. Great quantity of load
Deposition	Large boulders deposited	More deposition, especially on the inside bend of meanders	Deposition now the main fluvial process. Fine material is now deposited
Why does the long profile change?	The upper course is set in a landscape of high relief. The long profile starts at its source. Trickles begin to merge to form a single channel which flows down a steep gradient. The steep descent gives the river more potential energy. In places, there may be waterfalls and rapids.	The middle course is further downstream in an area of hilly rather than mountainous relief. Discharge has increased as the channel is deeper, and the volume of water has been increased by the many tributaries that have joined the main river. The river's energy results in less vertical erosion and causes lateral erosion at meanders. As vertical erosion reduces, the gradient of the long profile becomes concave.	The lower course is the section closest to the river mouth, where the surrounding land is low-lying. Erosion is now confined to lateral erosion at meanders. Lack of vertical erosion means the gradient is almost flat.

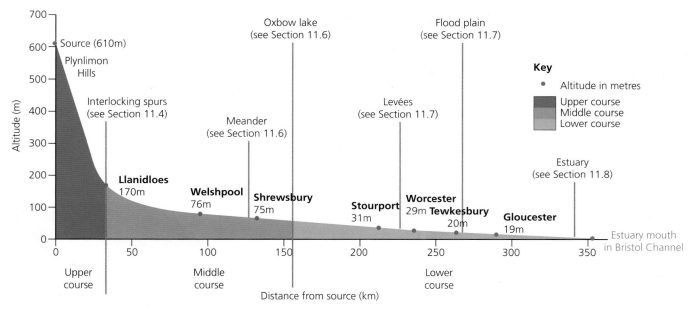

▲ Figure 11.7 Long profile of the River Severn, Wales

Why do discharge and velocity increase downstream?

Discharge is the volume of water passing through a given point on the river. It is measured in cubic metres per second (cumecs). Discharge = velocity × cross-sectional area. Discharge increases downstream as tributary streams join the main river and add their volume of water to it.

The average velocity of a river increases along its course. Despite the steep descent in the upper course, it is the lower course which has the greatest velocity. Velocity depends on how much water comes into contact with the channel's banks and bed. In the upper course, a small channel means there is much friction. So, despite the steep slope, velocity is low. Conversely, in the broad, deep channels of the lower course, less water is in contact

with the bed and banks, so velocity is much higher. This is because speed is boosted by the additional discharge from all the tributaries.

 Fieldwork: Get out there!

1 Assume you have been given the following equipment: a stop watch, two one metre rules, a tape measure and an orange. Explain how you could use this equipment to work out the velocity of a river in its upper course.

2 If you repeated the experiment downstream, how would you expect the velocity to change? Explain your answer.

→ Activities

1 Study Figure 11.5.

a) Name a town in the upper, middle and lower course of the River Severn.

b) From the list below, select the area that best describes the size of the drainage basin of the River Severn. Use the scale of the map to help you.

- 5,000 square kilometres
- 11,000 square kilometres
- 21,000 square kilometres

2 Study Figure 11.7.

a) Draw a labelled sketch to describe the three stages of a long profile.

b) Explain the shape of each stage.

3 a) Describe how the following things change along the course of a river: river width, depth, type of erosion, transport and deposition.

b) Explain why the changes occur.

Changing cross-profiles of a river

What is a cross-profile?

A **cross-profile** is a section taken sideways across a river channel and/or a valley.

- A channel cross-profile only includes the river.
- A valley cross-profile includes the channel, the valley floor and the slopes up the sides of the valley.

How and why does a channel cross-profile change downstream?

How does it change?

Changes downstream in the channel cross-profile are summarised in Figure 11.9.

Why does it change?

- In the upper course, the river erodes its bed by hydraulic and abrasive action. As the river travels downstream, it is joined by a number of tributaries. These increase the volume of water which gives the river kinetic energy, a higher velocity and thus more erosive power. This allows it to cut a much deeper channel with increased distance downstream.
- The channel becomes wider downstream because, as the gradient becomes less steep, there is less vertical erosion. By the time the river is in the middle course, lateral (sideways) erosion is dominant. This erodes the river banks, which makes the channel wider.

Steep V-shaped cross-profile

Interlocking spurs

River takes up most of the valley floor

▲ Figure 11.8 Valley cross-profile of the upper course of the River Severn

▼ Figure 11.9 How a channel cross section changes downstream. Note: The diagrams are for a straight section of a river. At a meander, the cross-profile will become asymmetrical (see page 158)

Upper course	Middle course	Lower course
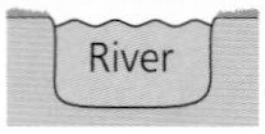 River	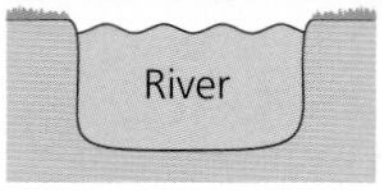 River	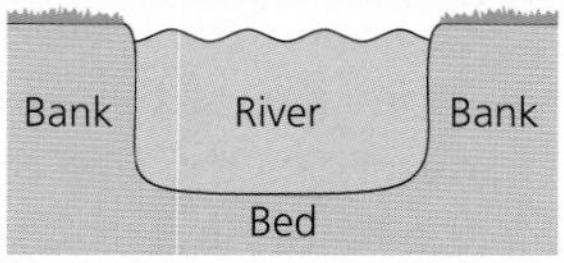 Bank River Bank / Bed
The channel is very narrow (only a few metres wide) and very shallow (less than 0.25 metres deep).	The channel becomes wider. It may be several metres wide. For many rivers, it will be over a metre deep.	The channel becomes wider still. A small river may only be five to ten metres wide, whereas the River Severn is 3.2 kilometres wide at its estuary mouth by the old Severn Bridge crossing. The channel is much deeper.

How and why does a valley cross-profile change downstream?

How does it change?

▼ Figure 11.10 How the valley cross-profile changes downstream

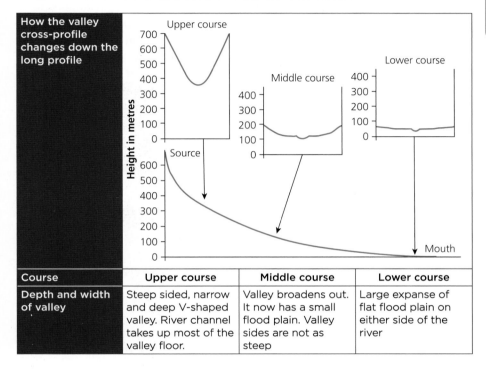

Course	Upper course	Middle course	Lower course
Depth and width of valley	Steep sided, narrow and deep V-shaped valley. River channel takes up most of the valley floor.	Valley broadens out. It now has a small flood plain. Valley sides are not as steep	Large expanse of flat flood plain on either side of the river

Why does it change?

In the upper course, there is a steep, V-shaped cross-profile. Vertical erosion by the river is the dominant process operating in the valley. This creates a slope that weathered material from the valley sides can fall down. On reaching the river, this material is removed. Rivers tend to have their source in upland areas, which means the rock is harder. The valley sides are therefore not broadened out much by weathering and erosion so slopes remain steep.

In the middle course, the river is flowing through lower country. The gradient is less steep, so the river begins to meander (bend) and erode laterally (sideways) into the valley sides. This broadens out the valley. In addition to this, the rate of weathering increases on the softer rocks of the valley sides. As the river uses more energy in lateral erosion, it is not able to remove all of the weathered material, so this builds up the valley floor to give it a more gentle profile.

In the lower course, the river is passing through low-lying country. Deposition from **floods** builds up the flood plain, and meanders migrate (see page 158). This builds up and widens the valley.

→ Activities

1. a) Draw a labelled sketch of Figure 11.8. Add to the labels to describe, in detail, a valley cross-profile in its upper course.
   b) Describe how this valley cross-profile would change in its middle course.
   c) Explain these changes. Include the following terms in your answer: vertical erosion, lateral erosion, deposition, meander, flood plain and weathering.
2. a) Draw a labelled diagram to show the channel cross-profile in the upper course.
   b) Describe how the channel cross-profile changes in the middle course.
   c) Explain the changes. Include the following terms: vertical erosion, lateral erosion, velocity, abrasion and hydraulic action.

Geographical skills

Describe how you might use an Ordnance Survey map to draw a valley cross section.

Interlocking spurs and rapids

Erosion processes result in the formation of **interlocking spurs** and rapids, on the upper course of a river.

What are interlocking spurs?

Interlocking spurs are projections of high land that alternate from either side of a valley and project into the valley floor (Figure 11.11). Interlocking spurs are valley landforms formed by fluvial erosion. They are found in the upper course of a river where rocks are hard, like in the Afon Dulas Valley, a tributary valley of the River Severn. Notice the zip-like interlocking nature of the hillsides, the very narrow valley floor with its winding river, and how the river takes up most of the valley floor. Figure 11.13 shows the OS map extract of the same area.

Characteristics of interlocking spurs:

- a steep gradient
- convex slopes
- project from alternate sides of the valley
- separated by a narrow valley floor mostly taken up by the river channel
- sometimes wooded
- may have scree slopes.

Figure 11.12 describes stages in the formation of interlocking spurs.

What are rapids?

Rapids are fast-flowing, turbulent sections of a river where the river bed has a relatively steep gradient. They are channel landforms of the upper course. The ability of a river to erode its bed depends on the river's energy. It also

▲ Figure 11.11 Interlocking spurs

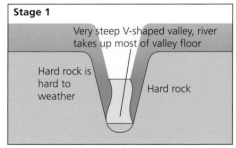

Stage 1

Very steep V-shaped valley, river takes up most of valley floor

Hard rock is hard to weather

Hard rock

In the upper course of a river, the river's water volume and discharge are low. The river uses most of its energy overcoming friction with the channel. What energy it has left over is used by hydraulic action to deepen the channel (vertical erosion).

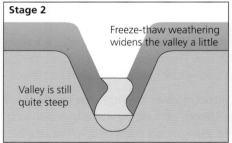

Stage 2

Freeze-thaw weathering widens the valley a little

Valley is still quite steep

In upland areas, the geology is composed of hard rock such as granite or slate. However, freeze-thaw weathering (when water gets into cracks in rocks and freezes at night and thaws by day) gradually broadens it out. This gives the valley a steep, V-shaped cross profile. Repeated weathering weakens the rock so fragments break loose and tumble down the hillside as scree, which the river then removes.

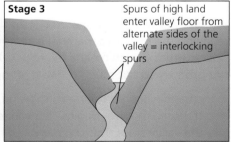

Stage 3

Spurs of high land enter valley floor from alternate sides of the valley = interlocking spurs

The winding path taken by the river is due to obstacles of harder rock in its path. The river takes the easiest route over the land. This results in projections of high land entering the valley from alternate sides. These projections are the interlocking spurs.

▲ Figure 11.12 Stages in the formation of interlocking spurs

depends on the degree of hardness of the bedrock. In some cases, there may be vertical bedding, whereby alternate bands of hard and soft rock cross the channel. Differential erosion will occur, as soft rock is more easily eroded than hard rock. This makes the river bed uneven and the river's flow becomes turbulent, resulting in 'white water' sections typical of rapids.

The River Severn has rapids just south of the Ironbridge Gorge. Characteristics:

- turbulent flow of water
- white water
- uneven river bed
- steep gradient.

> Notice the brown contour lines that show the height and slope of the land. They show high land projecting into the Afon Dulas Valley from alternating sides of the valley. The contours are close together, showing that the slopes of the interlocking spurs are steep.

Scale 1 : 50 000

▲ Figure 11.13 Map extract of the Afon Dulas Valley near Llanidloes

Key

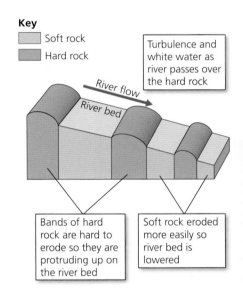

Soft rock
Hard rock

Turbulence and white water as river passes over the hard rock

River flow
River bed

Bands of hard rock are hard to erode so they are protruding up on the river bed

Soft rock eroded more easily so river bed is lowered

▲ Figure 11.14 Formation of rapids

▲ Figure 11.15 An example of rapids in Wales (River Llugwy)

→ Activities

1　Draw a sketch of Figure 11.11 and add labels to describe the characteristics of the interlocking spurs and the nature of the valley.

2　Suggest three reasons why few people live in the area in Figure 11.13.

Geographical skills

1　Study Figure 11.13, the OS map extract.

　a)　State the four-figure reference for Garreg y Gwynt.

　b)　What is the maximum height shown on the map? Give a six-figure reference for its location.

　c)　Draw an annotated sketch cross-profile from north to south along eastings 96. Label the Afon Dulas River and interlocking spurs.

Fieldwork: Get out there!

Imagine you were visiting Jackfield Rapids (Figure 11.15) to carry out fieldwork to determine the velocity of the river.

1　Identify two potential hazards of the site.

2　Suggest precautions that may be taken to reduce the risk.

Waterfalls and gorges

Waterfalls and **gorges** are channel landforms found in the upper course of a river.

What are the characteristics of a waterfall?

A waterfall is where water falls down a vertical drop in the channel, usually from a considerable height. The River Severn itself does not have a waterfall, although the Water Break-its-Neck waterfall on one of its tributaries is sometimes referred to as the River Severn Waterfall.

Stage 1
At the beginning

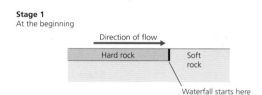

Rocks are laid down horizontally. A waterfall will form where there is a junction between a hard rock capping upstream and soft rock downstream. Differential erosion means the river erodes the softer rock and the water falls vertically from the hard rock to the soft rock below.

Stage 2
Becoming established

3. Continued undercutting means that the hard rock capping loses support and overhangs the drop

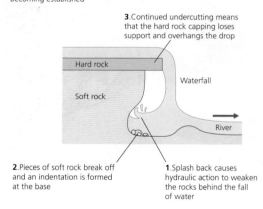

2. Pieces of soft rock break off and an indentation is formed at the base

1. Splash back causes hydraulic action to weaken the rocks behind the fall of water

Stage 3
Well established

4. The unsupported cap rock breaks off

6. Undercutting continues, creating a new overhang. As this process of headward erosion repeats, the waterfall retreats upstream leaving a gorge downstream

5. The fallen rock breaks up and is caught in the turbulent flow of water. Some rocks become trapped and drill into the bed, which creates a plunge pool

▲ Figure 11.16 Stages in the formation of a waterfall

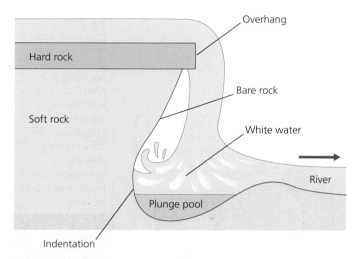

▲ Figure 11.17 Characteristics of a waterfall

What characterises a gorge?

A gorge is a narrow, steep-sided valley, with bare, rocky walls. A gorge of recession is a gorge found immediately downstream of a waterfall (see Figure 11.18). Gorges may be classed as either channel or valley landforms. Although the River Severn does have a gorge at Ironbridge, which displays some of these characteristics, it was primarily formed as a glacial overflow channel, so cannot be used as an example of fluvial (river) erosion.

Characteristics:

- very narrow valley
- very steep, high valley sides
- located immediately downstream of a waterfall
- river channel takes up most, if not all, of the valley floor
- turbulent, fast flowing white water
- many areas of bare rock on valley sides
- boulders litter the river bed.

How does a waterfall create a gorge?

By the end of stage 3 in the formation of a waterfall (Figure 11.16), the scene is set for a gorge to be formed. As the waterfall retreats upstream it leaves a steep-sided valley downstream which is called a gorge. Every time the overhanging cap rock breaks off, the gorge retreats and grows longer.

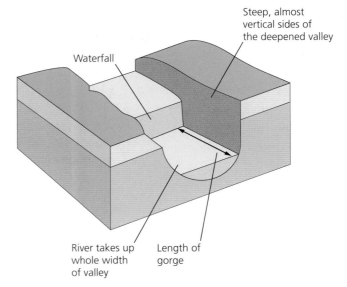

Steep, almost vertical sides of the deepened valley

Waterfall

River takes up whole width of valley

Length of gorge

▲ Figure 11.18 Block diagram of a gorge

→ Activities

▲ Figure 11.19 Water Break-its-Neck waterfall, Radnor Forest, Wales

▲ Figure 11.20 Fairy Glen, a gorge on the Conwy River near Betws-y-Coed, North Wales

1   Sketch Figure 11.19 and Figure 11.20 and add the correct labels to each one.

2   Explain how a waterfall can result in the formation of a gorge. In your explanation use the following terms: overhang, headward erosion, soft rock, hard rock capping, undercutting, hydraulic action, steep-sided.

3   a)   Suggest what it is about landscapes with waterfalls and gorges that attract tourists.

    b)   Based on what you know about river processes, and the information in Section 11.5, write a safety leaflet for people visiting a waterfall or gorge.

Meanders and oxbow lakes

A combination of fluvial erosion and deposition leads to the formation of meanders and oxbow lakes, mainly in the middle and lower courses of a river.

What are the characteristics of a meander?

A meander is a bend in a river. The characteristics of a meander are shown in Figure 11.21. The River Severn has many meanders some of which have carved out huge loops. The centre of Shrewsbury is inside one of these large loops.

Inside bank
Slip-off slopes on River Severn near Welshpool

- curved, beach-like feature on the inside bank
- very gentle, convex slope
- sediment consists of sand, gravel and pebbles that are smoothed and rounded by attrition
- vegetation begins to grow furthest from the water.

Outside bank
River cliff on River Severn near Buildwas

- a steep drop down into the river on the outside bend
- can be several metres high
- composed mostly of bare earth
- unconsolidated material at the base.

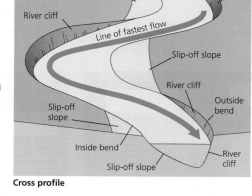

Cross profile

➤ Figure 11.21 Characteristics of a meander

Slow flowing water   **Fast flowing water**

How is a meander formed?

In the early stages of meander formation, water flows slowly over shallow areas (riffles) in the riverbed, and faster through deeper sections (pools). This eventually sets in motion a helicoidal flow that corkscrews across from one bank to another. This starts the erosion and deposition processes which continuously shape a meander.

▲ Figure 11.22 A meandering river and meander scars, which mark former positions of a meandering river

- Fast-flowing water on the outside bank causes lateral erosion through abrasion and hydraulic action, which undercuts the bank and forms a river cliff. (The point of maximum erosion is slightly downstream of the mid-point of the loop.)
- Helicoidal flow is a corkscrew movement. The top part of the flow hits the outside bank and erodes it. The flow then 'corkscrews' down to the next inside bend, where it deposits its load as friction slows the flow.
- Fast flow causes vertical erosion on the outside bend. This deepens the river bed, resulting in an asymmetrical cross-profile.
- Sand and pebbles are deposited on the inside bank where the current is slower, forming a gentle slip-off slope.

A sinuous river is one with many meanders. The loops increase in size as erosion continues on the outside bank and deposition continues on the inside bank. As meanders grow, they move or migrate over the flood plain. A river today may have been in a completely different part of the valley in the past (Figure 11.22).

What are the characteristics of an oxbow lake?

An **oxbow lake** is a small, horseshoe-shaped lake that is located several metres from a fairly straight stretch of river in its middle and lower courses. These landforms may be seen near Welshpool on the River Severn.

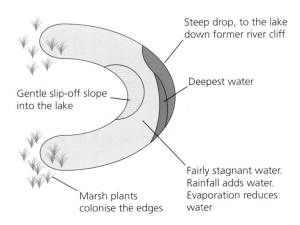

Steep drop, to the lake down former river cliff

Deepest water

Gentle slip-off slope into the lake

Fairly stagnant water. Rainfall adds water. Evaporation reduces water

Marsh plants colonise the edges

▲ Figure 11.23 Characteristics of an oxbow lake

→ Activities

1 Draw a cross-profile of a meander. On it, label the following: lateral erosion, vertical erosion, greatest velocity, slow-flowing water, shallow water, deposition, slip-off slope, river cliff, deepest water.

2 a) Suggest how the cross-profile you have drawn may differ from that of a straight section of a river.

   b) Explain why the cross-profiles would differ.

3 a) Sketch or trace the course of the river shown in Figure 11.22. In a different colour, sketch how you think this river would look in 200 years' time.

   b) Explain why you have given the river this new route.

4 Use Figure 11.24 to write down phrases in sequence to explain the formation of an oxbow lake. Make sort cards using these phrases. Shuffle these then test yourself to re-order them.

5 Without using a diagram, describe the characteristics of an oxbow lake.

How is an oxbow lake formed?

The formation of an oxbow lake is shown in Figure 11.24.

Stage 1

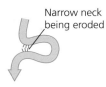

Narrow neck being eroded

- Meander loop becomes very large.
- Only a narrow strip of land separates the river channel (the meander neck).
- Continued lateral erosion
- Neck becomes increasingly narrow.

Stage 2

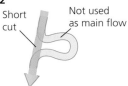

Short cut

Not used as main flow

- River floods, so main flow of water cuts straight across the neck.
- This 'shortcut' begins to break down the banks and carve a new channel.

Stage 3

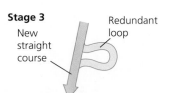

New straight course

Redundant loop

- Floods recede, so the river reverts to its normal meandering channel.
- Process is repeated over and over again with every flood event.
- This new channel becomes so established by the continued lateral and vertical erosion that it becomes the main channel.

Stage 4

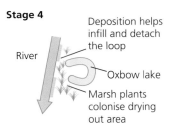

River

Deposition helps infill and detach the loop

Oxbow lake

Marsh plants colonise drying out area

- Loop of the old river channel is increasingly detached as it is no longer receiving river water.
- Subsequent flooding causes deposition on the new river banks. This aids the detachment of the old loop.
- Marsh plants colonise the area, which further widens the gap.
- In time, only the far end of the meander loop is left, sometimes several metres from the main channel. This is the oxbow lake.

▲ Figure 11.24 Stages in the formation of an oxbow lake

Fieldwork: Get out there!

Imagine you have been given a tape measure, ranging pole, chain or rope and a metre rule (see Chapter 27). Describe in detail how you would use this equipment to gather data from which to draw the cross-profile of a meander in a river's middle course.

Levées and flood plains

A period of prolonged heavy rain will cause an increase in a river's discharge, so water rises over its banks and floods over the surrounding land. Repeated annual flooding eventually builds up **levées** and a flood plain. These are landforms of fluvial deposition, found in the middle and lower courses of a river.

What are the characteristics of levées?

Levées are naturally raised river banks found on either or both sides of a river channel that is prone to flooding. The lower course of the River Severn has many levées such, as those at Minsterworth near Gloucester (Figure 11.26).

Characteristics:

▲ Figure 11.25 Levée on The River Severn at Minsterworth, near Gloucester

- raised river banks (about 2–8-metres high in the UK)
- composed of gravel, stones and alluvium
- grading of sediments with the coarsest closest to the river channel
- steep-sided, but steeper on the channel side than on the land side
- fairly flat top, naturally covered by grass often used as a footpath, e.g. the Severn Way.

How are levées formed?

When a river bursts its banks, friction with the land reduces velocity and causes deposition. Heavy sediment is deposited closest to the river. The size of sediment then becomes progressively smaller with increased distance from the river. With each successive flood, the banks are built up higher (Figure 11.26, stages 1 and 2).

Although it may seem that levées may make it more difficult for the river to flood next time, this is not the case. This is because over time the bed of the river develops a thicker layer of sediment, which raises the river in its channel (stage 3).

Stage 1 Before levée

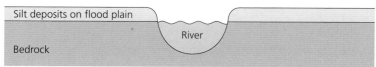

Silt deposits on flood plain

River

Bedrock

Stage 2 During a flood

Heaviest, most coarse sediment deposited closest to river as more difficult to transport

Finest particles of sediment carried further on to the flood plain

Silt deposits on flood plain

Bedrock

River bed deposits

Stage 3 After many floods

With every flood, the river banks are built a little higher

The raised banks either side of the river are natural levées

Silt deposits on flood plain

River

Bedrock

River bed builds up bed load deposits over time. This raises the level of the river so increases probability of flooding

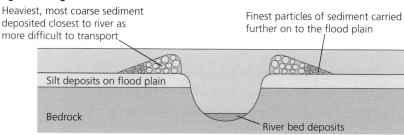

➤ Figure 11.26 Stages in the formation of levées

What are the characteristics of a flood plain?

A flood plain is a large area of flat land either side of a river that is prone to flooding. Figure 11.27 shows the River Severn in flood over part of its flood plain at Tewkesbury, Gloucestershire. Another settlement on the River Severn that is prone to flooding is Gloucester. The characteristics of a flood plain are shown in Figure 11.28.

▲ Figure 11.27 Flooding at Tewkesbury, January 2014

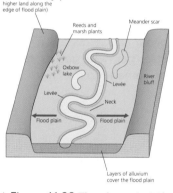

▲ Figure 11.28 The characteristics of a floodplain

How is a flood plain formed?

The width of the flood plain is due to meander migration (see page 158), where the outside bends erode laterally into the edges of the valley. Their position is also gradually moving downstream. Eventually, this cuts a wider valley. When floods have receded, the flood plain is slightly higher and more fertile due to the deposits of silt and alluvium caused by the river flooding. Alluvial deposits also infill old meander scars. A flood plain is built up over hundreds of years. Each flood makes the flood plain a little higher.

Geographical skills

1   Study Figure 11.29, the OS map.

   a)   Place a piece of tracing paper over the map. Trace its frame, the river and the first contour line either side of the river. Colour the river blue, the flood plain green and the area beyond the flood plain brown. Mark and label the levées.

   b)   Explain why levées make good footpaths.

   c)   Give two pieces of map evidence to show that Minsterworth Ham has poor natural drainage.

   d)   What is the width of the flood plain measured along northing 17? (On this map 2 cm represent 1 km)

→ Activities

1   Draw a sketch of the Minsterworth levée in Figure 11.25. Label it to show its characteristics.

2   Draw an annotated diagram or sequence of diagrams to show how the height of a flood plain is built up.

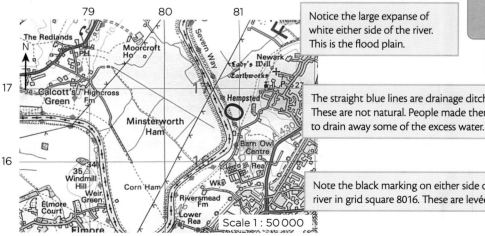

Notice the large expanse of white either side of the river. This is the flood plain.

The straight blue lines are drainage ditches. These are not natural. People made them to try to drain away some of the excess water.

Note the black marking on either side of the river in grid square 8016. These are levées.

Scale 1 : 50 000

▲ Figure 11.29 OS map of part of the west side of Gloucester

Example

⊛ KEY LEARNING

➤ The characteristics of an estuary

➤ How an estuary is formed

➤ How estuary mud flats are formed

The River Severn and its estuary

The River Severn completes its 354 kilometre journey in an estuary which enters the Bristol Channel. An estuary is the tidal part of a river where the channel broadens out as it reaches the sea.

What are the characteristics of an estuary?

■ An estuary is the tidal part of a river where freshwater from the river merges with salt water from the sea. It is therefore affected by both fluvial and marine processes.

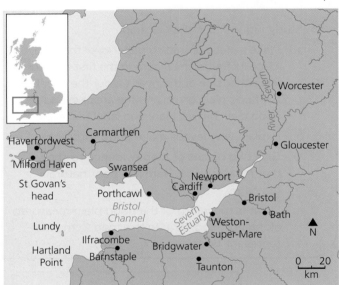

▲ Figure 11.30 The Severn Estuary

■ It may have a high tidal range. The River Severn has a tidal range of 15 metres, which is one of the highest in the world.

■ It may be very wide. The Severn Estuary is 3.2 kilometres wide at the old Severn Bridge Crossing.

■ It will have mudflats (Figure 11.33) that are visible at low tide and some of the mud will be covered by salt marshes.

■ It may have tidal bores, which are huge waves that funnel up the river. The Severn Estuary has a tidal bore which travels as far as Gloucester on very high spring tides. Large bores occur about 25 days a year. Bores travel at 8–21 kilometres per hour, getting faster upstream. They cause great damage to the river banks and vegetation.

How is an estuary formed?

Figure 11.31 shows how a valley was flooded by the post glacial rise in sea level to create an estuary.

Figure 11.32 shows how an estuary is influenced by both fluvial and marine processes. The river is flowing from east to west. Notice how:

■ the salinity (saltiness) increases towards the sea

■ there are two sources of sediment (from the river and from the sea)

■ the estuary is tidal, so fluvial and marine processes operate.

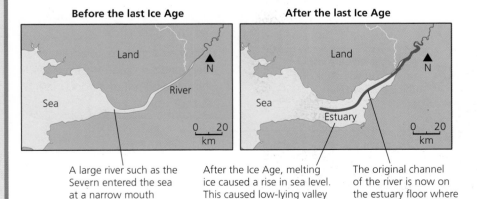

A large river such as the Severn entered the sea at a narrow mouth

After the Ice Age, melting ice caused a rise in sea level. This caused low-lying valley sides either side of the river to become flooded

The original channel of the river is now on the estuary floor where it provides a deep channel for shipping

▲ Figure 11.31 Estuary formation

How are estuary mudflats formed?

Mudflats form in sheltered areas where tidal water flows slowly. As a river transports alluvium down to the sea, an incoming tide transports sand and marine silt up the estuary. Figure 11.32 shows that just downstream of the tidal limit, fresh river water begins to mix with salty sea water. Where the waters meet, velocity is reduced, which causes deposition. This builds up layers of mud called mud flats (Figure 11.33). They will be covered at high tide, but exposed at low tide.

Within the mud flats there are many small streams (creeks). After a while, the mudflats may become colonised by salt-marsh vegetation such as cordgrass.

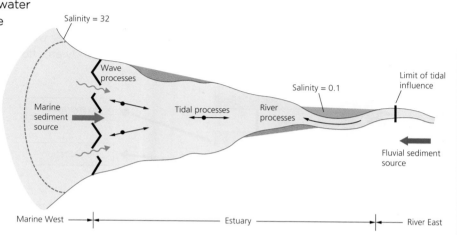

▲ Figure 11.32 Processes operating in an estuary

▲ Figure 11.33 Mudflats off Chepstow, Wales, with view to M4 Severn Bridge crossing

→ Activities

1 Draw a labelled diagram to show the characteristics of an estuary.

2 a) Explain what is meant by a tidal bore.
   b) Suggest what damage tidal bores may cause.

3 Study Figure 11.32.
   a) What is the origin of sediments in an estuary?
   b) What causes deposition to occur?

4 Study Figure 11.33.
   a) Describe the mudflats near Chepstow.
   b) Explain why these mudflats are not always visible.

Physical causes of flooding

How does water get into and out of a river?

The drainage basin hydrological cycle explains how precipitation falling in a catchment area gets into a river (Figure 11.34.).

Precipitation is any form of moisture reaching the ground. It includes snow, rain, sleet and hail. The risk of flooding depends on how quickly precipitation gets into a river channel. Surface runoff is the fastest route. If there is a lot of runoff then the discharge of a river will increase quickly. Other flows into a river are throughflow and groundwater flow (see Figure 11.34).

★ KEY LEARNING

➤ How water gets into a river
➤ How precipitation increases flood risk
➤ How geology and relief can increase flood risk

▼ Figure 11.34 The drainage basin hydrological cycle

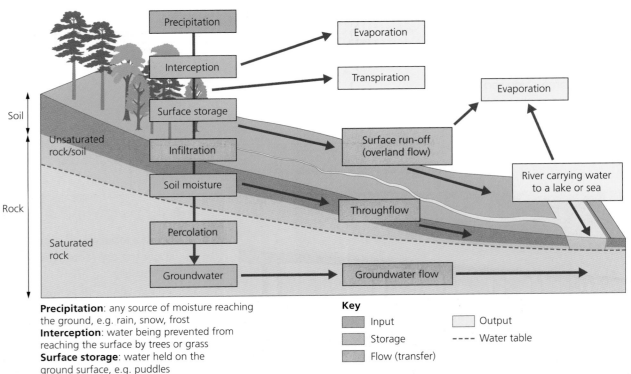

Precipitation: any source of moisture reaching the ground, e.g. rain, snow, frost
Interception: water being prevented from reaching the surface by trees or grass
Surface storage: water held on the ground surface, e.g. puddles
Infiltration: water sinking into soil/rock from the ground surface
Soil moisture: water held in the soil layer
Percolation: water seeping deeper below the surface
Groundwater: water stored in the rock
Transpiration: water lost through pores in vegetation
Evaporation: water lost from ground/vegetation surface
Surface run-off (overland flow): water flowing on top of the ground
Throughflow: water flowing through the soil layer parallel to the surface
Groundwater: water flowing through the rock layer parallel to the surface
Water table: current upper level of saturated rock/soil where no more water can be absorbed

Key

▢ Input	▢ Output
▢ Storage	- - - - Water table
▢ Flow (transfer)	

Flood risk is increased by:

- Bands of depressions passing over the UK at frequent intervals, especially in winter, result in continuous heavy rain, which may saturate the soil. The soil can no longer store water so surface runoff is increased. Rainwater will therefore enter the river quicker resulting in higher discharge and floods.

- Sudden bursts of heavy rain often results in the infiltration rate being too slow to cope. This may occur after a period of drought that has baked the soil hard. Surface runoff occurs, discharge increases quickly and flash floods occur.

- Prolonged light rainfall may cause floods if there has been a lot of previous (antecedent) rainfall that has saturated the soil.

- Sudden snow melt causes a release of stored water that flows over the ground as surface runoff.

How can geology and relief increase flood risk?

Geology:

- The type of rock found in mountains is usually impermeable rock such as slate, which does not allow water to pass through it. The rock is often bare, with thin soils and little vegetation to intercept the rain.
- Low lying areas often contain an impermeable clay soil. It is usually vegetated, but the soil is so compacted that it is difficult for infiltration to occur.
- Flooding is much less likely in areas of permeable rock such as chalk and limestone, as water passes through these rocks.

Relief is the height and slope of the land. Steep slopes mean that surface runoff occurs on mountainsides before rain has had time to infiltrate the soils. A valley floor with steep sides such as the Llanberis Pass in Snowdonia (Figure 11.35) therefore has a high flood risk.

Low-lying, flat flood plains also have a high flood risk as there is not enough gradient to remove the water. The flood risk is increased further by the impermeable clay soils. Notice how the flat, low-lying relief in Surrey is prone to flooding (Figure 11.36).

▲ Figure 11.35 Llanberis Pass, Snowdonia, North Wales

▲ Figure 11.36 Flooding in the low-lying, flat relief of Surrey

Finally, it must be noted that precipitation, geology and relief are interconnected. They combine to increase the risk of flooding. The high relief of Snowdonia causes air from onshore winds to rise to cross the mountains. As air rises, it cools and condenses to form rain. This is called relief rainfall. In Snowdonia, the rainfall runs over steep-sided, impermeable slate to flood the valleys below.

→ Activities

1. The definitions below have been mixed up. Can you sort them out?

Precipitation	Water flowing on top of the ground
Infiltration	Moisture reaching the ground e.g. rain and snow
Interception	Water lost through pores in vegetation
Percolation	Water flowing through the soil layer parallel to the surface
Transpiration	Water seeping deeper below the surface in the rock
Groundwater flow	Water being prevented from reaching the surface by trees and grass
Surface runoff	Water sinking into the soil from the ground surface
Throughflow	Water flowing through the rock layer parallel to the surface

2. Write these phrases in the correct order to show how rain falling on an impermeable surface may cause flooding: full river channel, heavy rain, increased discharge, flooded land, water gets into the channel quickly, much surface runoff, river bursts its banks.

3. a) Draw a labelled diagram to show how high relief increases precipitation.
   b) Place these types of precipitation in order to show how likely they are to cause flooding: a short, heavy rain storm, several days of drizzle, several days of snow. Explain your answer.

4. Explain how geology and relief have caused flooding in Figure 11.36.

Geographical skills

1. Draw a field sketch from Figure 11.35. Annotate it to show how geology and relief may cause flooding.

Human causes of flooding

How can urban land use increase flood risk?

⭐ KEY LEARNING

➤ How urban land use increases flood risk

➤ How rural land use increases flood risk

In the UK, urban land use is the main cause of increased flood risk. This is due to **urban sprawl** associated with urbanisation (see Section 15.9).

New infrastructure

Urbanisation leads to the growth of towns and cities. As the UK's population increases, new roads, shopping centres, schools and leisure centres are built. The greater the area covered by buildings and roads (with impermeable surfaces), the greater the potential flood risk.

New houses

With the increased demand for homes in the UK, thousands of new houses are built each year (see Chapter 15), many on **greenfield sites**. Between 2001 and 2011 there was an increase of 72 per cent in the average density of new dwellings in England. Property developers are squeezing several houses into a plot formerly occupied by only one house.

There are strict planning controls regarding building on flood plains, but even so, seven per cent of new dwellings in England in 2011 were built in areas of high flood risk. Large areas of flood plains are now covered with impermeable tarmac roads and concrete pavements. Cities therefore have few natural areas in which to store excess water. Water runs off quickly through gutters, drains and culverts and this leads to a speedy rise in a river's discharge – hence the increased flood risk.

Disappearing gardens

The growth of impermeable surfaces is becoming worse in our large cities, where people pave over back gardens to save mowing the lawn; this is often seen in rented accommodation. Parking is also a problem. A Direct Line survey in 2013 revealed that 47 per cent of UK households have two cars. With an absence of garages and little 'on road' parking in inner cities, many households have concreted over their front gardens to accommodate their cars (Figure 11.37).

How can rural land use increase flood risk?

The increased risk of flooding in rural areas is more localised and, in the UK, does not have much effect on areas downstream. However, changing land use and farming practices can increase flood risk.

Forestry

Felling (chopping down) trees reduces interception and roots no longer take water from the soil (Figure 11.38). The impact of felling could be considerable, as a dense forest uses up 40 per cent of any precipitation. After felling, the soil soon gets saturated, runoff occurs, river discharge increases quickly and so the risk of flooding increases. Felling trees also causes exposed soil to wash into rivers, building up their beds. This reduces the capacity of channels, so rivers are more likely to flood.

▲ Figure 11.37 Concreted front gardens create more impermeable surfaces

▲ Figure 11.38 Forestry reduces interception

Farming

Since the First World War, hedges have been ripped out to make way for huge fields that are more efficient for highly mechanised arable farming. Loss of hedges means less interception.

Farming has become more intensive and there has been a further increase in arable farming (crops) at the expense of pastoral farming (animals). Once crops have been harvested, they leave the soil bare in winter. This means there is no vegetation to intercept the rainfall.

Additionally, when fields are ploughed up and downhill, the furrows create channels for water to flow down easily (Figure 11.39). More soil is transported into rivers, raising their beds and so increasing the flood risk.

Disappearing fields

Fields intercept rainfall and soak up excess water through infiltration. Just as UK gardens are disappearing, so too are fields. Some fields near towns may be sold off to property developers, while others may be converted to riding stables. As large-scale factory farming increases, fields have also been replaced by huge sheds and concrete yards. Pastures have been over-grazed. This has compacted soil and degraded pastures, resulting in muddy runoff into rivers and an increased risk of flooding.

Additionally, to extend the growing season of fruit and salad crops, vast areas of polythene like those near Hereford in Figure 11.40 cover the fields of crops. While there may be some interception and evaporation from the polytunnels, the ability of the area to soak up water is reduced.

▲ Figure 11.39 An example of downhill plowing

▲ Figure 11.40 Fields of polytunnels reduce grassy areas

→ Activities

1 Explain how building a new housing estate on a flood plain increases the risk of flooding.

2 Explain why a change from pastoral to arable farming increases the risk of flooding.

3 Create a flow diagram to show how forestry can increase the risk of flooding.

4 'Urban land use causes a more significant risk of flooding than rural land use.' Do you agree with this statement? Justify your answer.

Fieldwork: Get out there!

For your house or the house of someone that you know:

1 Calculate the total area of the front garden

2 Calculate the percentage of the front garden taken up by impermeable surfaces.

3 Display this information as a pie chart or divided rectangle.

4 Considering both the front and back garden, write a report for the house owner suggesting how they might reduce the risk of flooding.

KEY LEARNING

➤ How precipitation links to discharge
➤ Hydrograph
➤ How hydrographs differ

How does precipitation link to discharge?

Rivers remove excess water from the land. The speed at which precipitation reaches a river is determined by physical and human factors. A river's discharge can vary depending on many factors, including the amount, type and intensity of precipitation. The flows to the river are shown in Figure 11.41. For detail, refer back to page 164.

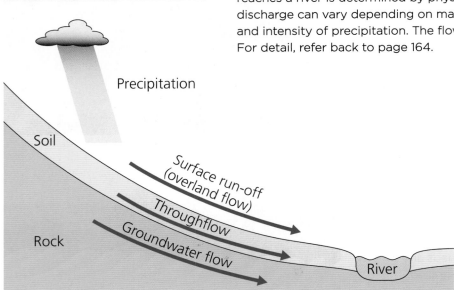

▲ Figure 11.41 How water gets into a river

What is a hydrograph?

A **hydrograph** shows how a river's discharge changes in response to a precipitation event. The vertical axis measures precipitation (usually rainfall) in millimetres and discharge in cubic metres per second (cumecs). The horizontal axis measures time, usually in hours or days. The bars represent rainfall and the line graph shows discharge. To understand how hydrographs work, it is worth taking time to understand the terminology.

Peak rainfall: the highest amount of rainfall per time unit (the highest bar)

Rising limb: shows how quickly the discharge rises after a rain storm (the first part of the line graph)

Peak discharge: the highest recorded discharge following a rainfall event (the top of the line graph)

Lag time: the time difference between peak rainfall and peak discharge (measure the horizontal distance between the top of the highest rainfall bar to the top of the discharge line and note the difference in hours)

Falling limb: shows the reduced discharge once the main effect of runoff has passed (the last part of the line graph which is going down)

Base flow: the normal flow of a river when its water level is being sustained by groundwater flow (usually shown on the hydrograph as a separate line)

Bankfull discharge: (does not always appear on hydrographs) will be drawn as a horizontal line marking the level of discharge above which flooding will occur as the river will burst its banks

▲ Figure 11.43 Hydrograph terminology

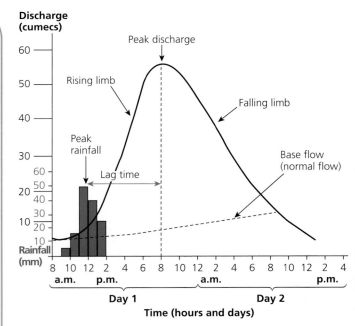

▲ Figure 11.42 A typical flashy response hydrograph

On any hydrograph, the rising limb will be steeper than its falling limb. The rising limb is fed by surface runoff, which reaches the river quickly over impermeable surfaces. The gentler slope of the falling limb reflects how discharge is steadily falling once surface runoff has stopped. Water is now reaching the river mostly through the soil as throughflow, which is slower than surface runoff. Eventually, this flow stops and the river returns to normal conditions, receiving water slowly through the rocks from groundwater (base flow).

How hydrographs differ

Hydrographs may be classified as having either a flashy response or a slow response.

Flashy response (storm) hydrograph

The flashy response hydrograph is associated with sudden flooding called 'flash floods'.

In Figure 11.42, the rising limb is steep because rainfall has occurred in conditions that have caused a lot of surface runoff. This means water gets into the channel quickly so there is a short lag time, giving the river a high peak discharge which puts it in danger of flooding.

Several conditions may result in a flashy hydrograph. For example:

- There may have been prolonged rainfall so the soil is saturated, or a long drought so soil is baked hard and cannot absorb the water.

- It may be a clay soil, which means water is unable to infiltrate.
- It may be on a steep-sided valley floor where water runs down the hillside, or on a flat flood plain where water cannot drain easily.
- It may have a small river basin so tributaries soon link with the main river to swell its discharge.
- It may have little vegetation, because of deforestation, to intercept precipitation.
- It may be an urban area, with large areas of impermeable tarmac and concrete.
- It may be in a rural area that has poor farming practices, like ploughing downhill.

Slow response hydrograph

On a slow response hydrograph an identical rainfall event will result in a less steep rising limb. The peak discharge is lower and the lag time longer. On this type of hydrograph the flood risk is low.

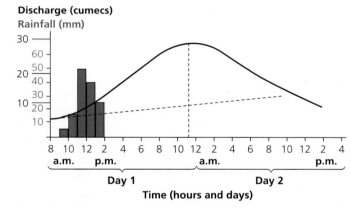

▲ Figure 11.44 Slow response hydrograph

→ Activities

1 Define the terms peak discharge, peak rainfall and lag time.

2 Study Figure 11.42, a flashy response (storm) hydrograph.
- What is the peak discharge?
- What is the peak rainfall?
- What was the total amount of rain that fell in this rainfall event?
- How many hours was the lag time?

3 Study Figure 11.44, a slow response hydrograph
- What is the peak discharge?

- How much lower is this peak discharge than that in Figure 11.42?
- What is the lag time?
- How much longer is the lag time than that in Figure 11.42?

4 Define the terms rising limb and falling limb
- How do these limbs differ between a flashy hydrograph and a slow response hydrograph?
- Describe conditions that may cause these differences.

KEY LEARNING

➤ Dams and reservoirs
➤ The benefits and costs of dams and reservoirs
➤ Channel straightening
➤ The benefits and costs of channel straightening

River management: hard engineering

Hard engineering uses heavy machinery to build artificial structures which work against nature to reduce the risk of flooding. **Dams and reservoirs** and **channel straightening** are two methods of hard engineering.

What are dams and reservoirs?

A dam is a large concrete barrier built across a river to impede its flow. This causes the valley behind the dam to flood, forming an artificial lake called a reservoir. This restricts the supply of water downstream. Water is released in a controlled manner through sluice gates in the dam. If water releases are carefully controlled and monitored, there should be no risk of flooding downstream.

What are the costs and benefits of dams and reservoirs?

Benefits

The benefits of the Kielder Dam and Reservoir in Northumberland, completed in 1981 can be seen in Figure 11.45.

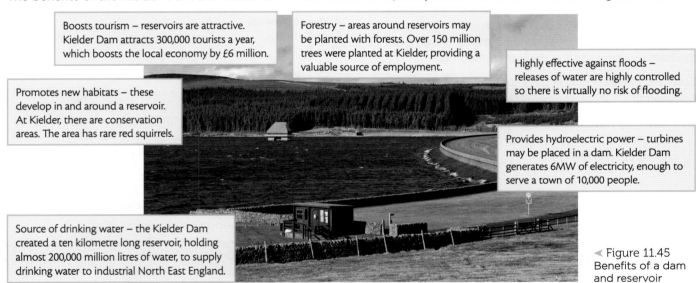

Boosts tourism – reservoirs are attractive. Kielder Dam attracts 300,000 tourists a year, which boosts the local economy by £6 million.

Forestry – areas around reservoirs may be planted with forests. Over 150 million trees were planted at Kielder, providing a valuable source of employment.

Highly effective against floods – releases of water are highly controlled so there is virtually no risk of flooding.

Promotes new habitats – these develop in and around a reservoir. At Kielder, there are conservation areas. The area has rare red squirrels.

Provides hydroelectric power – turbines may be placed in a dam. Kielder Dam generates 6MW of electricity, enough to serve a town of 10,000 people.

Source of drinking water – the Kielder Dam created a ten kilometre long reservoir, holding almost 200,000 million litres of water, to supply drinking water to industrial North East England.

◄ Figure 11.45 Benefits of a dam and reservoir

Costs

Social	Economic	Environmental
The flooding of a valley displaces people, usually farmers from their homes. At Kielder 58 families were displaced. This causes distress and breaks up communities.	Dams are expensive. Kielder Dam cost £167 million and may have been a waste of money. Loss of industry in North East England meant the demand for water and HEP was less than expected. Additionally, soils downstream can become less fertile through lack of sediment from floods, which reduces crop yields.	A concrete dam interferes with the path of migrating fish. Sediment is trapped behind the dam and this interferes with fish spawning grounds. Algae often collects behind a dam which deoxygenates the water. If there should be a sudden release of water through the sluice gates, this can cause river bank erosion downstream. The building of a dam may trigger an earthquake. Landslides often occur on the sides of a reservoir; this increases sediment and create shock waves which damage buildings. Reservoirs often flood areas of outstanding natural beauty. At Kielder, 1.5 million trees were lost along with 2,700 acres of farmland. This had a negative effect on habitats. New plantings are confined to Sitka Spruce.

What is channel straightening?

Channel straightening is when a meandering section of a river is engineered to create a widened, straightened and deepened course. This more efficient course improves navigation and reduces flood risk. In the nineteenth century, a new course was cut across a large meander loop on the River Tees to improve navigation. Centuries of straightening have also taken place on the River Parrett, to reduce flood risk in the low-lying Somerset Levels.

What are the benefits and costs of channel straightening?

Benefits

Social:

A straightened river reduces flood risk by moving water out of the area more quickly, as there is less friction with the bed and banks. The faster-flowing water also removes sediment that would otherwise build up the height of the river bed.

Economic:

- The historic cuts on the River Tees reduced the length of the river by 4.4 kilometres. This straightened course improved navigation considerably and increased trade at Stockton's port.
- Home owners gain confidence to invest in their property as they no longer expect to be flooded. Insurance costs also go down due to the lower flood risk.

Costs

Social:

Water flows through a straightened section quickly, but when it meets a meandering section downstream, such as at Burrowbridge in the Somerset Levels (2014), velocity is reduced. This causes sedimentation of the channel, so the river is more likely to flood, causing problems in another area.

Economic:

River straightening is expensive. Dredging a river to remove silt accumulation downstream is also expensive. After the 2014 flood damage, the EU authorised the £5.8 million dredging of a five-mile section of the Rivers Parrett and Tome near Burrowbridge. In some cases the impact of a straightened section downstream has been so severe that the river is restored to its original course. At Lewisham, London, £1.1 million was spent putting meanders back in the River Quaggy.

Environmental:

- The changes in hydrology and flooding downstream that can occur endanger animals and destroy habitats. The river's ecosystem is changed.
- A straightened river may have a concrete lining. This is visually unattractive and it deprives burrowing river bank animals of their habitat.
- In straightened sections, there is some evidence of increased pollution on the land from agro-chemicals, as runoff cannot drain into the river so easily.

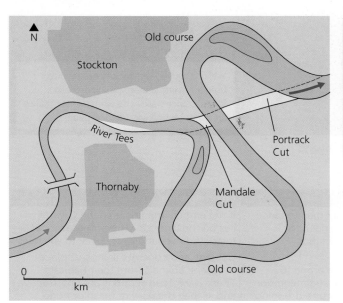

▲ Figure 11.46 Straightening the River Tees

→ Activities

1. Use information from Figure 11.45 to assess the social, economic and environmental benefits of dams and reservoirs.

2. Roleplay in a group of four. Take on the roles of an environmentalist, a local forestry worker, an elderly farmer living in the area to be flooded and a member of the local tourist board. Engage in a discussion about the proposal to dam the valley to make a reservoir.

3. Imagine you live in the valley that is to be flooded. Write a diary entry describing your thoughts on hearing of the proposal.

4. Draw an annotated diagram to explain how straightening a river reduces flood risk in an area.

River management: more hard engineering

Two other methods of hard engineering are **embankments** and **flood relief channels**.

What are embankments?

An embankment is an artificially raised river bank. In raising the banks, more water is contained in the channel. This reduces the flood risk. In Figure 11.47 the embankment at Bridge of Allan in Stirling, Scotland, is protecting the houses on the right that have a high flood risk. An embankment is made by bulldozers moving huge mounds of impermeable soil on to the river banks to build up their height. Some embankments are reinforced by gabions (wire cages filled with stones) or lined with concrete.

What are the benefits and costs of embankments?

Benefits

Safer from flooding – the channel now has an increased carrying capacity. It is less likely to burst its raised banks, so the risk of flooding to settlements behind the embankment is reduced.

Cheap – compared to other methods of hard engineering, the cost of building embankments is quite low.

Habitats – earthen embankments provide habitats for riverbank animals such as water voles, kingfishers and otters.

Walking routes – embankments are often used for riverside footpaths. Some follow long-distance walks. This embankment makes an attractive walkway for local people.

▲ Figure 11.47 The embankment protects houses in Bridge of Allan, Scotland

Costs

Social	Economic	Environmental
Embankments deprive people of easy access to the river for fishing and boating. Although they reduce the risk of flooding, embankments are not as reliable as other types of hard engineering. Their presence gives people a false sense of security, which means they may not be prepared for floods.	Embankments have higher maintenance costs than other hard engineering methods, as they need constant monitoring and repair. Earthen embankments are prone to erosion and this increases sedimentation downstream, which will incur a dredging cost if flooding is to be avoided.	If the embankment is breached, water lies on the land for a long time, as it has a restricted overland route back to the river. Gabions and concrete linings displace riverbank animals from their habitats. These reinforced sections are unattractive and, if they break, wire mesh or huge slabs of concrete litter the river bed.

What is a flood-relief channel?

A flood-relief channel is an artificially made channel that is designed as a backup channel for a river that frequently floods. It works like a bypass. The newly engineered channel runs roughly parallel to the main river (as seen with the Thames in Figure 11.48). The River Exe at Exeter has three relief channels, which were constructed at a cost of £8 million following devastating floods in 1960. The largest of these channels is the Exwick spillway. A gate has been built across the River Exe which automatically closes off the river in times of high discharge and diverts water along the Exwick spillway, thus reducing the risk of the River Exe flooding.

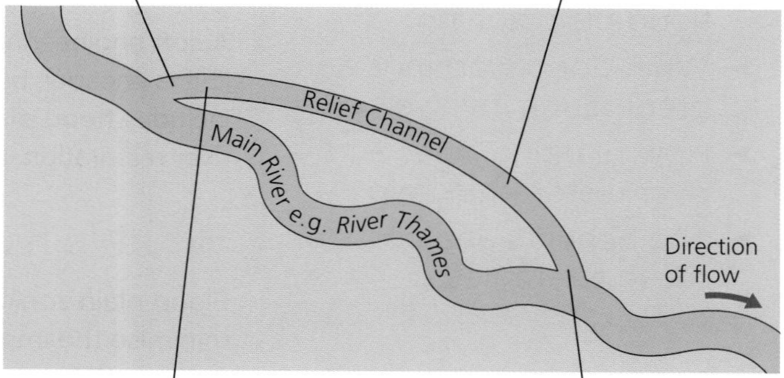

2. A cut is made into the river's bank to join up with the new channel.

1. A new channel is carved out of the land by heavy engineering equipment.

Relief Channel

Main River e.g. River Thames

Direction of flow

4. In times of high discharge, the relief channel takes the excess water thus preventing the main river from flooding.

3. Another cut is made into the main river to take back the diverted water.

▲ Figure 11.48 How a flood-relief channel works

Benefits

Social	A relief channel removes the risk of flooding from a designated area. Exeter's relief channels protect around 3,000 properties.
	Footpaths and cycle tracks are built along a new channel.
	Calm water provides areas for model boating and canoeing.
	Where reed beds have been included, birdwatching and nature reserves may be set up.
Economic	Insurance costs are lower in the vicinity.
	The value of homes increases and houses are easier to sell.
	There is a more secure environment for setting up business ventures.
Environmental	Some relief channels include artificial reed beds and grass-covered concrete sides. These provide new habitats.
	When full of water, they produce a tranquil setting.

Costs

Social	People living in the path of a relief channel have to be moved, causing disruption.
	Settlements downstream of a relief channel suffer from increased flooding, as the merging of water from the relief channel swells this part of the river. This raises the question of the ethics of protecting some settlements to the detriment of others.
Economic	Flood-relief channels are expensive.
	Sometimes, as in the case of the Jubilee River (see Section 11.15), they run out of funds. They also need to be maintained and repaired.
	The schemes take a long time to come into effect; Exeter's relief channels took twelve years to build.
Environmental	In the construction of relief channels, habitats are disturbed.
	The level of water in a relief channel varies considerably. This provides an unreliable habitat.
	Relief channels look unattractive in times of low flow, when vast expanses of concrete and gabions are exposed.

→ Activities

1 Study Figure 11.47.
   a) List the social, economic and environmental benefits of embankments.
   b) Consider the social, economic and environmental benefits and costs of embankments. In each case, do you think the cost or the benefit is greater?

2 Imagine you live in an area prone to flooding and the council is considering two proposals. One is to embank the river. The other is to build a relief channel.
   a) What four questions would you raise at the council meeting?
   (b) Which proposal are you more likely to favour, and why?

- ➤ The benefits and costs of flood plain zoning
- ➤ What flood warnings and preparation involve
- ➤ How planting trees and river restoration help
- ➤ The benefits and costs of those strategies

River management: soft engineering

A soft engineering strategy involves adapting to a river and learning to live with it. It is cheaper, but often less effective than hard engineering. Soft engineering includes **flood plain zoning**, **flood warnings**, preparation, planting trees and river restoration.

What is flood plain zoning?

Flood plain zoning is where land in a river valley is used in such a way as to minimise the impact of flooding. In England and Wales, the Environment Agency (EA) categorises land into four flood-risk zones and issues flood risk maps.

Local authorities are required to use these maps to produce flood-risk assessments and to guide decisions regarding new building applications. The land use zones shown in Figure 11.49 show how land can be used sustainably by having different land uses parallel to the river. More permanent structures can be installed further away from the river without any substantial economic or social costs, should flooding occur.

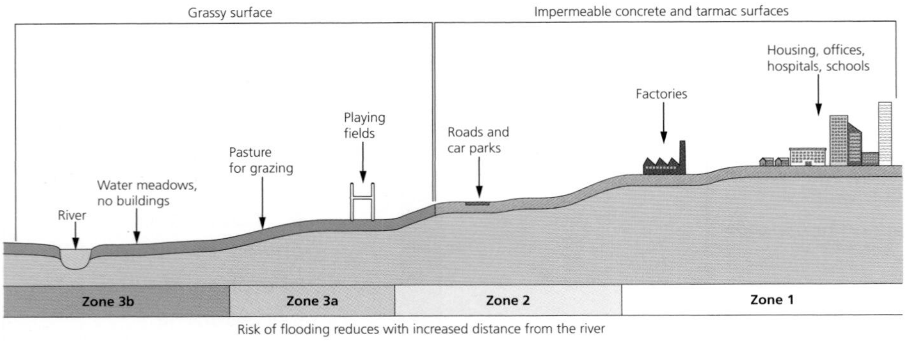

▲ Figure 11.49 Flood plain zoning

How do flood warnings and preparation work?

The EA and other agencies, such as district councils, and the Water and Highways Authorities co-ordinate efforts to devise and carry out action plans for areas at risk. Distinct roles are identified for the emergency services, the armed forces and voluntary groups such as the Royal National Lifeboat Institution.

The meteorological office analyses data from its 200 automated weather stations and passes this to the EA, who uses it, along with river level data, to provide updated flood alert information. The media, and occasionally sirens or loudspeakers, publicise this information. The EA provides a flood map website, a three day flood forecast, and personalised warnings. The EA also provides information on how to prepare oneself for a flood. This is summarised in Figure 11.50.

How does planting trees help?

Planting shelter belts of trees across slopes and woodland in floodplains (rewilding) reduces the risk of flooding, as trees intercept water by taking it up through their roots. Wales plans to plant ten million trees over the next five years.

How does river restoration help?

River restoration is when a river that has previously been hard engineered is restored to a natural channel. For example, near Sutcliffe Park in Greenwich, River Quaggy had previously been re-routed through underground drains, but by 2007 it was brought back to the surface and restored close to its original meandering course. Part of the floodplain was lowered to create a floodwater storage area, and wildflower meadows and avenues of trees were planted.

What are the benefits and costs of these strategies?

Strategy	Benefits	Costs
Flood plain zoning	• By restricting building on the active flood plain, local authorities do not increase impermeable surfaces, so the risk of flooding is reduced • It is low-cost: only administration costs are involved. • Traditional water meadows by a river (Figure 11.49, zone 3b) are protected from development. In some meadows, cows may graze there when it is not flooded. • By conserving the flood plain, planners provide a welcome green space in UK towns.	• This approach has limited impact as many UK cities have already sprawled over the active flood plain. • It is very difficult to get planning permission to extend or rebuild homes in the flood plain. • There is a housing shortage in the UK. Restricting building makes the problem worse. Restricted supply will inflate house prices. • Habitats are destroyed due to increased building on other greenfield sites.
Flood warnings and preparation	• This is a very cheap way of protecting people and their property, as it is largely dependent on communication via the internet. The EA's personalised flood warning option makes people feel more secure and more in control • If people are warned in advance of a flood, then they protect their valuables earlier • It is a way of ensuring people's safety without having to invest in high-cost hard engineering	• Flood warnings are only effective if people listen and take action. Education is needed: not everyone listens to, or has access to, the media or the internet. • It does not help people living in areas prone to flooding. The clear-up operation is distressing, people may have to move to temporary accommodation, their insurance costs will increase and their houses will be difficult to sell
Planting trees	• Reduces water flowing downstream as shelter belts of broad-leaf trees can reduce surface runoff • More carbon dioxide is absorbed • Adds variety to the landscape and new habitats. Increases biodiversity • Relatively inexpensive as there are EU grants available	• Changed appearance: countryside wooded rather than open grass, arguably artificial looking and less aesthetically pleasing • Loss of potential grazing land
River restoration	• Creates new wetland habitats and increases biodiversity, e.g. damselflies • Increased water storage areas reduce risk of flooding downstream – the River Quaggy scheme has protected 600 homes and businesses from flooding • Aesthetically pleasing – visitor numbers to Sutcliffe Park have increased.	• Possible loss of agricultural land and flooding of crops near the river • Can be expensive: the initial cost of restoring River Quaggy was estimated at £1.1 million • Not always the most effective or practical strategy

▼ Figure 11.50 Environment Agency Flood Warnings

FLOOD ALERT

Flooding is possible.
Be prepared

FLOOD WARNING

Flooding is expected.
Immediate action required

SEVERE FLOOD WARNING

Warning
Severe flooding
Danger to life

→ Activities

1. Explain the land use zoning shown in Figure 11.49.
2. Draw a flow chart to show how information is gathered and communicated to the public to warn them about a risk of flooding.
3. Go to the website www.environment-agency.gov.uk/flood. Read the leaflet. Assume you live by a river. Make a flood action plan for your family.
4. Evaluate the UK government's policy of not building on flood plains. Consider the costs and benefits of flood plain zoning. Then write a short report (300 words) to give your conclusions.

Fieldwork: Get out there!

Select a 50 metre stretch of river that runs through a settlement and which has a path alongside it.

1. With an adult, walk along the path and gather information to demonstrate how land use is zoned parallel to the river (how it changes on both sides as you look further away from the river). Take a photo.
2. Draw an annotated sketch map and annotate your photo to present your findings.

⭐ KEY LEARNING

➤ The characteristics of the scheme

➤ Why the scheme was needed

➤ The issues that arose from the scheme

Jubilee River flood-relief channel

The Jubilee River is a relief channel for the River Thames in south-east England. The relief channel runs through Berkshire and Buckinghamshire, flowing roughly parallel to the River Thames. It starts to the south-east of Maidenhead and flows in south-easterly direction passing just to the north of Eton. Once it has passed Eton, it re-joins the River Thames (Figure 11.51).

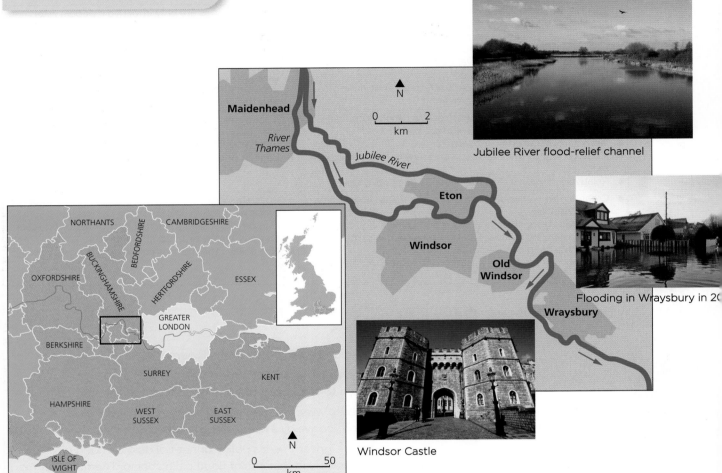

Jubilee River flood-relief channel

Flooding in Wraysbury in 20[...]

Windsor Castle

▲ Figure 11.51 Location of the Jubilee River

What are the characteristics of the scheme?

It was funded by the Environment Agency (EA), and cost £110 million. It opened in 2002, and at 11.7 kilometres long and 50 metres wide, it is the UK's largest artificial channel. The channel was designed to look like a natural river, so it has meanders and shallow reed beds, and a nature reserve with bird hides has been created in the area. It has five weirs, or large dams, along its course.

Only two of the weirs are navigable by paddle craft. Under normal conditions, the level of water in the river is low but when discharge is high, the Jubilee River effectively diverts water from the River Thames, thus preventing the River Thames from overflowing its banks. This reduces the flood risk in southeast Maidenhead, Eton and Windsor.

Why was the scheme required?

This area of the Thames flood plain is low-lying and prone to flooding. It contains the royal settlement of Windsor, which attracts many international visitors, as well as Eton, home of a prestigious public school. The impermeable surfaces of the built-up areas have historically resulted in flooding following high rainfall events. Given the high-value property in this area, the EA decided to increase the level of flood protection.

What measures were taken?

The Jubilee River was created to take overflow water from the River Thames in times of high discharge following heavy rainfall.

What issues arose from the scheme?

Social

- Is it ethical to protect some properties at the expense of others? Three thousand properties were protected in affluent Eton and Windsor, but to the detriment of the less wealthy settlements of Old Windsor and Wraysbury downstream. The Thames at Old Windsor now suffers from a much higher discharge due to the merging of the two channels just upstream. The scale of the problem came to light in 2014, when the area suffered its worst floods since 1947.
- Paddle boaters had been promised a navigable river. However, on two weirs they have to carry their boats around them, and Taplow Weir is considered too dangerous to cross.

Economic

- The Jubilee River scheme is the most expensive flood-relief scheme in the UK. Yet, one year after its completion, the weirs were damaged by floods. The initial repair bill for Slough Weir alone was £680,000. Maintaining the channel is a huge economic burden.
- At a projected cost of £330 million, the Jubilee River was one of four flood-relief channels planned for the lower course of the Thames. However, the EA ran short of funds. If further engineering is to alleviate flooding downstream, local councils and businesses will have to make up a £110 million shortfall. Is this fair, when Windsor and Eton residents did not have to pay?
- Until a solution is found, small businesses such as shops stand to lose money, as they cannot open when their premises are flooded. Insurance costs are high. Business insurance costs for Wraysbury alone were around £500 million in 2014. This will cause future insurance premiums to increase.

Environmental

- In 2014, there was extensive flooding immediately downstream from where the flood-relief channel rejoined the Thames. The built environment suffered from flooded roads and buildings. Fields were inundated and habitats were disturbed.
- The concrete weirs are rather ugly, especially under normal flow conditions, when more concrete is exposed. Ongoing repair work has made the matter worse, such as at Manor Farm Weir (Figure 11.52).
- There is also the problem of algae collecting behind the weirs. This disrupts the natural ecosystem.

▲ Figure 11.52 Unsightly repairs on Manor Farm Weir

→ Activities

1. Assess the social, economic and environmental costs and benefits of the Jubilee River scheme. Try to limit each cost and benefit to six words.

2. a) Group work: debate the success of the Jubilee River scheme. Possible roles: Windsor resident, Wraysbury resident, insurance company manager, Wraysbury shopkeeper, environmentalist and a canoeist.
   b) Write a report summarising the views expressed.
   c) In your view, was the scheme worthwhile? Justify your answer.

12 Glacial landscapes

The power of ice in shaping the UK

Ice was a powerful force in shaping the physical landscape of the UK. Today, there is no permanent ice cover in the UK. It was a very different story 20,000 years ago (see Chapter 4), when ice covered most of the UK, as part of a vastly expanded Arctic ice sheet. In places it was three kilometres thick (Figure 12.1). Gravity caused large bodies of ice called glaciers to flow slowly from highland into lowland areas. Around 10,000 years ago, Earth's climate warmed again. As the ice age ended, the ice melted and retreated, revealing a transformed upland landscape, chiselled into steep peaks and sharp ridges.

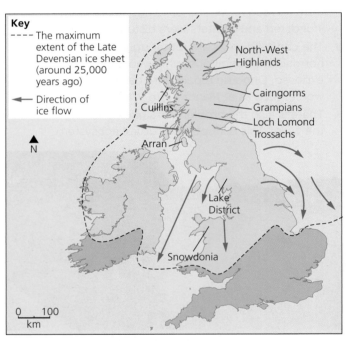

▲ Figure 12.1 Map of extent of ice coverage

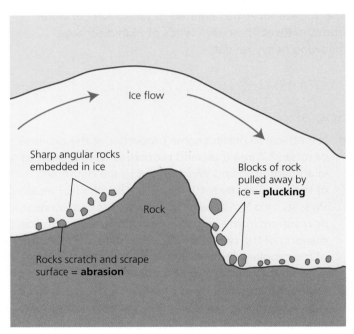

▲ Figure 12.2 The ice erosion processes of plucking and abrasion

What are the main processes of ice erosion?

Chapters 10 and 11 explain erosion in relation to the moving forces of water. In this chapter, the moving, eroding force is ice. When gravity causes ice to move down a mountainside, there are two main ways in which the rock below becomes eroded (Figure 12.2):

■ **plucking** – as the ice moves over the rock surface below, meltwater freezes around loose sections, pulling them away. Plucking is especially effective when the rock contains many joints (cracks) which the water can seep into. One reason why meltwater is present under a glacier is the sheer weight of ice above. Ice at the base of the glacier melts because of the great pressure it is under (this is called pressure melting). Also,

meltwater has travelled from the surface of the glacier to its base through crevasses (giant cracks) in the ice.

■ **abrasion** – erosion is caused by rocks and boulders embedded in the base of the glacier. These act like sandpaper, scratching and scraping the rocks below. Very large boulders can do enormous damage this way, scarring the landscape with features called striations. These are still visible in the UK today.

When a lot of plucking has taken place, large numbers of rocks and boulders become embedded in the ice. This increases the rate of abrasion.

How does freeze-thaw weathering take place?

Erosion requires a moving force like ice to break apart rock and carry it away. In contrast, weathering describes the destruction of rock that takes place in a particular place. The remains of the rock do not move; they remain in situ. Weathering is caused by temperature and moisture changes, along with any chemical processes caused by mild acids in rainwater. You can see the effects of weathering all around you on a daily basis on roads and buildings.

In glaciated areas, freeze-thaw weathering (or frost shattering) takes place on rock surfaces above the surface of the ice and at its margins. This is a physical weathering process and does not involve chemical changes:

- Water seeps into cracks in a rock face (this may be water from summer rainfall or snowmelt).
- The temperature falls at night, causing the water to freeze.

- Water expands by ten per cent when it turns to ice. This expansion puts pressure on the rock either side of the crack, prising it apart and causing the crack to tear wider open.
- During the next 24-hour cycle, the ice melts, sinks deeper into the crack, and then freezes again.
- Over time, large blocks of rock can be shattered apart by repeated cycles of this weathering process.

The evidence for freeze-thaw weathering is seen in landscape features called scree slopes and blockfields. These are piles of rock debris that blanket large upland areas in the UK. Some debris dates from the last ice age, but some is more recent. Freeze-thaw weathering is still an important process in areas where many repeated freeze-thaw cycles take place during the winter months. It can also be classified as a weather hazard because of the damage that can be done to houses.

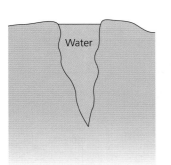
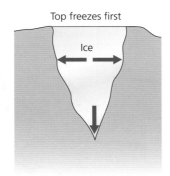
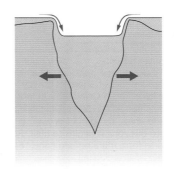

Top freezes first

Water

Ice

If the ice thaws the next day the resulting water will not fill the crack, which is now both wider and deeper because of its 9% expansion. Dew or rainfall on the rock surface can refill the crack.

▲ Figure 12.3 How freeze-thaw weathering operates in cycles

→ Activities

1. Explain the difference between weathering and erosion.
2. a) How does plucking work?
   b) Suggest how the rate of ice erosion can be affected by the type and angle of the rock a glacier is moving over.
3. Many people live in areas where freeze-thaw takes place regularly. This can cause their water pipes to burst if they get very cold. Write an information leaflet aimed at homeowners in an at-risk area. You should explain what the risk is and suggest steps that people can take to protect their homes.
4. The process of freeze-thaw weathering works very slowly in the world's coldest places. In contrast, it works very quickly in some parts of the UK. Suggest why this is the case.

→ Going Further

What happens underneath a glacier?

The way ice at the base melts under pressure is an interesting idea to explore further. You can try out your own experiments on an ice cube (but take care not to simply melt the ice with the warmth of your hand). You can also research the different types of active glacier in the world today.

Glacial movement and sediments

How do glaciers move?

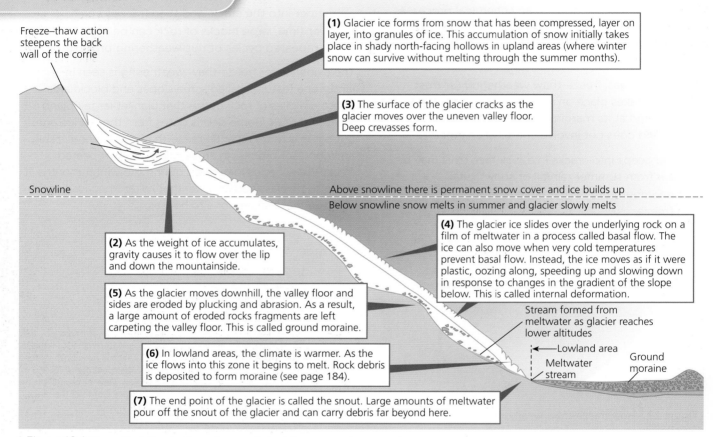

Freeze–thaw action steepens the back wall of the corrie

(1) Glacier ice forms from snow that has been compressed, layer on layer, into granules of ice. This accumulation of snow initially takes place in shady north-facing hollows in upland areas (where winter snow can survive without melting through the summer months).

(3) The surface of the glacier cracks as the glacier moves over the uneven valley floor. Deep crevasses form.

Snowline

Above snowline there is permanent snow cover and ice builds up

Below snowline snow melts in summer and glacier slowly melts

(4) The glacier ice slides over the underlying rock on a film of meltwater in a process called basal flow. The ice can also move when very cold temperatures prevent basal flow. Instead, the ice moves as if it were plastic, oozing along, speeding up and slowing down in response to changes in the gradient of the slope below. This is called internal deformation.

(2) As the weight of ice accumulates, gravity causes it to flow over the lip and down the mountainside.

(5) As the glacier moves downhill, the valley floor and sides are eroded by plucking and abrasion. As a result, a large amount of eroded rocks fragments are left carpeting the valley floor. This is called ground moraine.

(6) In lowland areas, the climate is warmer. As the ice flows into this zone it begins to melt. Rock debris is deposited to form moraine (see page 184).

Stream formed from meltwater as glacier reaches lower altitudes

Lowland area

Meltwater stream

Ground moraine

(7) The end point of the glacier is called the snout. Large amounts of meltwater pour off the snout of the glacier and can carry debris far beyond here.

▲ Figure 12.4 Characteristics and processes of glacial movement

▲ Figure 12.5 The snout of a glacier

How does a glacier transport material?

The front of a glacier is called its snout (Figure 12.5). As ice from upland areas descends into lowland areas, the snout **bulldozes** material. Soil, rocks and boulders are shoved forwards by the sheer force of the moving ice. Material is also carried on the surface of the glacier (you can see this in Figure 12.6 on page 181). Freeze-thaw weathering takes place on mountainsides above the glacier. This causes blocks of rock to become detached and fall onto the ice below. Some material is carried inside the glacier too for two main reasons:

- Plucking has torn away rock at the bed of the glacier that is now embedded in the base of the moving ice.
- Some rocks fall into crevasses at the surface of the glacier. Some crevasses reach deep into the glacier, resulting in a build-up of material inside the moving ice.

Why does glacial deposition take place?

Glaciers can carry ice far from the regions where the snow falls. As glaciers move from upland to lowland areas they enter a warmer climatic zone. During summer months, meltwater pours off the snout of some glaciers. Meltwater rivers can transport vast quantities of water from glaciers and ice sheets into the oceans. These rivers carry large amounts of sediment called glacial **outwash**. Because it has been carried by water, outwash material has been rounded and reduced in size by attrition. It has also been deposited sequentially and sorted, with fine material carried furthest from the glacier.

Under normal conditions, the snout of a glacier does not actually retreat, even though constant melting is happening. This is because new ice continually flows down from upland areas to replace and balance the meltwater loss. The following examples illustrate this point:

- A glacier moves forwards at four metres a day. Old photographs show its snout has neither advanced nor retreated. This means that a four metre length of the glacier is being melted each day.
- Another glacier moves forwards at ten metres a day. Old photographs show its snout has retreated (Figure 12.6). This means that more than a ten metre length of the glacier is being melted every day.

The behaviour of a glacier is similar to that of an escalator. Rocks and boulders are constantly moved down the mountain side. Eventually, the section of ice they are carried on reaches the warmer lowland areas. The same section of ice temporarily becomes the snout of the glacier – and is promptly melted away too! Deposition occurs then.

Constant transport of new, debris-laden ice into lowland areas results in the widespread deposition of all of the eroded and weathered material from the uplands that the glacier has carried. The dumped material is called glacial **till** (Figure 12.6).

▲ Figure 12.6 A retreating glacier and the till (deposition) it leaves behind, Greenland

→ Activities

1 Describe how snow is converted into glacier ice over time.

2 Compare how material is transported by the top and the underside of a glacier.

3 Explain how the processes of basal flow and internal deformation help a glacier to move.

4 Look at Figure 12.4.

   a) Where is the snowline found and what does it show?

   b) Draw a table with two columns, labelled 'Landforms and features above the snowline' and 'Landforms and features below the snowline'. Add as many landforms and features as you can from Figure 12.4 to this table, together with a brief description of each one.

5 The snout of a glacier is observed to be retreating. Yet material on top of the glacier is still being moved forwards. Explain why.

➤ How upland areas are affected by ice erosion
➤ How upland river valleys are modified by ice erosion

Landforms resulting from ice erosion

How are upland areas affected by ice erosion?

Glaciated places in the world are home to unique landforms that give them a special character that is not found elsewhere. These special features were revealed in the UK at the end of the last ice age in the regions shown in Figure 12.1 (see page 178).

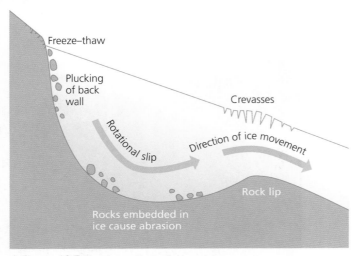

▲ Figure 12.7 The formation of a corrie

One such landform is a **corrie**. This is a deep armchair-shaped hollow found on the flank of a mountainside, where glaciers begin. A corrie is formed by a snow patch in a hollow, deepening and compressing to become a dense mass of ice. Freeze-thaw weathering (see page 179) increases its density by removing air. As it pulls away from the walls of the hollow, blocks of rock are plucked. Once embedded in the base of the ice, these blocks abrade the hollow, causing it to get wider, deeper and steeper (Figure 12.7).

Freeze-thaw weathering takes place on the back wall of the hollow, above the surface of the ice, which soon becomes covered with fallen, weathered rock. Over time, the back wall retreats backwards, cutting deep into the side of the mountain.

When two corries develop side by side or back to back, an **arête** (a freeze-thawed ridge) develops between them. When three or more corries grow in hollows on all sides of a mountain, a **pyramidal peak** is produced. As the corries erode the mountain behind them, the remaining rock is weathered into a sharp point (Figure 12.8).

The deepest corries in the UK are found on the northeast side of mountains where least sunlight is received. When the ice disappeared at the end of the last ice age, some deep corries filled with water to create corrie lakes.

How are upland river valleys modified by ice erosion?

At the start of the ice age, small corrie glaciers developed in the UK's mountainous regions. As the climate grew colder, ice began to spill out of the corries and flowed into the numerous river valleys that are a feature of upland areas. Because ice flows far less quickly than water, the river valleys were soon filled entirely with slow-moving but powerful ice.

▲ Figure 12.8 Pyramidal peak, arête and corrie

The result was that the shape and appearance of these river valleys changed completely.

- Before glaciation, a river valley would have been V-shaped in profile. River tributaries flowed down the gentle valley sides to reach a meandering valley floor.
- During glaciation, the rock in the valley sides is torn away by a combination of plucking and abrasion. The result is a U-shaped valley or **glacial trough**. It's very steep, almost vertical sides lead down to a straight and wide valley floor.
- The tributaries that used to flow down the river valley sides now exit abruptly through a gap in the new cliff-shaped valley wall. The water cascades down from a high altitude and creates a waterfall. The portion of the original tributary valley that remains is now called a **hanging valley**.

- While rivers meander around spurs of land, ice has the erosive power to remove any obstacles in its path. **Truncated spurs** can usually be identified along the sides of a glacial trough.

These changes to the river valley are clearest to see in an area that has only recently been de-glaciated. The UK's glaciated landscapes have continued to change in the 10,000 years since the ice retreated. Figure 12.9 shows a typical U-shaped valley today. Parts of the valley floor that were over-deepened by plucking have filled with water to create a ribbon lake.

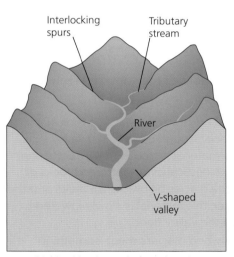

A highland landscape before glaciation

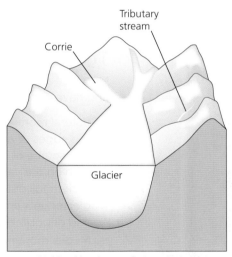

A highland landscape during glaciation

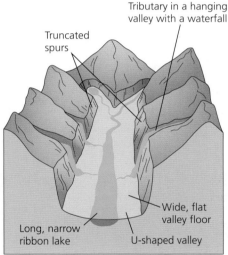

A valley after glaciation

⌃ Figure 12.9 How an upland valley and its landforms are modified by glaciation

→ Activities

1. Sketch Figure 12.8 and label the three features (corries, arêtes and peaks).

2. What is a hanging valley? How is it formed?

3. a) What is the difference between a V-shaped valley and a glacial trough?
   b) Explain how the movement of ice is responsible for changing a V-shaped valley into a glacial trough.

4. Draw a picture of a corrie filled with ice. Add labels to show the characteristics and formation of this feature. Use two different coloured pens to make these annotations (one for the characteristics and a different one for the formation). You can base this task on Figure 12.7 but aim to add additional annotations that use more detailed information drawn from the text.

5. Draw a table with two columns, labelled 'Before glaciation' and 'After glaciation'. Fill it with as many matched pairs of landforms as you can, for example, a V-shaped valley and a glacial trough.

Landforms resulting from ice transport and deposition

Which landforms result from moving or melting ice?

Depositional landforms are produced when a glacier loses the ability to carry material.

- When a melting ice mass reduces in size, the material it is carrying drops to the ground.
- Material is sometimes deposited underneath a moving glacier. Plucking sometimes results in very large amounts of rock fragments being carried along in the base of the ice. Later, some of this debris gets dropped back onto the valley floor. This could be because it gets lodged, or stuck, behind obstacles the ice is flowing over, such as bands of hard rock.

Much of the material that gets deposited is till (see page 181). Where ice has retreated, we can see **moraines** and **drumlins**.

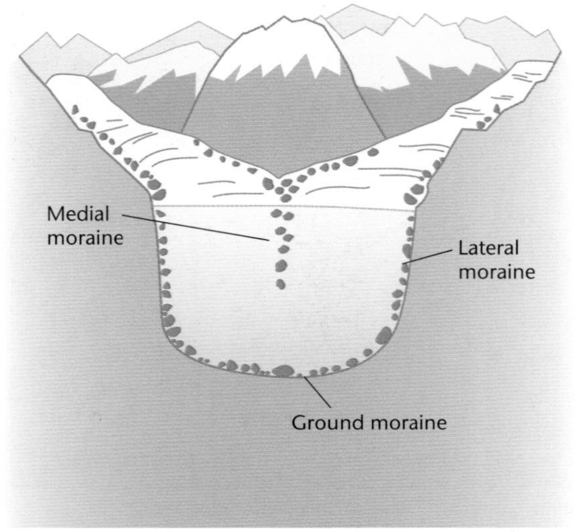

The different types of moraine

Moraines are accumulations of rock debris (Figure 12.10) and have distinct shapes:

- Lateral moraine – a ridge of material that runs along the edges of a glacial trough close to the valley side. The source of the material is freeze-thaw weathering, high on the valley sides, causing shattered blocks of rock to fall onto the glacier below. As ice melts and the glacier gets smaller, this material is slowly lowered to and deposited on the valley floor.
- Medial moraine – when glaciers meet, something very interesting happens. Two lateral moraines merge together to form a very large ridge of rock debris: the medial moraine. In Figure 12.11, you can see this by the thick dark stripe running straight down the middle of the glacier below the point where the two glacial tributaries have met.
- Ground moraine – the material that gets lodged and deposited underneath the glacier is simply called ground moraine. Vast amounts of ground moraine can be produced when glaciers disappear entirely in response to climate change. Ground moraine covers large areas of the UK as a legacy of the last ice age. Your house might even be built on ground moraine!
- Terminal moraine – this is the enormous ridge of material that gets bulldozed by the snout of the glacier (see page 180). These features can still be identified in the landscape today. They allow us to work out how far the ice advanced during the last ice age.

Medial moraine

Lateral moraine

Ground moraine

▲ Figure 12.10 The formation of different types of glacial moraine

▲ Figure 12.11 An aerial view of a glacier showing how the medial moraine is produced

Drumlins

Drumlins are egg-shaped hills composed of mounds of till. Two processes are involved in their formation:

- First, material is deposited underneath a glacier as ground moraine.
- Second, this ground moraine is sculpted to form drumlin shapes by further ice movements (think of the ridges that fingers leave when they move through sand).

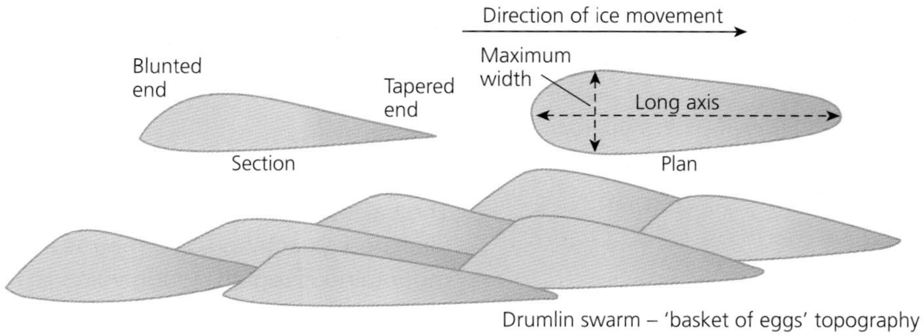

▲ Figure 12.12 The characteristics of drumlins

Drumlins show the direction of movement in the past. For landscape detectives, they are an important clue! They can be 100 metres or more in length. A large group is said to resemble a 'basket of eggs'. Some drumlins may have a very large rock fragment at their core, which could help explain why they formed in some places and not others.

What do 'glacial erratics' tell us about past ice movements?

Glacial **erratics** are another brilliant clue for landscape detectives. We have learned a great deal about ice movements in the past from the study of erratics in the UK. An erratic is a large boulder that stands out like a sore thumb in the landscape. This is because it is composed of a rock type that is nowhere else to be seen.

Figure 12.13 shows granite erratics resting on a sandstone platform in Arran, Scotland. This site is some distance away from the nearest outcrop of granite. The boulders provide clear evidence that a glacier flowed here in the past.

▲ Figure 12.13 Glacial erratics on the Isle of Arran

Fieldwork: Get out there!

'A glacier used to be here.' How could this hypothesis be investigated for a lowland area?

- Make a list of landforms that you might be able to identify and any measurements you could take.
- What difficulties might you encounter while carrying out fieldwork to identify landscape features such as moraine?
- See Figure 12.16 on page 187 for an example of an OS map. How could maps provide additional evidence to help with your investigation?

→ Activities

1 Look at Figure 12.12. (a) What is a drumlin? (b) Describe the characteristics of a 'swarm' of drumlins.

2 Use an annotated diagram to help show how medial moraines are formed.

3 What is a glacial erratic? Give two characteristics of a boulder that could help identify it as being a glacial erratic.

4 Explain how different ice processes contribute to the formation of different types of moraine. Make sure you refer to all the key words below (you can use them more than once). You can present your answers in a table with the column headings: Ice process; Types of moraine; How the process contributes to the formation.

- plucking
- abrasion
- freeze-thaw
- bulldozing
- lateral moraine
- medial moraine
- terminal moraine
- ground moraine

> ⭐ KEY LEARNING
>
> ➤ Major landforms of ice erosion
> ➤ How ice deposition has affected the Lake District

Glacial landforms in England's Lake District

The Lake District is one of numerous upland areas that were previously glaciated. Reaching 1,000 metres at Scafell Pike, it is England's highest mountainous region. Tough volcanic rocks were scoured and re-shaped by glaciers to produce the landforms you now see.

Which major landforms of ice erosion can be seen in the Lake District?

The Lake District is well-known for its mountains and ribbon lakes.

▲ Figure 12.15 Striding Edge with Red Tarn to the left, Helvellyn, Cumbria

Its many mountains and hills were thoroughly researched by the walker and writer Alfred Wainwright in the 1950s. Wainwright counted 214 significant peaks. The Lake District lacks a truly good example of a pyramidal peak, but nonetheless has some spectacular arêtes and corries. Figure 12.15 shows a corrie lake and arête called Red Tarn and Striding Edge respectively. They are located just to the east of Lake Thirlmere in the map extract shown in Figure 12.16. There is a hanging valley at Grisedale.

Red Tarn provides us with landscape evidence of how **rotational slip** in a corrie eroded deep into the mountain side. You can see the steep back wall that forms part of the arête. The lake is testament to the erosion that once took place there. It has left behind a deep, wide hollow that has filled with water in the post-glacial period. Immediately after the ice first retreated, the corrie's sides would have been even steeper than they are today. Over time, the Lake District's features have softened. This is because rain and running water, rather than ice, are now the main influences on landscape development.

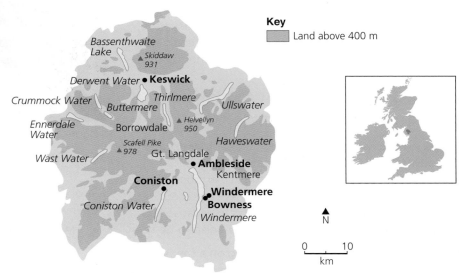

▲ Figure 12.14 Glaciated upland areas and ribbon lakes in the Lake District

There are many ribbon lakes in the Lake District. They mark the position of over-deepened glacial troughs. Like Scotland's lochs, these lakes are much deeper than you might expect from looking at a photograph. In the 1930s, Lake Coniston gained notoriety when Donald Campbell was killed there. He was attempting to break a world water speed record in his high-powered boat, the Bluebird. Sadly, the boat overturned. Campbell had chosen Coniston because it was long, straight and deep.

In other parts of the Lake District, settlements have developed on dry, flat sections of the area's wide and U-shaped valley bottoms. Keswick is situated on the floor of a glacial trough that the River Derwent now flows into.

How did ice deposition affect the Lake District in the past?

Ice deposition features are still visible in the landscape even after 10,000 years. Farming in lowland areas takes place on ground moraine, though you cannot see it now that vegetation is present. Some relief features are visible, however.

■ Fields in Borrowdale use terminal moraines as boundaries. You may think you are looking at an artificially created earth embankment, but it is entirely natural.

■ Swarms of drumlins can be seen in some places, such as Swindale in the northeast Lake District (Figure 12.16).

■ Glacial erratics are strewn across low-lying areas of the Lake District. Deposited by melting ice between 10,000 and 30,000 years ago, some glacial erratics may have travelled all the way from Scotland!

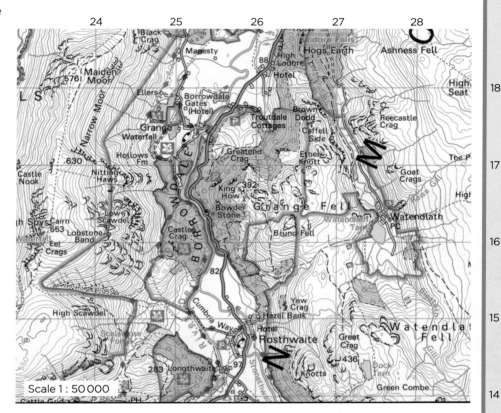

▲ Figure 12.16 Ordnance Survey map showing the part of the Lake District

Geographical skills

Map work

The OS map (Figure 12.16) has a scale of 1:50,000. This means that two centimetres on the map represents one kilometre.

1 How large is the map area in square kilometres?

2 Identify (a) the highest area shown on the map, (b) the lowest area and (c) the flattest area where slope angle is least.

3 Give the four-figure grid reference of a corrie lake.

4 What features are found at square 2415?

→ Activities

1 Make a table with two columns. In one column, list all the erosion and deposition features found in the Lake District. In the second column, write the name of where the feature can be seen.

2 a) Identify three uses that humans have found for glacial landforms in the Lake District.
   b) Explain what the characteristics of these landforms are that make them useful.

3 Design an advertisement for tourism in the Lake District. Who would be your target audience?

4 Look at Figure 12.14.
   a) Based on landscape evidence, where do you think the deepest ice was found in the Lake District during the ice age?
   b) Can you identify directions in which particular glaciers may have flowed? Give reasons for your answers.

Economic activities in glaciated upland areas in the UK

Why are farming and forestry important land uses in glaciated upland areas?

Glaciated upland areas can be extreme environments. For rural landowners, there are very few ways to earn a living from the land. It is difficult to farm crops because of the:

- steep slopes, due to past ice erosion, which makes using machinery difficult.
- thin soils with limited fertility, due to resistant underlying rocks and steep slope gradients.
- low temperatures at high altitudes, resulting in a short growing season.
- heavy relief rainfall, especially in western areas that are most exposed to weather fronts, bringing waterlogging to flat sites and soil erosion on slopes.

▲ Figure 12.17 Stag on the Highlands

Extensive agriculture such as animal grazing is well suited to glaciated upland areas. In the Scottish Highlands, sheep were introduced to many upland estates during the 1800s, when there was a growing demand for wool from textile factories in the UK's growing industrial cities. More recently, some landowners have introduced deer for venison meat, and Highland cattle for speciality beef (Figure 12.17). Some farmers have even brought more exotic species, such as ostriches.

Another competing land use is commercial forestry. Coniferous woodland occupies two million hectares of land in the UK, much of it in upland areas. Around half is managed by the Forestry Commission, which was established in 1919 to ensure that the UK would never run short of timber supplies. In the 1980s, forest cover increased further thanks to private investors. Many upland areas of Scotland, such as the Isle of Arran, were soon carpeted with fast-growing pine and spruce trees. Wood is used as timber for furniture and building construction, and to make wood chips for gardens and biofuel.

▲ Figure 12.18 OS map extract showing upland land uses in Scotland's Isle of Arran

How important is tourism for glaciated upland areas?

Tourism is a major draw in many upland areas (Figure 12.20). It employs more people than any other industry in these places and is often the most important source of an upland area's income. Glaciated landscapes attract walkers who appreciate the often breath-taking views. For mountain climbers, there are plenty of exciting challenges. Some glaciated upland areas in the UK fare especially well from tourism on account of their location. The Lake District is highly accessible, for instance. It also has a milder climate than more northerly upland areas. The Lake District's glacial features are explored on pages 186–7. The benefits of tourism for Scotland's Isle of Arran are covered in detail on pages 192–3.

Why does quarrying take place in some glaciated upland areas?

Why are rocks quarried in upland areas, far from the towns and cities where they are used for construction? It is because the geology of upland areas tends to be very different from lowland areas. By their very nature, upland areas are composed of tough, resistant rocks that are not found in lowland areas. A lack of population means that there are fewer dangers (and objections!) when explosives are used to shatter rock into blocks that can be easily transported elsewhere. The rock types shown in Figure 12.19 all have a high economic value.

▼ Figure 12.19 Quarrying in glaciated upland areas

Lake District slate	This distinctive blue-grey rock is used around the world as a roofing and decorative material. The Lake District has thirteen active quarries.
Pennines limestone	Limestone is a widely used building material. Limestone fragments and gravel are a popular landscaping material for gardens.
Highlands granite	This tough, resistant rock has a range of uses, from pavement edges to kitchen work surfaces. Granite from the glaciated Scottish island of Ailsa Craig is used in the sport of curling due to its unusually uniform hardness.

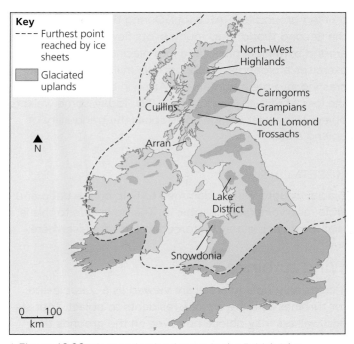

▲ Figure 12.20 Glaciated upland areas in the British Isles (UK and Eire)

→ Activities

1. a) Describe what is meant by 'extensive' agriculture.
   b) Explain three physical factors that mean only extensive agriculture can be carried out in glaciated upland areas.
2. Why has the amount of forest cover in upland areas increased in the last century?
3. Look at Figure 12.20. Suggest why some glaciated upland areas may receive more tourists than others.
4. Outline possible costs and benefits of carrying out quarrying in upland areas. In your answer, think about different types of costs and benefits, such as the economic or environmental impacts of quarrying.

Geographical skills

Identifying land uses

Figure 12.18 shows part of the Isle of Arran in Scotland (also see pages 192–3).

1. Identify the land uses shown in grid squares 0038 and 9843.
2. Give the grid reference for the quarry.
3. Identify a grid square likely to be popular with mountain climbers. Use map evidence to explain your choice.
4. Suggest two reasons why forest has not been planted in some places.

> **KEY LEARNING**
> - Why land use conflicts sometimes develop
> - How development and conservation needs can be balanced

Land use conflicts in glaciated upland areas in the UK

Why do land use conflicts sometimes develop?

A **land use conflict** develops when the activities of two different groups of people are incompatible. Figure 12.21 shows a conflict matrix for glaciated upland areas. Conflict can develop in relation to all the following land uses, in addition to tourism (page 188).

Military training

The UK's mountainous terrain provides an excellent training ground for RAF pilots. During their training in the Second World War, young pilots flew at high speeds around pyramidal peaks and into glacial troughs. The exercises are dangerous though: six military aircraft crashed into the mountains of Arran in the 1940s, with the loss of 51 lives (see Figure 12.22). Today, some walkers object to how jet engine noise spoils the tranquillity of upland areas.

Reservoirs

Glacial troughs can be dammed to help create deep and wide reservoirs (see Chapter 11). The downside to this is that local people may be forced to relocate elsewhere.

Wind turbines

Increasingly, upland areas are viewed as a suitable site for turbines as there are few residents to object. Tourists and walkers may object, however, on the grounds that turbines ruin the landscape.

Hunting

Many Scottish upland areas are privately owned land where hunting is allowed. On the Isle of Jura, visitors pay £475 to shoot a stag. Some hunters visit upland areas in the hope of achieving a 'Macnab'. This means that they stalk a red deer, shoot a grouse and catch a salmon all within the same day. Many nature lovers are opposed to hunting.

Key
- No conflict
- Some conflict
- Strong conflict

		Development/exploitation						Conservation/recreation				
		Sheep and deer farming	Forestry	Quarrying	Reservoirs	Military training	Wind turbines	Riding	Walking and climbing	Hunting and shooting	Photography and filming	Wildlife conservation
Conservation/recreation	Wildlife conservation	No conflict	No conflict	Some conflict	Strong conflict	Some conflict	Some conflict	No conflict	Some conflict	No conflict	No conflict	No conflict
Conservation/recreation	Photography and filming	No conflict	No conflict	No conflict	No conflict	Strong conflict	No conflict	No conflict	No conflict	No conflict		
Conservation/recreation	Hunting and shooting	Some conflict	Strong conflict	Strong conflict	Strong conflict	Some conflict	Some conflict	No conflict	Some conflict			
Conservation/recreation	Walking and climbing	No conflict	No conflict	No conflict	No conflict	No conflict	No conflict	No conflict				
Conservation/recreation	Riding	No conflict	Some conflict	Some conflict	Strong conflict	Some conflict	No conflict					
Development/exploitation	Wind turbines	No conflict	Some conflict	No conflict	Strong conflict	Some conflict						
Development/exploitation	Military training	No conflict	No conflict	No conflict	No conflict							
Development/exploitation	Reservoirs	Strong conflict	Strong conflict	Strong conflict								
Development/exploitation	Quarrying	Some conflict	Some conflict	No conflict								
Development/exploitation	Forestry	Strong conflict	No conflict									
Development/exploitation	Sheep and deer farming	No conflict										

▲ Figure 12.21 A matrix of possible land use conflicts in glaciated upland areas

Forestry

Conifer plantations block out the view for tourists and acidify the soil where they grow. Often they are planted very close together and light cannot reach the forest floor. As a result, very few animal or bird species live there.

After the forest has been cut down, there is no vegetation left to intercept rainfall. Worse still, machinery used by the loggers compacts the soil so water cannot soak in. This can result in localised flooding.

▲ Figure 12.22 A present-day clue that RAF pilots used to train for war here

▲ Figure 12.23 The Hogwarts Express is filmed passing through a glaciated upland area

How can development and conservation needs be balanced?

Many people who live in upland areas would like more types of employment, yet others do not want exploitation of the environment. It is possible to compromise by allowing some development to take place, but adopting strict management to make sure ecosystems do not become permanently damaged, and landscapes are not spoilt by too much activity and noise. Common management measures include:

- maximum visitor numbers – it is important that visitor numbers do not exceed the carrying capacity.
- signing – signs can be used to show people areas that are permanently or temporarily out of bounds.
- seasonal closure – in some places, visitor attractions are closed in winter months to give a site chance to recover.
- restricted activities – landowners are allowed to ban the use of motorbikes or horses. Camping is restricted to designated campsites.

An excellent strategy that balances development and conservation needs is film-making. Glaciated upland areas, both in the UK and elsewhere, are very popular locations for television and movie making. The Harry Potter film series used upland glaciated areas in Yorkshire, Northumberland and Scotland (Figure 12.23).

Places that feature in films benefit in several ways. Actors and production staff may stay in the area for a long time, generating revenues for local hotels and businesses. Landowners may charge a fee for the use of the location. Once the film is released, it may capture the interest of the public, causing more people to visit.

→ Activities

1 a) State two ways in which upland glaciated landscapes can be used to help generate electricity.
   b) Suggest why upland areas are especially well-suited to these types of energy production.

2 Explain two advantages and two disadvantages of bringing forestry to upland glaciated areas.

3 Make a list of any television programmes or films you have watched that feature mountainous upland areas. Do you know where any of the film locations were?

4 Design a year-long management plan to help restore an upland area that has been heavily damaged by erosion from walkers' boots. Think about possible strategies and how long they would need to stay in place.

Geographical skills

Conflict matrix

Study the conflict matrix in Figure 12.21.

1 Identify five pairs of strongly conflicting activities and five pairs of activities where there is no conflict.

2 Suggest reasons for the conflicts you have identified. How might these conflicts be resolved?

✪ KEY LEARNING

➤ The tourist attractions in Arran's glaciated upland areas

➤ The varied impacts of tourism in Arran

Tourism in the upland Isle of Arran, Scotland

What are the tourist attractions in Arran's glaciated upland areas?

The glaciated Isle of Arran is reached easily from the Scottish coast, making it a popular destination for day-trippers (Figure 12.24). The following impressive glacial features provide striking views and pose an active challenge for those who like to spend their **leisure** time outdoors:

- Goatfell – Arran's highest mountain and most popular natural visitor attraction. From the 874 metre summit of its pyramidal peak, visitors can see all of the way to Ireland on a clear day. Goatfell is flanked by many more dramatically-shaped mountains.

- A'Chir ridge – a knife-edged glacial arête that divides two corries. It is Arran's most exciting and challenging ridge. Climbers must be extremely careful as there is a long, vertical drop on either side.

- Glacial troughs – the past action of the ice has left Arran with several deep and wide, U-shaped valleys. They include Glen Rosa, Glen Sannox (Figure 12.25) and Glen Catacol. Glacial striations (see page 178) and polished rock surfaces can still be seen in these valleys.

Walkers of all ages take in the views at their own pace. Younger adults may also be interested in sporting activities. Among the island's many visitors are people who have come with the intention of climbing, running, biking, paragliding and abseiling. Helicopter tours of the mountains are also available, for those who can afford it.

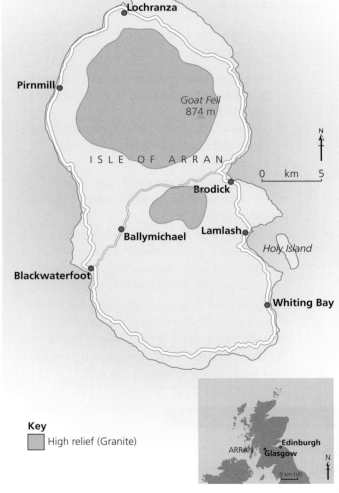

Key

◼ High relief (Granite)

▲ Figure 12.24 Map of Scotland and the Isle of Arran

Figure 12.25 A view of Arran's mountains from a glacial trough, North Glen Sannox

What are the impacts of tourism in Arran?

The carrying capacity of popular visitor sites has been exceeded, resulting in footpath erosion and other problems (Figure 12.26). Additionally, Arran's upland areas experience heavy frontal and relief rainfall for much of the year, resulting in a markedly seasonal pattern of tourism. Many tourist workers are left unemployed in winter.

▼ Figure 12.26 Costs and benefits of tourism in Arran's upland areas

Negative impacts	Positive impacts
Footpath erosion occurs on popular paths at 'honeypot' sites. One of the worst affected routes is the North Goatfell Ridge. The problem is made worse by steep slope angles in the valleys and corries. High altitude brings relief rain and more soil erosion.	The island has grown to rely on tourism as a large part of its income. Visitor numbers reach 200,000 annually. Many used to cater for themselves or camp outdoors, but more are now using the island's hotels and restaurants.
There is a lot of congestion, especially at popular times of the year. Arran's upland roads are narrow single-track roads due to the difficulties of construction in upland areas.	A range of new visitor attractions have helped to create more jobs, generating around £30 million annually. Recently, a new whisky visitor centre was opened at the village of Lochranza.
Injuries and fatalities are frequent. Many people have died walking and climbing Arran's mountains after underestimating the challenge. This is upsetting for the islanders and costly too. Arran's Mountain Rescue Service is staffed by volunteers but sometimes requires the help of expensive RAF helicopters.	Tourism is helping to tackle the island's ageing population problem. Visitors often fall in love with the island's dramatic upland landscape and decide to move there permanently. This brings young families and children to the island. In turn, this ensures that the island's schools have a sustainable future.

➜ Activities

1. a) What is meant by the 'carrying capacity' of a tourist area?
   b) Identify and explain three signs that carrying capacity has been exceeded in a place.

2. Write a postcard home from the Isle of Arran. Make reference to the island's interesting physical features and its visitor attractions.

3. Draw a table with three columns, labelled 'Social impacts', 'Economic impacts' and 'Environmental impacts'. Rearrange the information given in Figure 12.26 to fill the three columns.

4. Draw a simple portrait of the following people (they can be stick figures):

   - a day-tripper from Glasgow who is keen on outdoor adventures, but does not have a lot of money to spend
   - two married geography teachers and their two teenage children, who are spending one week at a hotel on the island
   - a very wealthy 70-year-old who used to enjoy walking in the mountains when they were younger.

   For each picture, list underneath the activities they might participate in while visiting a glaciated upland area.

5. Using an atlas, identify the area that day-trippers to Arran are likely to be drawn from. The sea crossing takes 45 minutes on a ferry which carries cars and foot passengers. How long in total do you think people would be prepared to travel for?

⭐ KEY LEARNING

➤ How the impacts of tourism in Arran are managed

➤ Judging the success of the management strategies

Evaluating tourism management in the Isle of Arran

How are the impacts of tourism in Arran being managed?

There are three main concerns in Arran: footpath erosion in tourist 'hot spots', climbing accidents, and the seasonal character of the island's tourist industry.

Footpath erosion

The National Trust for Scotland owns some of the worst-affected parts of the island. They have established a mountain path team to restore mountain paths, and have raised money through the Footpath Fund Appeal. Now, the National Trust is carrying out maintenance and small-scale restoration work on the mountain paths. Some paths are stabilised using paving stones to create steps. This stops soil erosion and mud flows, which allows natural regeneration of vegetation to occur.

The National Trust's aim is to maintain mountain paths for future generations. The techniques they use have a low environmental impact. Wherever possible, they use locally-sourced natural materials, instead of cement, so that the area still looks visually pleasing. However, the work costs up to £140 per metre.

▲ Figure 12.27 Tackling footpath erosion

Climbing and walking accidents

The Arran Mountain Rescue Team was formed in 1964. It is funded by a combination of grant aid and public donations. The team provides search and rescue assistance to walkers and climbers on the Isle of Arran. All members are experienced hill-walkers or mountaineers who know Arran's glaciated upland areas like the back of their hands. They are on call 365 days a year. All are unpaid volunteers who give their time freely to help people in need. The team have rescued many people, although fatalities still occur, most recently in 2013 and 2015. The Arran Outdoor Education Centre gives safety talks to visiting school students and staff.

▲ Figure 12.28 Mountain rescue over Goatfell

Seasonal tourism and visitor numbers

Purpose-built visitor attractions, such as the Balmichael Centre (Figure 12.29) and Auchrannie Resort, encourage people to visit the island in winter, when walking in the upland areas becomes difficult. This means more local people have permanent full-time tourism jobs. In addition, local businesses collaborate to create and maintain a new web site called *VisitArran*. It promotes Arran as an ideal short break or one-day holiday destination.

Each year, around 200,000 visitors now pay between £30 and £35 million into the island economy. Almost four out of every five tourists visit again, which shows that people often fall in love with Arran's upland scenery and want to return. The growth in recent years of online bookings has helped, too. Many Americans and Canadians are descended from Scottish migrants and they are keen to visit Arran and see the landscape of their ancestors.

Have the management strategies in Arran been a success?

Although the costs of footpath repairs are high, there have been tremendous improvements overall. Special attention has been paid to making sure that water drains from the paths quickly, meaning that the repairs are intended to be lasting and sustainable. The popular coastal path, the land owned by the National Trust (including Goat Fell), and the Forestry Commission estates have all benefited the most. In contrast, some privately-owned upland land still suffers from eroded footpaths.

Additionally, visitor numbers and spending are still rising. Some local people are unimpressed, however. From their perspective, it was better when there were fewer visitors and cars, and there was more peace and quiet!

Here's where it gets really interesting, though. Arran's three tourist issues – footpaths, accidents and visitor numbers

▲ Figure 12.29 Tourist activities in an upland area of Arran (the Balmichael Centre)

– are interrelated with one another. Precisely because more visitors than ever are using improved footpaths, more walkers may be getting into danger than in the past. In 2015, there were a record number of call-outs for the mountain rescue. Therefore, one unexpected result of successful management of the tourist industry seems to be that more people are becoming exposed to risk in upland areas.

Geographical skills

Evaluating success

The focus of an evaluation will often be on the level of success for a management strategy or action. There are three steps to take when carrying out an evaluation:

- Work out what information is needed.
- Decide which facts can be categorised negatively as 'costs' ('failures' or 'challenges'), and which facts are better described positively as 'benefits' ('successes' or 'opportunities').
- Weigh up the positives and the negatives in order to arrive at an overall judgement.

On balance, do you feel the people responsible for the management have done a good job, or not?

For Section 12.9, try answering this question: 'Evaluate the success of strategies used to manage the impacts of tourism in a glaciated upland area.'

1 What were the strategies?

2 In each case, what worked and what didn't work? Have some strategies worked better than others?

3 On balance, is it your judgement that the area as a whole has been successfully managed? What, if anything, might you advise is done differently?

→ Activities

1 Take a look at the example on pages 266–67, which shows the effects of the tourism industry on Tunisia. To what extent are the effects shown also applicable to upland areas of the UK, such as the Isle of Arran?

2 Use Figure 12.29 as the basis for a fictional account of a mountain climbing accident. Use the maps of Arran provided in this chapter to plot a course for your mountain climbers.

- Where did they go?
- What features did they hope to see, explore, or climb?
- Where might the accident have happened and why?
- What challenges might the rescue team and helicopter have encountered (think about the weather and the slope angles to help you)?

Question Practice

Unit 1 Section C
Coastal landscapes

1 'On a _____ coastline, the geology is the same along the edge of the coast.' Choose **one** word to complete the sentence.

Concordant ☐   Discordant ☐   [1 mark]

2 Suggest **one** reason for the uneven shape of the coastline shown in Figure 10.10.

Key
- ☐ Bagshot beds (soft)
- ☐ Chalk (tertiary)(hard)
- ☐ Wealdon clays and greensands (soft)
- ☐ Purbeck limestone (hard)
- ☐ Portland limestone (hard)
- ☐ Kimmeridge clay (soft)

ISLE OF PURBECK

N

◄ Figure 10.10 Isle of Purbeck, Dorset   [1 mark]

3 Name **one** landform usually located at a headland.   [1 mark]

4 State **one** way in which waves can erode coastlines.   [1 mark]

5 Study Figure 10.18 (page 128). Name one process of erosion that has affected these cliffs.   [1 mark]

6 Study Figures 10.39 (page 138) and 10.45 (page 142). Suggest how these engineering defences protect coastlines.   [4 marks]

7 Using Figure 10.13 (page 125) and your own knowledge, explain how these landforms may have been created.   [6 marks]

River landscapes

1 Study Figure 11.5 (page 150). State one reason why the gradient becomes concave.   [1 mark]

2 (a) Study Figure 11.7 (page 151). Describe the shape of the river's long profile.   [1 mark]

 (b) Suggest one reason why the size of sediment would be larger in the upper course.   [1 mark]

3 Study Figure 11.13 (page 155).

 (a) State the four-figure reference for Garreg y Gwynt.   [1 mark]

 (b) Name one landform you would expect to find at this location.   [1 mark]

4 Explain the environmental effects of river flooding on the area shown in Figure 11.27 (page 161).   [4 marks]

5 Study Figure 11.20 (page 157). Explain the processes that have taken place to create the landforms shown.   [6 marks]

Glacial landscapes

1   Study Figure 12.15. Name one glacial landform shown in this photo.

◄ Figure 12.15
Striding Edge with
Red Tarn to the left,
Helvellyn, Cumbria   [1 mark]

2   Study Figure 12.16 (page 187). Name the feature found at grid square 2415.   [1 mark]

3   Study Figure 12.18 (page 188).

   (a) Identify the glacial landform at grid reference 999426. Select **one**
       letter only.

       A Corrie   B Arête   C Drumlin   [1 mark]

   (b) Name one feature found at square 9938.   [1 mark]

   (c) Identify the grid square that shows the highest area of the land.   [1 mark]

4   Suggest how the tourist activities shown in Figure 12.29 (page 195) may   [4 mark]
    affect the environment.

5   Using Figures 12.10. and 12.11, explain how depositional landforms are
    produced in glacial environments.   [6 marks]

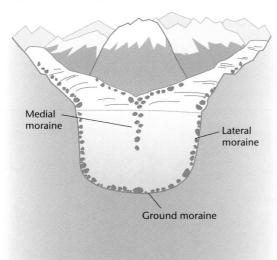

▲ Figure 12.10 The formation of different types of
glacial moraine

▲ Figure 12.11 An aerial view of a glacier

13 The global pattern of urban change

⭐ **KEY LEARNING**

➤ How the world's urban population is growing

➤ Where the world's largest cities are

➤ How rates of urbanisation differ between continents

Urban trends

How is the world's urban population changing?

Over half the world's population now lives in cities. By 2030, it is expected that 60 per cent of the world's population will live in urban areas, and by 2050 it will be 70 per cent. This process is known as **urbanisation**.

Where are the world's largest cities?

The world's largest cities, with populations over ten million, are known as **megacities**. In 1975 there were only four megacities – Tokyo, New York, Mexico City and São Paolo. Today, there are over twenty (Figure 13.1) and the number is rising year by year. You may notice that London, the UK's largest city, is not among the world's megacities. Its population is not predicted to reach ten million until 2030. That might give you an idea of how big megacities really are.

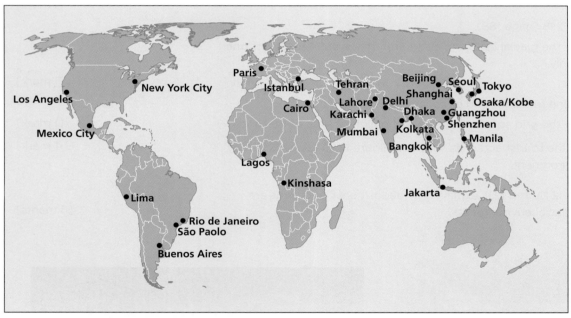

▲ Figure 13.1 The world's megacities in 2015

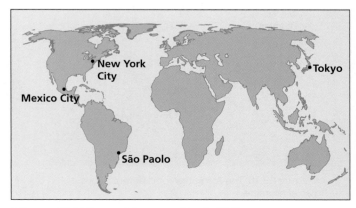

▲ Figure 13.2 The world's megacities in 1975

How do rates of urbanisation vary around the world?

Rates of urbanisation differ between continents (see Figure 13.3). The highest rates of urbanisation are in poorer, low-income countries (LICs) in Asia and Africa. In most of these countries, a majority of the population still live in rural areas and the rate of **rural–urban migration** is high. The population of cities is younger, so the rate of **natural increase** is also high.

There are lower rates of urbanisation in richer, high-income countries (HICs) in Europe, North America and Oceania. In these countries, urbanisation has slowed down as the majority of the population already live, in cities. The urban population is ageing, so the rate of natural increase has also slowed down.

One exception to this pattern is South America, with many **newly emerging economies** like Brazil. Here, urbanisation happened earlier and has slowed down, even though these countries are not yet among the richer, high-income countries.

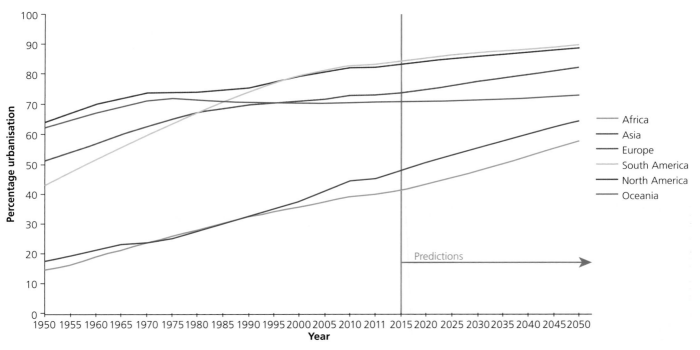

▲ Figure 13.3 The urbanisation of the world's continents, 1950–2050

→ Activities

1   Look at Figure 13.1. Describe the distribution of the world's megacities by continent. Which continent has most megacities? How many megacities are in each continent?

2   Compare Figures 13.1 and 13.2.

   a)   How has the number of megacities changed since 1975?

   b)   How has the distribution changed?

3   Make a list of the world's megacities in Figure 13.1, and name the countries in which each one is found. Try to do this without an atlas. Then check the atlas to see if you named the correct countries.

Geographical skills

1   Look at Figure 13.3.

   a)   Which continents had the highest rates of urbanisation in 1950?

   b)   Which continents had the highest rates of urbanisation by 2015?

   c)   Explain why your answers to (a) and (b) are different.

⊛ KEY LEARNING

➤ Factors affecting population growth

➤ Factors affecting the rate of urbanisation

➤ How the world's megacities have grown

➤ Where the world's megacities will be in future

How urbanisation happens

What factors affect population growth?

Population growth is the difference between **birth rate** and **death rate**. When the birth rate is higher than the death rate, population grows. This is natural increase. When the birth rate is lower than the death rate, population falls. This is **natural decrease**.

World population grows because on average, globally, birth rates have been higher than death rates. However, there are important differences in birth rates and death rates between countries.

What factors affect the rate of urbanisation?

A large proportion of the world's megacities are in Asia. There are two main reasons:

■ Asia is where over half the world's population lives. China and India both have more than a billion people.

■ The majority of Asia's population is still rural, although this is changing as people move to cities. Over 50 per cent of China's population now live in cities compared to just 20 per cent in 1980.

There are a number of factors that have led to urbanisation, not just in Asia, but in other parts of the world too.

▼ Figure 13.4 Factors affecting growth of megacities. The photo is of Shanghai, a megacity in China

Migration

Rural–urban **migration** is the main driver of urbanisation. Most of these migrants are young. They migrate from the countryside to cities because of pull factors, like jobs and a better education.

Natural increase

The young population in many cities leads to high rates of natural increase. Cities also tend to have better health care than rural areas, so death rates are lower and life expectancy is higher.

Location

Historically, cities have grown on rivers, coasts and other busy transport routes where trade can thrive. Even today, many of the world's megacities are ports, which are a good location for trade.

Economic development

Cities that trade are also a good place for business, so they grow economically. It is economic growth that creates jobs, which attract people, and it is people who bring the ideas and enterprise on which cities thrive.

How have the world's megacities grown?

Cities do not grow at a constant rate. Some cities that grew rapidly in the twentieth century, such as Tokyo in Japan, have now slowed down. Meanwhile, other cities that grew slowly in the twentieth century, such as Lagos in Nigeria, are now urbanising rapidly (Figure 13.5).

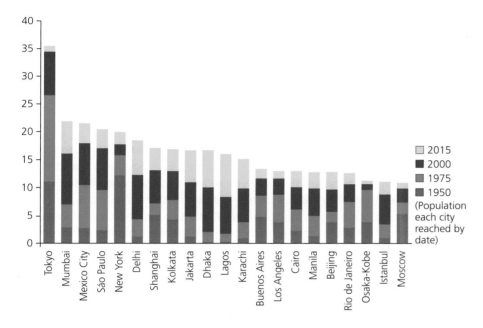

Legend:
- 2015
- 2000
- 1975
- 1950

(Population each city reached by date)

▲ Figure 13.5 Growth of the world's megacities

Where will the world's megacities be in future?

By 2050, the world's largest megacity is likely to be one that does not even exist at the moment! At least, it does not have a name. China has plans to merge cities in the Pearl River Delta to create one large megacity with a population of 120 million. The existing cities of Hong Kong, Shenzhen and Guangzhou would merge to form an urban area twenty times the size of London, with a population twelve times bigger. Most of the new megacities in future are likely to be in Asia, particularly in China and India.

→ Going further

1 Find out more about a city you know. It could be a city in the UK, e.g. London or Birmingham. Answer these questions for your city to explain how urbanisation happened.

■ Why did the city first grow?

■ When did its most rapid growth occur? What caused this?

■ What is the present day population?

■ How is the population changing now, and why?

→ Activities

1 Look at Figure 13.4. Explain how each of these factors affects the rate of urbanisation. Write a sentence for each one.

a) migration
b) natural increase
c) location
d) economic development.

2 Look at Figure 13.5. By which date – 1950, 1975, 2000 or 2015 – did each of these cities reach a population of ten million?

a) Tokyo
b) Mexico City
c) Shanghai
d) Lagos

3 Suggest why some megacities are growing faster than others in the twenty-first century. For example, why is Lagos growing faster than Tokyo? (You can find out more about Lagos in Chapter 14.)

4 a) How many of the megacities in Figure 13.5 have a coastal location? Make a list, then check in an atlas or the map in Figure 13.1 on page 198.

b) Why is coastal location an important factor in urban growth?

14 Urban growth in Nigeria

⊙ KEY LEARNING

➤ How Lagos compares with your image of Africa
➤ Where Lagos is located
➤ The importance of Lagos to Nigeria and Africa

Welcome to Lagos

What is Lagos like?

Lagos is Africa's biggest city and one of the fastest-growing cities in the world. In this chapter you will get to know it well. You will discover an exciting, vibrant, crowded, enterprising megacity – sometimes shocking, but never boring!

From above, Lagos could be any modern city. The city centre is dominated by modern, high-rise offices, surrounded by miles of sprawling suburbs, linked by busy roads (Figure 14.1). But, get down to street level and the noise around you will be unlike anything you have heard before.

In the background is the constant drone of generators that power the city (Lagos does not have a reliable electricity supply). Even louder is the roar of traffic. Gangs of motorcycles, fleets of yellow minibus taxis and old trucks belching smoke from their exhaust are gridlocked with cars, all honking their horns in a cacophony of noise. Above it all rises the chorus of street vendors selling their wares on every corner, loudspeakers blaring a mixture of traditional Nigerian music and modern afrobeat, and the wailing call to prayer from the city's mosques. Welcome to Lagos!

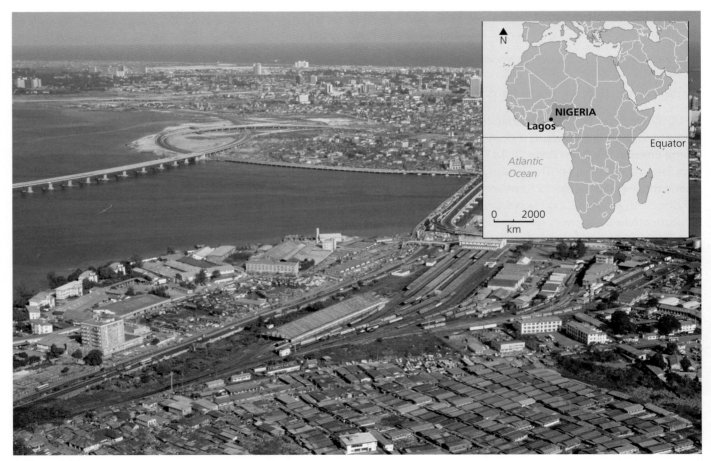

▲ Figure 14.1 Lagos, Nigeria.

Where is Lagos located?

Lagos is the largest city in Nigeria, itself the most populous country in Africa. It lies in the southwest of the country, on the coast of the Gulf of Guinea, close to the border with Benin (Figure 14.2).

Until the fifteenth century it was a small fishing village on an island. In 1472, Portuguese settlers gave it the name 'Lagos' (or 'lakes' in Portuguese) after the water that surrounds it. In the early twentieth century, by then under British rule, Lagos was made the capital of Nigeria. It remained the capital after independence from Britain in 1960.

What is the importance of Lagos?

In 1991, the Nigerian government moved to Abuja, which became the new capital of Nigeria, though Lagos retained its importance as the country's centre of trade and commerce. About 80 per cent of Nigeria's industry is based in and around Lagos and it is now the main financial centre in West Africa. The city also has a major international airport and a busy seaport. The population of Lagos continues to grow (Figure 14.3).

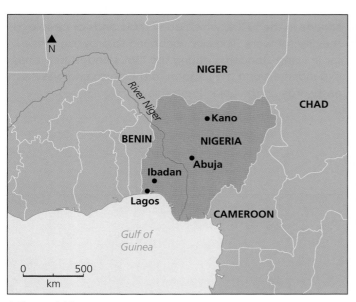

▲ Figure 14.2 Lagos' location in Nigeria

Key facts

Population: 15 million (2015 estimate)

Population growth rate: 600,000 per year

Area: 1,000 km²

Average earnings: £670 per year (2012)

Gross domestic product: £18 billion per year (total value of goods and services)

Waste: 10,000 tonnes per day

Reported murders: 1 per cent of households have reported the murder of a family member

Religion: 68 per cent describe themselves as Christian. There is a large Muslim minority

▲ Figure 14.3 Lagos fact file

→ Activities

1. Close your eyes for a minute. What images of Africa come into your mind? Write a list.

2. Look at Figure 14.1.
   a) Describe Lagos from above. Mention its size, the buildings (their age and height), the roads and its surroundings.
   b) Compare the photo with your images of Africa. What differences do you see?

3. Look at the maps in Figures 14.1 and 14.2. Describe the location of Lagos (a) within Africa and (b) Nigeria. Mention the Equator, the Atlantic Ocean, the Gulf of Guinea and Benin.

4. Look at Figure 14.2.
   a) Why do you think Abuja was chosen as the new capital of Nigeria?
   b) Was this a good decision for Lagos? Give reasons for your answer.

5. Study Figure 14.3. What does the data tell you about:
   a) the population of Lagos?
   b) the wealth of Lagos?
   c) life in Lagos?

6. Compare Lagos with a city you know in the UK. Mention at least three similarities and three differences.

✪ KEY LEARNING

➤ How fast Lagos is growing
➤ Causes of population growth in Lagos
➤ How push and pull factors lead to rural–urban migration

Growing Lagos

How fast is Lagos growing?

Lagos is growing so fast that no one can agree what its population really is! In 1960, the city still had a population of less than a million people, but this grew to about four million by 1990 and about fifteen million by 2015 (Figure 14.4). Some estimates are higher than this and, if you include the surrounding area, the population is over twenty million.

As Lagos' population has grown, so has the area of the city (Figure 14.5). The original site was on Lagos Island, surrounded by Lagos Lagoon. By 1960 the city had expanded northwards onto the mainland, following the line of the main railway.

Lagos' expansion really took off during the oil boom in Nigeria in the 1970s, which drew thousands of people to the city for work. The city continued to grow despite a fall in living standards during the 1980s and 1990s. It has expanded around the Lagoon to the north and west, and eastwards on the Lekki Peninsula.

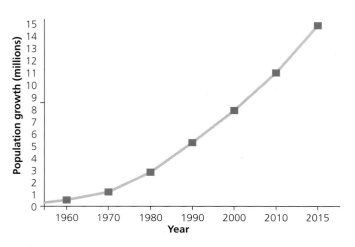

▲ Figure 14.4 Population growth in Lagos

▲ Figure 14.5 The growth of Lagos

What causes population growth in Lagos?

The main driver of growth in Lagos over the past 50 years has been rural–urban migration. People are encouraged to leave the countryside by **push factors** such as the lack of job opportunities and low wages (Figure 14.6). They are brought to the city by pull factors such as the prospect of well-paid work and the attraction of an urban lifestyle.

Another reason for Lagos' population growth is the high rate of natural increase in the city's population. This is due to the city's youthful population, since most migrants to the city are young.

Nigeria is becoming an increasingly urbanised country. By 2015, just over half the population was still living in rural areas, but as rural–urban migration continues, the majority will be urban within the next few years.

How do push and pull factors lead to rural–urban migration?

Education and health services are poor in rural areas.

There is a land shortage due to population growth. Despite urbanisation, rural population continues to grow.

Changing climate is making the weather less predictable. Droughts and floods occur more often, now.

Farming pays low wages but requires a lot of hard work.

Land is degraded due to farming and other activities. Land in the Niger Delta region is polluted by the oil industry.

Few job opportunities exist other than farming.

Political unrest creates insecurity. The terrorist group, Boko Haram, is active in the north of Nigeria.

▲ Figure 14.6 Push factors that lead to rural–urban migration in Nigeria

→ Activities

1 Look at Figure 14.4. Describe how Lagos' population has grown since 1960. Use data from the graph in your description.

2 Look at Figure 14.6. Read the list of push factors for rural–urban migration in Nigeria. Write an equivalent list of pull factors that would attract people to Lagos, such as more job opportunities (you might get more ideas in Section 14.3). Draw a table like this to list the push and pull factors for Lagos.

Push factors for Nigeria	Pull factors for Lagos
Few job opportunities	More job opportunities

3 Imagine you are one of the people in the photo in Figure 14.6. Send a text message to your cousin in Lagos to explain why you want to move there. (Keep a copy of your message until Section 14.8, so you can see how your expectations of Lagos might have changed after you have lived there for a year.)

Geographical skills

1 Look at Figure 14.5.

   a) Describe how Lagos has grown in size since 1960. Use the scale on the map to give a rough idea of its size.

   b) Explain how the growth of the city has been affected by its location and physical geography.

Case study

⊛ KEY LEARNING

➤ Opportunities in Lagos
➤ Why urbanisation has helped Nigeria to develop
➤ How Lagos' growth creates more inequality

Lagos: a city of opportunity

What are the opportunities in Lagos?

At first sight, Lagos might not look like the sort of place where you would choose to live. The city is straining under the pressure of a growing population. Congested roads, high crime rates, violent clashes among street gangs, electricity in short supply, a sewage system that hardly works – these are problems that Lagos residents regularly have to contend with. Yet, despite the problems, there are plenty of **economic opportunities** and **social opportunities**.

Employment

More jobs are available in Lagos than anywhere else in Nigeria. Even if you can't find work in the **formal economy**, paying tax, it is possible to work in the **informal economy**, for example as a street vendor or recycling waste (see Section 14.4), paying no tax.

Education

There are also more schools and universities in Lagos than you find outside the city. If you are educated, you are more likely to find work in one of Lagos, growing industries, like finance, film or fashion.

Health care

Although it is not always free, at least health care is available in Lagos. The nearest clinic or hospital is a lot closer than if you live in a village, though you might have to queue if you cannot afford to pay.

Why has urbanisation helped Nigeria to develop?

As the largest and wealthiest city in Nigeria, Lagos has made a big contribution to the country's development. In the twenty-first century, Nigeria's life expectancy, years spent in school, and wealth have all improved significantly (Figure 14.7). The graphs show average increases across the whole country, but Lagos would be above average. That is because health care, education and employment are all better in Lagos than in rural areas. Urbanisation has been an important factor in Nigeria's recent development.

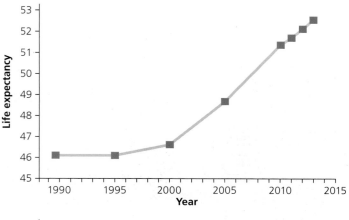

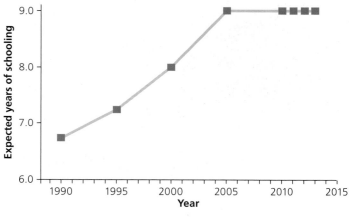

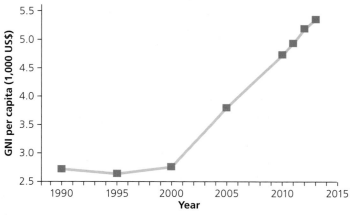

▲ Figure 14.7 Development in Nigeria

How does Lagos' growth create more inequality?

One consequence of urbanisation in Lagos is the widening gap between rich and poor. Lagos' economy is booming, leading to fabulous wealth for some. Victoria Island, close to the city centre, has become a wealthy neighbourhood (Figure 14.8).

Meanwhile, over 60 per cent of Lagos' population live in **squatter settlements**, or slums, like Makoko (Figure 14.9).

Most of the people here work in the informal economy and live on less than $1.25 (roughly £1) a day.

Lagos' rapid growth means that the city's infrastructure has struggled to keep up with the number of people. However, there are signs that it is beginning to improve.

▲ Figure 14.8 Housing for the rich on Victoria Island

▲ Figure 14.9 Housing for the poor in Makoko

Transport

The first stage of a planned rapid transit network in the city has already been built (read more in Section 14.7).

Electricity

Two new power stations are planned to reduce the city's shortage of electricity and to light the streets at night. Most wealthy households and businesses rely on generators to provide power when the network fails.

Water supply

Only the wealthiest homes have a piped water supply. Others use public taps and boreholes or buy their water from street vendors (read more in Section 14.6).

Crime reduction

In order to tackle high levels of armed muggings, burglaries and carjackings, the city has bought three helicopters for police to spot criminal activities.

→ Activities

1 Look at Figure 14.7. Describe how (a) life expectancy, (b) education and (c) wealth have improved in Nigeria since 1990.

2 Compare Figure 14.7 with Figure 14.4 in section 14.2.

   a) Compare Lagos' population growth with development in Nigeria.
   b) Explain how urbanisation could be linked with development.

3 Look at Figures 14.8 and 14.9. In which area of Lagos – Victoria Island or Makoko – would you be most likely to:

   a) drive a car?
   b) use a kerosene lamp?
   c) buy water from a street vendor?
   d) pay for health care when you are ill?

   In each case, explain your answer.

4 If you were a newly arrived immigrant in Lagos, what would:

   a) please you?   b) disappoint you?

   In each case, mention at least three things.

⊗ KEY LEARNING

➤ The advantages Lagos' location has for industry
➤ The contribution Lagos makes to the Nigerian economy
➤ The benefits and problems of the informal economy

Economic opportunities and challenges

What are the advantages of Lagos' location for industry?

Like many other megacities around the world, Lagos has a coastal location. Over the centuries, it has transformed from a small fishing village into a busy seaport. Lagos Lagoon provides a good, sheltered harbour for shipping. In addition to its port, Lagos now has a major international airport, which is the main arrival point for 80 per cent of flights to West Africa (Figure 14.10).

Good transport connections have helped Lagos to develop into a major industrial centre. Its growing population also provides a large market for goods and services, encouraging more industry to locate there. With many schools and universities, Lagos also has a well-educated and skilled workforce, attracting more companies.

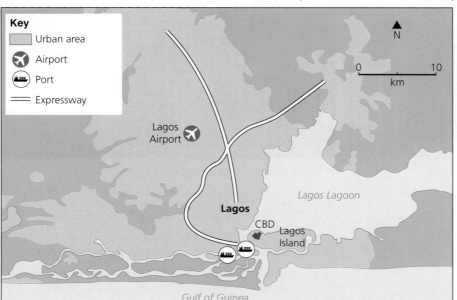

▲ Figure 14.10 Lagos' transport connections

What contribution does Lagos make to the Nigerian economy?

Lagos is the major contributor to the Nigerian economy. With about 10 per cent of the country's population, it contributes about 30 per cent of its **gross domestic product (GDP)**. Most of Nigeria's manufacturing industry is based in Lagos, as well as many of the new service industries, like finance in the **central business district (CBD)** (Figure 14.10).

Now, Lagos is building a new city on the coast called Eko Atlantic, destined to be the new financial hub of West Africa. Inspired by Hong Kong, or perhaps Canary Wharf in London (see Section 15.5), it is a joint project between the city government of Lagos and international private investors. It will be home to a quarter of a million people and employ 150,000 more.

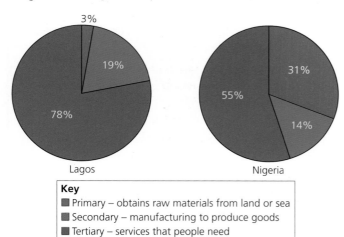

Key
■ Primary – obtains raw materials from land or sea
■ Secondary – manufacturing to produce goods
■ Tertiary – services that people need

▲ Figure 14.11 Employment structure for Lagos and Nigeria

What are the benefits and problems of the informal economy?

Unemployment in Lagos is much lower than in the rest of Nigeria at a rate of 9.9 per cent in 2015 (not much higher than London). However, unlike the UK, there is no unemployment benefit for those without work. Although a 2016 Employment Trust Fund bill has been passed to grant loans to unemployed people and help them become self-employed, most people need to find work in the informal economy in order to survive. They do jobs like street vending, car washing, shoe shining or **waste recycling**; 90 per cent of new jobs created in Lagos are in the informal sector.

Although the formal economy in Lagos is growing, there is a limit to the number of jobs it can create. About 40 per cent of the workforce work in the informal economy. It plays a vital role in both providing employment and helping the city to function. Lagos has earned the reputation of being an enterprising city where people are prepared to do anything to make a living – even sorting rubbish (Figure 14.12).

Olususun is a huge landfill site near the heart of Lagos. The city has grown up around the site.

Around 500 people work at the dump. Each day they sort 3,000 tonnes of waste by hand, picking out valuable items to sell.

Workers even live at the dump, building their homes out of discarded materials.

There are also shops, restaurants, bars, cinemas and a mosque at the dump.

Municipal governments collect just around 40% of the 10,000 tonnes of waste produced in Lagos every day. This 40% is taken to landfill sites like Olusosun. Only 13% of this waste is recycled. It would be better if people could separate their own waste for recycling.

Rubbish can be turned into energy by harnessing methane gas, emitted from rotten waste. A new project by the Lagos State Waste Management Authority is planned to produce 25MW of electricity. This is enough to power a town, though not a city the size of Lagos.

Electric waste, imported to Nigeria, is also brought to the site. It is treated with chemicals to extract the reusable materials, but toxic fumes are released.

Without the dump, a lot of reusable items would go to waste. People in Lagos can save money by buying recycled goods.

Natural gases build up under the decomposing waste, especially when it is dry. This often leads to fires that are hard to extinguish.

Olusosun is an example of the way people in Lagos find solutions to problems – seeing an opportunity where others might just see junk.

▲ Figure 14.12 The Olusosun rubbish dump in Lagos

→ Activities

1 Look at Figure 14.10.
   a) Draw a sketch map to show Lagos' location.
   b) Annotate your map to list the main advantages for industry. Mention at least four factors.

2 Look at Figure 14.11. Compare the employment structure for Lagos with the rest of Nigeria. Can you explain the differences?

3 Classify each of these jobs into primary, secondary or tertiary employment. In each case, give a reason for your classification:

- Recycling at Olusosun
- Selling mobile phones
- Working in a car factory
- Fishing at Makoko
- Driving a minibus taxi
- Working for a bank

4 Look at Figure 14.12.
   a) Draw a table to list the benefits and problems of working at Olusosun dump.
   b) Do you think that Olusosun is an effective way of managing waste disposal in Lagos? Give reasons.

⊗ KEY LEARNING

➤ Where squatter settlements are found

➤ The problems of living in squatter settlements

➤ Whether squatter settlements should be demolished or improved

Squatter settlements

Where are squatter settlements found in Lagos?

The lack of properly built homes in Lagos has forced millions of people to build their own homes on land (or even water!) they do not own (Figure 14.15, opposite). These so-called squatter settlements are found all over the city, particularly on marshy, poorly drained land where no one else wants to build (Figure 14.13).

Squatter settlements are not unique to Lagos. They are found in cities in low-income countries and newly emerging countries all around the world. They can also be called informal settlements, shanty towns or slums.

▼ Figure 14.14 Housing conditions in squatter settlements in Lagos

Condition	%
Housing density	
Households living in more than one room	25
Households living in one room	75
Household facilities	
Kitchen, bath and toilet	10
Lacking either kitchen, bath or toilet	52
No kitchen, bath or toilet	38
Water supply	
Piped water	11
Public tap	14
Well or borehole	55
River	4
Water vendor	16
Toilets	
Septic tank (underground tank in which sewage collects)	10
Pit latrine (sewage soaks straight into the ground)	55
Pail latrine (sewage is poured into a drain or river)	33
Bush (i.e. no toilet!)	2

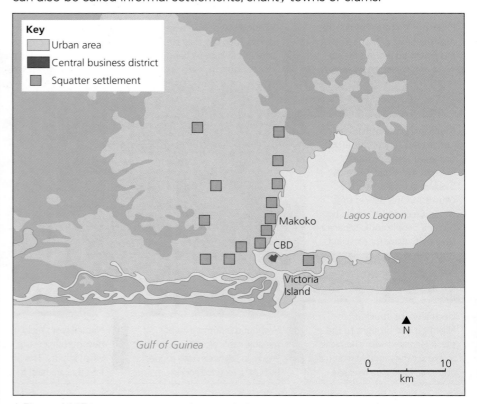

▲ Figure 14.13 Large squatter settlements in Lagos

What are the problems of living in a squatter settlement?

Squatter settlements in Lagos are densely populated due to the shortage of available land. In the case of Makoko, on the edge of Lagos Lagoon, homes extend into the water, built on stilts (Figure 14.15).

The homes are usually makeshift shelters built from materials like tin sheets and wooden planks. They lack basic facilities and good **sanitation** (Figure 14.14). The population of Makoko is estimated at up to a quarter of a million people. Most of them make a living in the informal economy and by fishing. This goes back to Makoko's origins as a fishing village outside Lagos. As the city grew it was swallowed up in the urban area.

Should Makoko be demolished or improved?

For all its problems, people in Makoko are fiercely protective of their homes. The authorities in Lagos are keen to demolish the area, but residents have nowhere else to go (Figure 14.16).

▲ Figure 14.15 Makoko

The demolition of Makoko

In the shadow of the Third Mainland Bridge in Lagos, fragile wooden huts have stood for decades on stilts above the water like long-legged birds. Beneath steaming traffic jams, the people of Makoko drift across the muddy water in canoes, casting nets for fish.

But after giving residents 72 hours to leave their homes, state authorities began demolishing the shanty town a week ago.

This is not the first attempt to wipe out Makoko. Lagos authorities call the shanty town 'unwholesome' and out of keeping with Lagos' 'megacity status'. Lagos governor said there were plans to build something much grander. 'We have a plan to turn that place into the Venice of Africa', he told protestors from Makoko.

◄ Figure 14.16 A news article about the demolition of Makoko, July 2012

→ Activities

1   Look at Figure 14.13.

   a) Describe the distribution of squatter settlements in Lagos. Mention their location in relation to the Lagos Lagoon.

   b) Try to explain the distribution. (Hint: think about what the land near the lagoon would be like.)

2   Study Figure 14.14. Identify the main problems of living in a squatter settlement in Lagos. Mention at least four.

3   Look at Figure 14.15. Draw an annotated sketch of the photo. In your labels, describe some of the problems faced by people in Makoko, including:

   ■ its location

   ■ building construction

   ■ housing density

   ■ sanitation.

4   Read the article in Figure 14.16.

   a) With a partner, discuss whether Makoko should be demolished or improved. One of you should play the role of a Makoko resident and the other should play the role of the Lagos governor.

   b) Together, decide how the growth of squatter settlements should be managed. Should Makoko be demolished or improved? Write a short report to give reasons for your decision.

Case study

⊛ KEY LEARNING

➤ How Lagos obtains its water supply

➤ How water pollution affects these supplies

➤ Why sea level rise could be a long-term threat to Lagos

▲ Figure 14.17 A water vendor in Lagos

Water supply and pollution

How does Lagos obtain its water supply?

Among the common sights in Lagos are water vendors on the street selling water in containers (see Figure 14.17). It is often difficult to obtain drinking water from any other source, even though the city is surrounded by water!

Only ten per cent of the population in Lagos have a piped water supply that has been treated and purified. The rest of the population either rely on water vendors or dig their own wells or sink boreholes to reach **groundwater** supplies that lie below the water table (see Figure 14.18). Water in the lagoon or tidal creeks around the city is not suitable for drinking because it is salty (not to mention polluted!).

In 2012, the newly formed Lagos State Water Regulatory Commission began the job of regulating the water supply and water vendors, and issuing licences for boreholes. It is responsible for ensuring a safe water supply at a reasonable price for consumers. This will be a big job.

▼ Figure 14.18 Water supply around Lagos

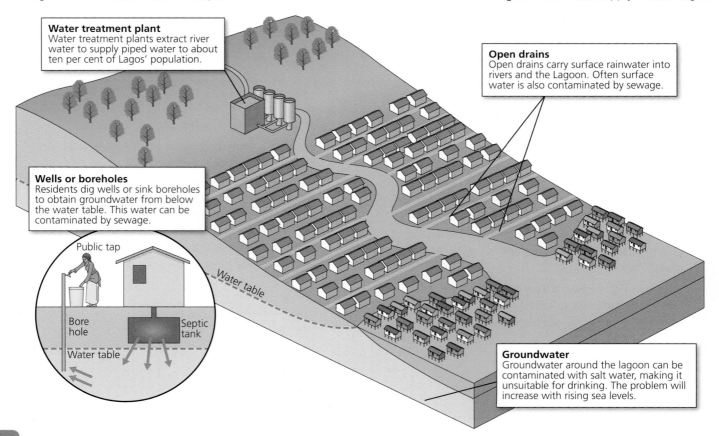

Water treatment plant
Water treatment plants extract river water to supply piped water to about ten per cent of Lagos' population.

Open drains
Open drains carry surface rainwater into rivers and the Lagoon. Often surface water is also contaminated by sewage.

Wells or boreholes
Residents dig wells or sink boreholes to obtain groundwater from below the water table. This water can be contaminated by sewage.

Public tap

Bore hole

Septic tank

Water table

Groundwater
Groundwater around the lagoon can be contaminated with salt water, making it unsuitable for drinking. The problem will increase with rising sea levels.

How does pollution affect Lagos' water supply?

Drinking water in Lagos often contains bacterial or chemical **pollution** that can lead to diarrhoea. The number of cases of diseases like cholera and dysentery has increased. One of the main causes of pollution is the lack of a proper sewage system in the city. Sewage is sometimes disposed of with rainwater through open drains. It is carried into rivers and the Lagoon, which also become polluted.

Sewage may also soak into the ground from pit latrines or leaking septic tanks. Here it can find its way into the water supply through wells and boreholes. Even water from vendors can be contaminated because they also obtain water from the same sources.

Why could rising sea level be a threat to Lagos?

Most of Lagos lies less than two metres above sea level. The predicted rise in sea level, of up to one metre in the twenty-first century due to global warming, is a severe threat to the city. Flooding could increase and groundwater could become contaminated by salt.

Already this century, there have been a number of serious floods in Lagos, caused by intense tropical rain. Roads quickly turn into rivers and drains overflow, flooding streets and homes with sewage.

There are several reasons why the impact of sea-level rise in Lagos could be more severe than in other cities, such as:

- Lagos' coastal location, making it vulnerable to the sea
- flat, low-lying land that is quick to flood but slow to drain
- a wet, tropical climate with over 2,000 millimetres of annual rainfall
- rapid urbanisation that has covered the land with buildings and concrete
- squatter settlements, built without any proper drainage
- land reclamation that has reduced the area of water for floodwater to drain into.

→ Activities

1   Explain why Lagos has a shortage of drinking water, despite being surrounded by water.

2   Look at Figure 14.18. Describe how water is supplied from: (a) rivers, (b) ground water.

3   Look at Figures 14.17 and 14.18. Compare three sources of water in Lagos – water vendors, piped water and wells or boreholes.

   a)   Draw a large table like this. Complete the boxes in the table.

Water source	Availability	Purity	Sustainability
Water vendors			
Piped water			
Wells or boreholes			

   b)   Which source of water would you prefer to use? Give your reasons.

4   Explain how rising sea levels in Lagos could lead to:

   a)   more flooding

   b)   less drinking water.

Case study

⊗ KEY LEARNING

➤ The impacts traffic congestion has on people in Lagos
➤ The efforts made to reduce congestion
➤ How a transport master plan could help Lagos

Traffic congestion

What impact does traffic congestion have on people in Lagos?

The average Lagosian commuter spends over three hours in traffic every day. It makes Lagos one of the most congested cities in the world. Perhaps that is not surprising when you consider that 40 per cent of new cars in Nigeria are registered in Lagos, which occupies just one per cent of the country's total area.

Traffic congestion causes many problems for people in Lagos. It is not just a matter of the inconvenience, though that is considerable (see Figure 14.19). The fatal accident rate in Lagos is 28 per 100,000 people – three times higher than the rate in European cities. Air pollution rates in Lagos are five times higher than the internationally recommended limit.

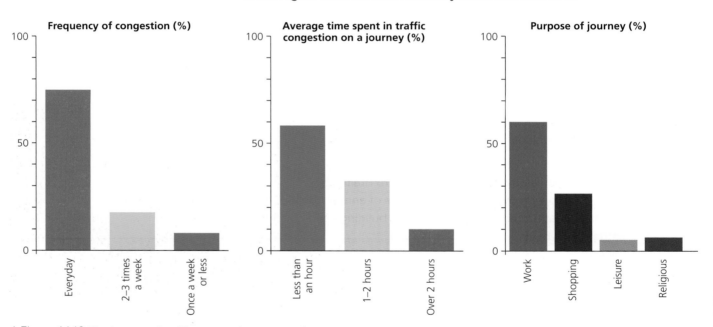

▲ Figure 14.19 The impact of traffic congestion on people

What efforts have been made to reduce traffic congestion?

In 2003, the Lagos state government set up the Lagos Metropolitan Area Transport Authority (LAMATA) to improve transport in the city. One of its first achievements was to introduce a bus rapid transit (BRT) system on a north–south route from the suburbs to the CBD on Lagos Island (see Figure 14.20). It provides a separate lane for buses to reduce travel times. 200,000 people use the service each day – a quarter of all commuters in Lagos.

However, a single BRT route is inadequate in a city the size of Lagos. The public transport system has to be supplemented by a large fleet of minibus taxis, known as 'danfos'. They are designed to carry ten to fifteen passengers, but demand is so high that they often carry twenty to thirty.

How could a transport master plan help Lagos?

Another scheme, due to open in 2016, is a new light railway on a west–east route into the CBD, designed to carry seven times as many passengers as the BRT. Eventually, there are plans for a network of seven new rail lines, known as Lagos Rail Mass Transit (LRMT).

This is part of a wider Strategic Transport Master Plan for Lagos which includes:

■ an **integrated transport system** where road, rail and waterway networks link together to make journeys easier

■ a new waterway network of ferries to make better transport use of the water areas around Lagos

■ a more efficient road network with separate bus lanes, and without obstacles like markets and street vendors, to speed traffic flow

■ better urban planning with mixed-use developments, (e.g. residential and commercial), to reduce the number of journeys people need to make

■ a new airport on the Lekki Peninsula, further from the congested urban area

■ better walking and cycling facilities (like pavements for pedestrians!).

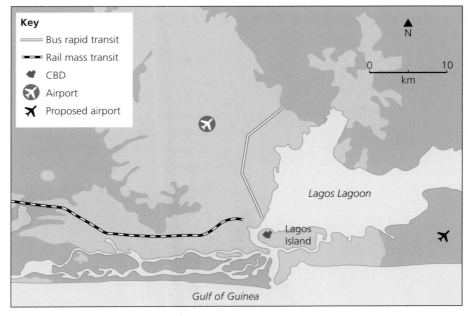

▲ Figure 14.20 New and planned transport developments in Lagos

▲ Figure 14.21 Minibus taxis, or danfos, in Lagos

→ Activities

1 Why might each of these people be concerned about traffic congestion in Lagos?

a) A Makoko resident

b) A danfo driver

c) A business owner in the CBD

d) The governor of Lagos

2 Look at the graphs in Figure 14.19.

a) Describe the impacts of congestion on people in Lagos.

b) How does this compare with a city you know?

3 Read the six ideas for the Strategic Transport Master Plan in Lagos. Explain how each of these ideas would help to reduce congestion in the city.

→ Going further

1 Look at Figure 14.20.

a) On a copy of the map, create an integrated transport plan for Lagos. Draw at least:

■ two more BRT routes

■ two more LRMT routes

■ two waterway routes.

Try to integrate the routes on your map.

b) Write a short report of up to 200 words to support your plans. You can compare your plans with the real thing at www.lamata-ng.com.

⭐ KEY LEARNING

➤ How urban planning can improve life in Lagos

➤ How planning can have environmental benefits

➤ The challenges facing Lagos in the twenty-first century

Urban planning in Lagos

How can urban planning improve life in Lagos?

How can you begin to plan a city like Lagos? Its population is growing by about 600,000 people each year. One idea is to create new floating communities, using the vast area of water that surrounds Lagos, to house the growing population (Figure 14.24). The idea is not as far-fetched as it may seem. Already several squatter settlements like Makoko are built on stilts at the edge of Lagos Lagoon. Effectively, they are communities on water.

In 2014, the Makoko Floating School was built (Figure 14.22). It has classrooms that can host lessons for up to 60 children at a time and it is also used as a community centre when not being used as a school.

How can planning also have environmental benefits?

The floating school does not just help to meet educational needs in Makoko though, of course, these are important. It is also a prototype for the sort of structures that could help to house the population of Makoko and other Lagos communities in the future.

The school is environmentally sustainable, and with its floating design, it would help communities to withstand the impact of rising sea levels as a result of climate change (Figure 14.23).

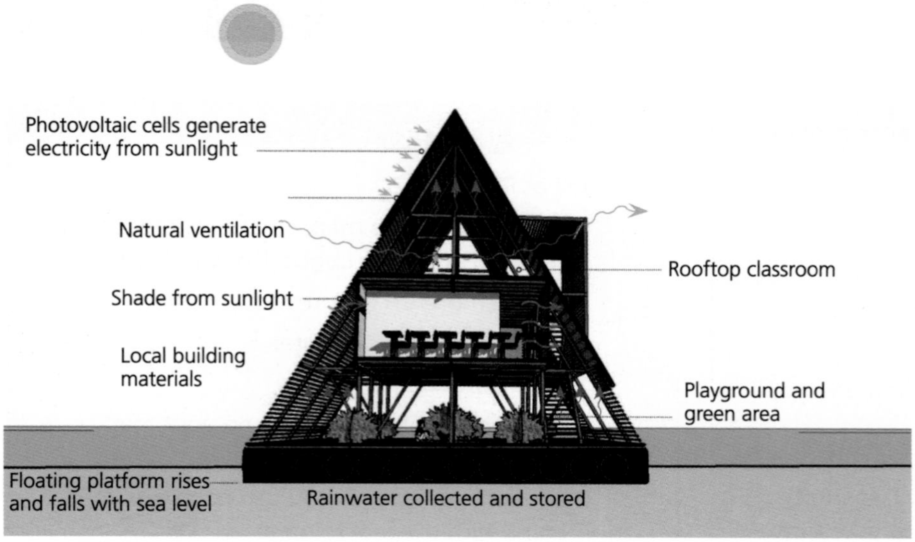

▲ Figure 14.22 The floating school in Makoko

Photovoltaic cells generate electricity from sunlight

Natural ventilation

Shade from sunlight

Local building materials

Rooftop classroom

Playground and green area

Floating platform rises and falls with sea level

Rainwater collected and stored

▲ Figure 14.23 Design for the floating school

What challenges does Lagos face in the twenty-first century?

Lagos faces a number of major challenges.

- Growing population – Lagos is predicted to reach a population of 40 million by 2035. This would put it among the world's top three megacities.
- Population density – there is only limited space to fit the growing number of people. Lagos is already four times more crowded than London.
- Rising sea level – sea level rises by about three millimetres every year. It is projected to rise by about one metre in the twenty-first century. This is a threat to low-lying coastal cities like Lagos.

- Water supply – Lagos has the advantage of being in a tropical area with high annual rainfall, but it lacks a system for storing water and piping it around the city.
- Power supply – Lagos' electricity supply is notoriously unreliable. Most wealthy people have their own generators, while the poor resort to other forms of energy.

▲ Figure 14.24 An artist's impression of a floating community in Lagos

➡ Activities

1   Look at Figures 14.22 and 14.23.

   a) What features of the floating school help to make it sustainable? List at least five.

   b) For each feature, explain how it makes the school more sustainable.

2   Look at Figure 14.24. Explain how floating communities could help Lagos to overcome the five major challenges the city faces in the twenty-first century. Write a short paragraph about each challenge.

3   Think about what you have learned about Lagos in this chapter and how that compares with your expectations at the start.

   a) Look back at the text message you sent from a person thinking of moving to Lagos in Activity 3 in Section 14.2. How might you feel now, having lived in Lagos for a year? Make a list of the good points and bad points.

   b) Send another text message back to your parents in the village you came from to describe your experiences in Lagos.

15 Urban challenges in the UK

⭐ KEY LEARNING

➤ How the population of the UK is distributed

➤ Where cities in the UK are located

➤ How UK cities are growing

Cities in the UK

How is the UK's population distributed?

The UK is one of the most urbanised countries in the world, with 82 per cent of our population living in cities. This is typical of most high-income countries (HICs) that went through the process of urbanisation during the nineteenth and twentieth centuries. It is different in low-income countries (LICs), such as Nigeria, which are still rapidly urbanising today (see Chapter 14).

If you live in London (Figure 15.1) or another UK city, you might get the impression that we are an overcrowded country. But when you look at a map of the UK's population distribution (Figure 15.2), you can see that people are unevenly distributed. While some areas such as South East England are densely populated, other areas, such as Northern Scotland, are sparsely populated.

Overall, the UK's population density is 260 people per square kilometre (km²), ranging from about 5,000/km² in London to less than 10/km² in northern Scotland. This makes us one of the more densely populated countries in Europe – more crowded than France, for example, but less crowded than the Netherlands.

▲ Figure 15.1 London is the UK's largest city. It contains the centre of the finance industry, often called the 'City of London'.

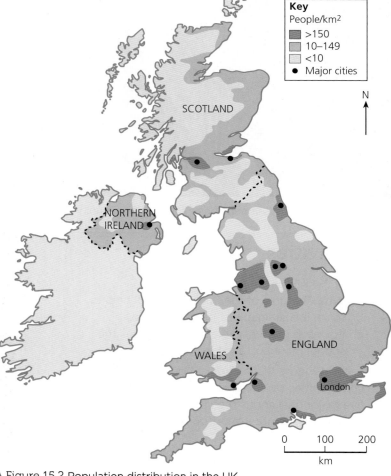

Key
People/km²
▢ >150
▢ 10–149
▢ <10
● Major cities

N

SCOTLAND

NORTHERN IRELAND

WALES

ENGLAND

London

0   100   200
km

▲ Figure 15.2 Population distribution in the UK

Where are cities in the UK located?

The UK's cities are found in the most densely populated areas (Figures 15.2 and 15.3). They tend to be located in flat, low-lying parts of the country, particularly on the coast or near major rivers. Historically, this is where many cities grew, supported by farming, **trade** and industry.

How are UK cities growing?

Today, the fastest-growing cities are in South East England, which is the region with the fastest-growing economy (read more in Chapter 20). By far the biggest growth so far in the twenty-first century has been in London, with over a million new people. At the other end of the scale, Sunderland, in the North East of England, is the only major UK city where population has fallen. This was due to the decline of industry and loss of jobs, forcing people to move away to find work.

→ Activities

1. Study Figure 15.1.
   a) If you live in London, why might you think the UK is overcrowded?
   b) Would you be right or wrong? Explain your answer.
2. Study Figure 15.2. Describe the distribution of population in the UK. Include each part of the UK in your description, e.g. South East England, northern Scotland.

Geographical skills

1. Compare Figures 15.2 and 15.3. Explain the connection between population distribution and the location of cities in the UK.
2. Study Figure 15.3.
   a. Where are the fastest-growing and slowest-growing cities in the UK?
   b. How can you explain the pattern? (Figures 15.1 and 15.4 should help you.)

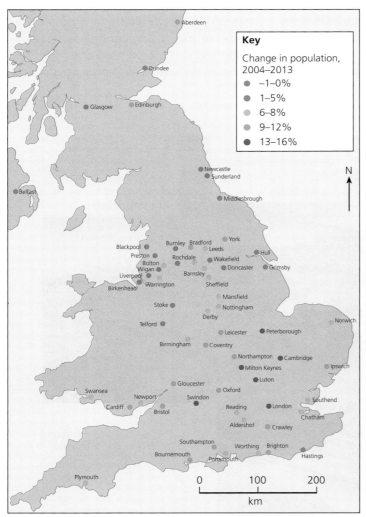

Key

Change in population, 2004–2013
- −1–0%
- 1–5%
- 6–8%
- 9–12%
- 13–16%

▲ Figure 15.3 Major towns and cities in the UK, showing how their populations are changing

▲ Figure 15.4 Sunderland, a city where ships were once built and coal was mined. The shipyards have long closed. The football stadium was built on the site of an old coal mine.

⊕ KEY LEARNING

➤ Where London is located
➤ Why London has grown
➤ The national and international importance of London

London on the map

Where is London located?

London is located in South East England on the River Thames. It is the site chosen by the Romans when they conquered the South of England in 43AD. They built a walled settlement on the north bank of the Thames to defend themselves against the defeated Britons. They called the settlement Londinium and it became the capital of the Roman colony in Britain.

Two factors were important in London's success as a city:

■ The Thames is a tidal river. At high tide, ships were able to navigate up the river to London and the city became a port.
■ London was built at the lowest bridging point on the Thames – the widest point on the river where it was possible to build a bridge.

Why did London grow?

Two thousand years after it was built by the Romans, London is still the capital city of the UK. From the eighteenth century onwards, new docks built along the river increased the number of ships using London as a port. London's importance as a centre of trade and commerce grew and new manufacturing industries developed. This, in turn, attracted more people and its population increased (see Section 15.3).

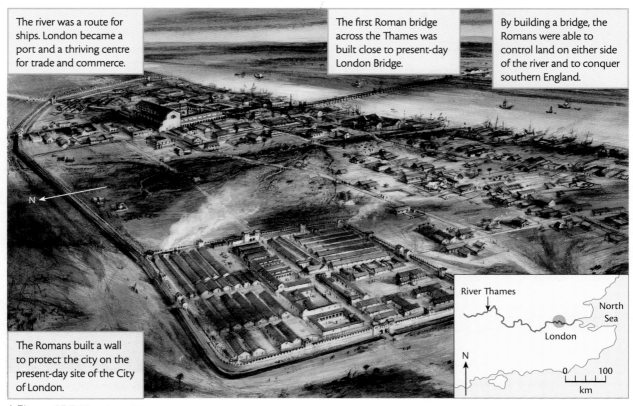

The river was a route for ships. London became a port and a thriving centre for trade and commerce.

The first Roman bridge across the Thames was built close to present-day London Bridge.

By building a bridge, the Romans were able to control land on either side of the river and to conquer southern England.

The Romans built a wall to protect the city on the present-day site of the City of London.

River Thames

North Sea

London

N

0   100
km

▲ Figure 15.5 The site of Roman London

London's role as a port declined towards the end of the twentieth century, with the opening of new docks on the coast. However, it remains the main hub for the UK transport network. Both the UK's road and rail networks focus on London (Figure 15.6). The UK's two busiest airports – Heathrow and Gatwick – are both close to London. They help to maintain London's global connections and its importance as a tourist destination.

What is London's national and international importance?

London is not just the UK's capital; it is also by far the UK's largest and wealthiest city. The gap between London and the rest of the UK has widened in the twenty-first century, as both earnings and house prices have risen faster in London than elsewhere (Figure 15.7).

An indication of London's modern-day importance is its status as a world city. A world city's influence is not just national, but also global. Along with New York, London is one of the two most important financial centres in the world. The headquarters of many large international companies, as well as most major British companies, are based there. London is also a national and international centre for:

- media and communications networks
- education, including renowned universities and research
- legal and medical facilities
- culture, entertainment and tourism.

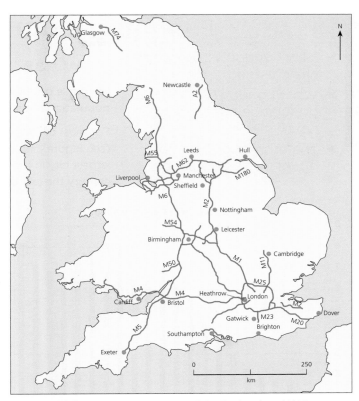

▲ Figure 15.6 The motorway network in England and Wales

The city attracts investment and people from all around the world. Many of London's iconic buildings, like the Shard (not to mention football teams!), are owned by foreign investors. Many migrants come to London to work in high-paid as well as low-paid jobs.

▼ Figure 15.7 London and the UK compared (2015)

Categories	London	UK
Population	8,630,000	64,100,000
Average male life expectancy	79	78
Average female life expectancy	83	82
Proportion of ethnic minorities in the population	37%	14%
Average earnings	£34,473	£22,044
Average house price	£514,000	£272,000
Unemployment rate	7.5%	6.6%
Educational achievement (pupils achieving good GCSE grades)	62%	55%
Murder rate	1.1/100,000	1.4/100,000

Geographical skills

Study Figure 15.5. Draw a sketch map, based on the drawing, to show the site of Roman London. Annotate your map to explain why this was a good site for a city.

→ Activities

1. Study Figure 15.6.
   a) Plan journeys from three other cities in the UK to London. In each case, list the motorways you would travel on.
   b) Describe the pattern of motorways in the UK in relation to London.

2. Study Figure 15.7. Compare London with the rest of the UK. Write three short paragraphs to compare (a) the population, (b) distribution of wealth and (c) quality of life.

3. What do you think are (a) the advantages and (b) the disadvantages of living in London? Give at least three examples of each.

⊗ KEY LEARNING

➤ How London's population has changed
➤ What London's population structure is now
➤ The ethnic composition of London's population

London's growing population

How has London's population changed?

London's population is higher now than it has ever been. In 2015, London's population reached 8.6 million, overtaking the peak it last reached in 1939.

For most of the past two hundred years, London's population has been growing (see Figure 15.8). In 1801, with just over a million people, it was already the largest city in the world. During the Industrial Revolution in the nineteenth century the city grew as it attracted migrants, mainly from other parts of the UK.

London's population reached its previous peak at the start of the Second World War. The city was badly bombed during the war and its population fell after 1939. Numbers continued to decline after the war as housing was demolished and people moved out. During the twentieth century many cities in other countries grew bigger than London.

London's population has been climbing again since 1991. It is likely to continue growing and is predicted to reach 10 million by 2030, which will make London one of the world's megacities.

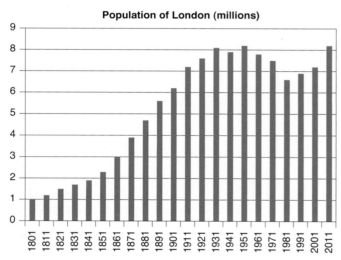

▲ Figure 15.8 London's population growth since 1801

How old is London's population?

London has a much larger population than any other UK city. It also has a younger population (see Figure 15.9). This helps to explain why its population is growing.

■ Young people in their 20s and 30s, especially university graduates, move to London for work. They are attracted by more job opportunities, higher pay and the perception of an exciting social life in London.

■ Younger people, particularly in the 20–30 age group, are more likely to have children. That leads to a higher rate of natural population increase in London.

■ Migrants from around the world add to London's population. At the same time as people arrive, others leave. The balance between the two groups is net migration.

■ Although net migration into London is quite low, most immigrants are young while most people leaving are older. This reduces the average age of the population and leads to greater natural increase.

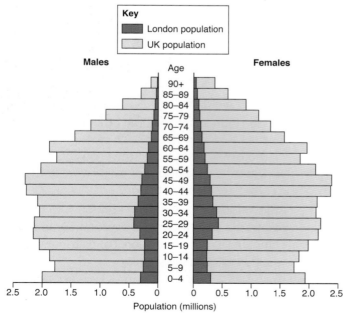

▲ Figure 15.9 London's population structure, compared with the UK (2011)

What ethnic groups make up London's population?

London is also the most diverse city in the UK. Less than half of London's population are of white British origin, while 37 per cent were born outside the UK (see Figure 15.10).

Migration to London goes back to Roman times. Later, Saxons and Normans also settled in London. In the seventeenth century, French Huguenot (Protestant) refugees arrived and settled outside the city walls to the east in Spitalfields. Later, in the nineteenth century, came Jewish refugees from Eastern Europe and more recently economic migrants from Bangladesh. Each group of migrants has helped to change the character of Spitalfields (see Figure 15.11).

Today, London's population comes from every part of the world. The largest numbers are from countries like India, Nigeria and Jamaica, each once part of the British Empire. Since 2007, more migrants have come from Eastern Europe, with the free movement of people in the European Union (EU).

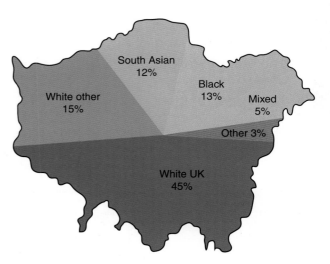

South Asian 12%

Black 13%

Mixed 5%

White other 15%

Other 3%

White UK 45%

▲ Figure 15.10 The ethnic composition of London's population

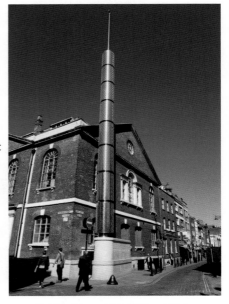

➤ Figure 15.11 Brick Lane Mosque: used by Bangladeshis in Spitalfields. It was first built as a Huguenot (Protestant) church and later used as a Jewish synagogue

→ Activities

1. Study Figure 15.8. Describe changes in London's population since 1801. Mention key figures and dates from the graph. Be as accurate as you can.

2. Study Figure 15.9. Compare the structure of London's population with the UK's. Which age groups make up a larger proportion of a) London's population and b) the UK's population?

3. a Use the information in Figure 15.9 to predict how London's population is likely to change in the future.

   b Explain how you made your prediction.

4. Study Figure 15.10. Turn the information into a bar chart to show the percentage of each ethnic group in London's population.

5. Study Figure 15.11. Buildings show one way in which ethnic diversity has helped to change the character of London. Think of at least five other ways in which ethnic diversity can change the character of an area.

Fieldwork: Get out there!

'Areas with a more diverse population also have more diverse shops and services.'

- Devise at least one fieldwork method you could use to test this hypothesis.

- Suggest two areas you know where you could carry out your fieldwork, one with a more diverse population than the other.

- Predict what results you would expect to get from your fieldwork and why.

⭐ KEY LEARNING

➤ How an old area of London has changed

➤ The cultural mix found there now

➤ The opportunities for recreation and entertainment

The new face of London

How has an old area of London changed?

One old area of London close to the city centre is Shoreditch (see Figure 15.12). It typifies the sort of changes that have happened around London, and in some other UK cities.

Just 30 years ago, Shoreditch was still a run-down inner-city area, with many old factories and warehouses. Most industries had closed down and people were moving out of the area. In their place newcomers were moving in, particularly Bangladeshi immigrants around Brick Lane (see 15.3).

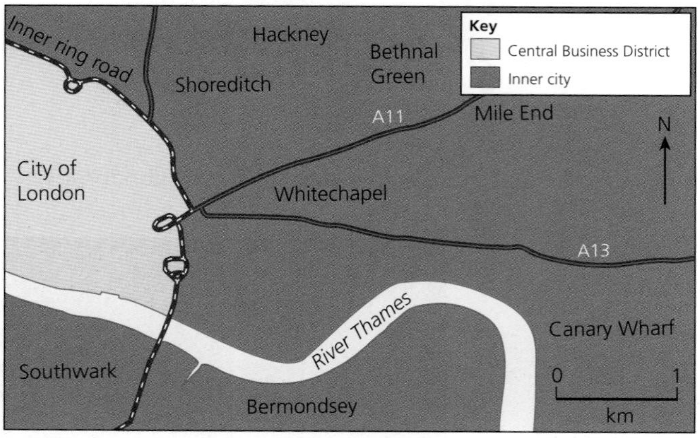

What is the cultural mix found in Shoreditch?

Shoreditch today is almost unrecognisable from 30 years ago. Old industrial buildings have been converted into flats and offices. Pubs and bars have been brought back into life as restaurants and art galleries. Jobs have been created in new creative industries, such as web design, film-making and art.

One focus for employment is around the Old Street roundabout. So many new hi-tech companies have appeared that it is nicknamed 'Silicon Roundabout' after Silicon Valley, the centre of the hi-tech industry in California where companies like Microsoft and Apple grew.

▲ Figure 15.12 The location of Shoreditch in London

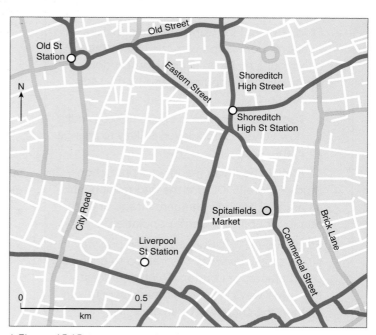

▲ Figure 15.15 Street map of Shoreditch

▲ Figure 15.13 New, hi-tech companies around Old Street roundabout

▲ Figure 15.14 Shoreditch is now well-known for its nightlife and themed cafes

What opportunities for recreation and entertainment are there?

The population of Shoreditch has changed too. Many older residents and Bangladeshi families are moving away, as rents and property prices go up. In their place, young professional workers, many in the finance and creative industries, are moving in (Figure 15.14).

This process of rising property prices and changing population is known as gentrification. With the new, younger population have come new forms of recreation and entertainment. Shoreditch is now one of the most vibrant parts of London, day or night!

◄ Figure 15.16 Street art or graffiti in Shoreditch? You decide!

➤ Figure 15.17 Spitalfields Market, once a fruit and veg market, is now a fashionable place to shop

Flat to rent, Shoreditch, E2, £330/week

A unique studio flat set in a recent warehouse conversion, close to Shoreditch station. The property comprises a surprisingly spacious studio room fitted with original hardwood floors, an integrated kitchen equipped with fully functional appliances and a modern bathroom.

The property also benefits from superior travel links. You'll be a 15-minute commute to Canada Water station (serving the Jubilee Line), 10 minutes to Liverpool Street and 5 to Dalston Junction, with 24-hour bus services reaching out to almost every corner of London.

Amenities:

| ☑ Balcony | ☒ Garage | ☒ Patio |
| ☑ Parking | ☒ Garden | ☒ Disabled access |

▲ Figure 15.18 An advert for accommodation in Shoreditch

→ Activities

1 Study Figure 15.12. Explain why Shoreditch is a good location in London to:
   a) start a new business
   b) live. (You could refer to an Underground map of London to help you.)

2 Study the photos and map in Figures 15.13–15.17.
   a) Identify different types of recreation and entertainment in the photos and list them.
   b) Write an advert for one of the places in the photos. Think about the type of person your advert would be aimed at. Use the map to give directions to the place.

3 Read the advert in Figure 15.18. What type of household is this advert aimed at? Think about:
   ■ their ages
   ■ their income
   ■ their occupations
   ■ whether they have children or not.

4 How would the following be affected by the changes in Shoreditch?
   ■ A student in London
   ■ A Bangladeshi family on a low income
   ■ A young couple working in a hi-tech industry
   ■ The owner of an old warehouse.
   Explain your ideas.

⊗ KEY LEARNING

➤ Why the docks in London declined

➤ Why new industries, like finance, have grown

➤ How employment patterns in London have changed

London rising

Why did the docks in London decline?

London has been a port since Roman times (see Section 15.2). Later, the docks were built to handle the huge volume of goods and raw materials brought to London by ship (see Figure 15.19). Around the docks, industries such as sugar refineries, flour mills and timber yards grew to process the materials.

By the 1970s, the docks were in decline. New container ships were being used and the docks were no longer large enough to hold them. One by one the docks closed, until by 1980 they were lying empty, with many of the industries gone too.

➤ Figure 15.19 Docks on the Isle of Dogs, still working in the 1960s (facing northeast)

▲ Figure 15.20 Canary Wharf on the Isle of Dogs today (from the east)

Why have new industries, like finance, grown?

In 1981, the government set up a new body, the London Docklands Development Corporation (LDDC), to plan the **regeneration** of the docks. It was given the task of finding new ways to use the land around the docks by attracting **private investment**. It was hoped this would create new economic opportunities and jobs to replace those lost when the docks closed down.

What happened next became a model for other regeneration projects around the UK. At the heart of Docklands lies Canary Wharf, dominated by high-rise office blocks that are now home to many international banks. Over 100,000 people work there and, together with the City of London, Docklands has helped establish London as one of the world's leading financial centres.

How have employment patterns in London changed?

The number of jobs in London has been rising almost continuously since 1994 (Figure 15.21). Even the recession after 2007 did not really slow the rise. So, what are all these new jobs?

The biggest growth in jobs was in services, especially 'Professional, real estate and business services' (Figure 15.22). This includes work in company head offices, management consultancy, law and accountancy, estate agents, advertising and market research. The biggest decline in jobs over the same period was in manufacturing. London has very few factories left.

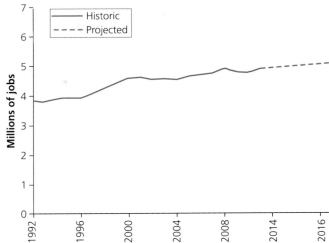

▲ Figure 15.21 London's historic and projected employment from 1992 to 2016 (calculated in 2013).

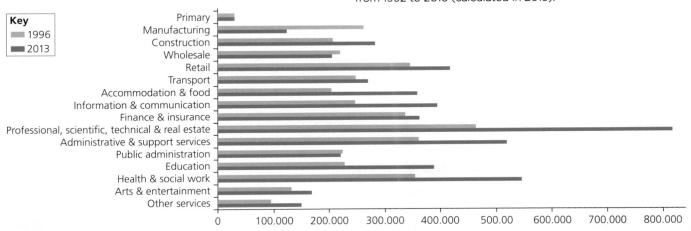

▲ Figure 15.22 Jobs in London in 1996 and 2013.

→ Activities

1   Study Figures 15.19 and 15.20. Turn them into 'living photos' by adding speech bubbles to copies or sketches of the photos.

a)   Choose quotes for your speech bubbles for each photo from the ones below:

- 'I don't think there's much of a future in this job.'
- 'I expect a bonus this year to pay for our holiday.'
- 'I like to go to the company gym at lunchtime.'
- 'It's freezing doing this job in winter.'
- 'I hope to get overtime this week to pay the rent.'
- 'I'd like to be chief executive by the time I'm 40.'

b)   Summarise the way in which employment in Docklands has changed since the 1960s.

Geographical skills

1   Study Figure 15.21.

a)   Describe how London's total workforce has changed since 1994. Mention dates and figures.

b)   Calculate the percentage growth in the workforce from 1998 to 2011.

2   Study Figure 15.22.

a)   List the activities in which the number of jobs declined from 1996 to 2013.

b)   List four activities with the largest increase of jobs from 1996 to 2013. Give one example of a job in each activity.

c)   Overall, say how employment patterns in London have changed.

Case study

⭐ KEY LEARNING

➤ Why there is a need for improved transport in London

➤ The transport improvements planned

➤ How Crossrail could impact on London

Keep the city moving

Why is there a need for improved transport in London?

London has a well-integrated transport system, but it is struggling to cope with the increase in passenger numbers (Figure 15.23). As the population grows and work opportunities increase, more people are using public transport to commute to work. Driving a car is not a sensible option for most Londoners, with limited space to park and traffic congestion causing long journeys.

In 2014 roughly 75 million passengers used underground trains (the Tube) and buses in London each week – 25 million on the Underground and 50 million on buses. The number is growing every year (see Figure 15.24).

What transport improvements are planned?

The demand for public transport in London is predicted to grow by 60 per cent by 2050, when the population will be much higher. Though 2050 may seem like a long time away, improvements to transport need long-term planning and investment.

Crossrail is a new, east–west rail route across London due to open in 2018, linking Shenfield and Abbey Wood in the east with Reading and Heathrow in the west (Figure 15.25). It will tunnel under the city centre, reducing journey times and increasing the total number of passenger journeys in London. Already, Crossrail 2 is being planned for 2030. It would be a similar project on a north–south route across London.

▲ Figure 15.23 A crowded London Underground train

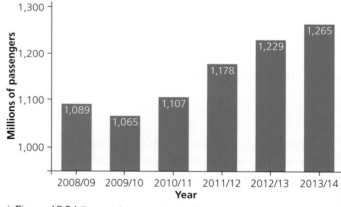

▲ Figure 15.24 Recent increase in Underground passenger journeys

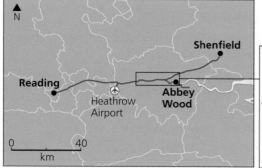

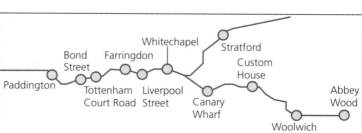

▲ Figure 15.25 The Crossrail route in London

What could the impacts of Crossrail be on London?

Crossrail is one of the largest infrastructure projects London has ever seen. It is expected to bring a number of benefits. It will:

■ reduce journey times – for example, the journey time from Liverpool Street to Heathrow will fall from over an hour to 35 minutes

■ increase the number of rail passenger journeys in London by ten per cent, or an extra 200 million journeys a year

■ bring an extra 1.5 million people within a 45-minute journey of central London, increasing the number of people who can commute to work in London

■ improve the integrated transport system in London by providing more interchanges with the Underground network

■ raise property values by about 25 per cent around stations along the Crossrail route

■ encourage further regeneration across London, providing access to thousands more jobs

■ improve access for disabled people to new stations, with no steps from platform to street level.

▲ Figure 15.26 Canary Wharf is one of London's main financial hubs (see Section 15.5). Crossrail will make commuting quicker and easier. There are plans for a new phase of Docklands regeneration nearby at Wood Wharf, with thousands of jobs in creative industries.

▲ Figure 15.27 Custom House is one stop from Canary Wharf on Crossrail, but a world apart. It is one of the poorest parts of London, with a high proportion of social housing and people on low income. On the other side of the Crossrail track is ExCel London, London's largest exhibition centre.

→ Activities

1 Study Figure 15.24.

   a) Describe how the number of passenger journeys changed from 2008 to 2014.

   b) Explain why this change happened and predict future changes.

2 Study Figure 15.25. Explain how Crossrail will:

   a) reduce journey times in London.

   b) increase the number of passenger journeys.

   c) improve integrated transport in London (you could refer to an Underground map of London to help you).

3 a) Rank the benefits of Crossrail for London in order of importance.

   b) Explain why you ranked your first benefit as the most important.

4 Study Figures 15.26 and 15.27.

   a) Which Crossrail station do you think will bring the biggest benefits – Canary Wharf or Custom House? Consider local residents, commuters, regeneration and property values.

   b) Give reasons for your decision.

⊛ KEY LEARNING

➤ How much of London is green

➤ The benefits of green cities

➤ What strategies could make London greener

Urban greening

How much of London is green?

London is one of the world's greenest cities. Almost half the city – 47 per cent – is green space, including parks, woodlands, cemeteries and gardens (Figure 15.28). The percentage might have been higher, but in recent years many people have paved over their gardens to create patios or make space to park their cars.

London's large area of green space is a result of the way in which the city has developed.

- Central London parks – London has more big parks than many cities. They include royal parks, such as Hyde Park, which once belonged to royalty. Now everyone can enjoy them.

- Local parks – many parts of inner and outer London have municipal parks run by the local council. They date back to the nineteenth century where there was concern about public hygiene in London and the need for people to have fresh air.

- Suburban growth – the expansion of London in the early twentieth century led to the development of suburbs. They were built on farmland, providing millions of new homes with gardens for Londoners.

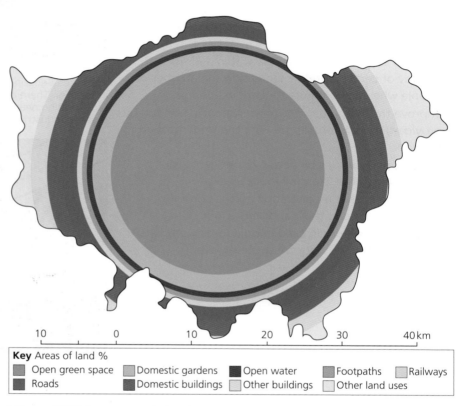

Key Areas of land %
- Open green space
- Roads
- Domestic gardens
- Domestic buildings
- Open water
- Other buildings
- Footpaths
- Other land uses
- Railways

▲ Figure 15.28 Green space in London

Why is it good to have green cities?

There are many reasons for cities to be green, but not all of these were known when London was developing.

- Trees produce oxygen, clean the air and help to reduce global warming by using carbon dioxide. There are 8.1 million trees in London – almost one per person!

- Trees and green open space reduce the danger of flooding by slowing down the rate at which rainwater drains from the land.

- Parks, woodlands and even domestic gardens provide a habitat for wildlife, including birds, insects and mammals. There are 13,000 wildlife species in London.

- People enjoy green open spaces and they help us to keep healthy. We use these spaces for walking, running, cycling and for sport.

- People also use green spaces for growing food. There are 30,000 allotments in London: shared open spaces where people grow their own food.

What strategies can be used to make London greener?

Urban greening is about how we increase and protect the green spaces we have in cities. London is already a 'green city', so urban greening here is more about protection.

- On a small-scale, this is about individual actions, like encouraging people to feed birds in winter or not paving over gardens.

- On a larger-scale, it could be about connecting the green spaces we already have to help species to migrate naturally. London now has a 'green grid' to link open spaces.

One new project is the Garden Bridge across the River Thames in central London, planned to open in 2018 (Figure 15.29). It is hoped it will bring environmental and economic benefits to the city, and become another link in the green grid.

▲ Figure 15.29 The proposed Garden Bridge in London

→ Activities

1   Study Figure 15.28.

   a) Estimate the percentage area of each land use in London. For example, open green space is 47%. Your percentage figures should add up to 100%. List them in order of the amount of space they occupy

   b) Does anything surprise you about the graph and percentages? Explain why you do or don't find it surprising.

2   a) What are the benefits you can think of for having green open space in a city? Make a list and include some of your own ideas.

   b) Write a letter to your council to persuade them to provide more green space, or at least, to keep the green space that is already there. (You could use the findings from your fieldwork below to help you to make your case.)

3   Study Figure 15.29. The cost of the Garden Bridge in London will be £175 million. Is this money well spent, do you think? Give your reasons.

Fieldwork: Get out there!

How much green space is there in your area?

To answer this question, you will have to do some land-use mapping in your area.

- Walk around the area with a large-scale outline map. Shade all the spaces on your map in different colours to show how the land is being used.

- Calculate the percentage of the total area for each land use. You could do this by placing a grid with 100 squares over your map and counting the squares for each land use. Is there more or less green open space than you predicted?

Case study

> ⭐ **KEY LEARNING**
> ➤ Social deprivation
> ➤ How deprivation varies between areas of London
> ➤ Why inequality is still a challenge in London

Urban inequalities

What is social deprivation?

You might not think of London as being a deprived city. After all it is the wealthiest city in the UK. But **social deprivation** is a major problem in London, with over two million people living in poverty. Social deprivation is the degree to which a person or a community lacks the things that are essential for a decent life, including work, money, housing and services.

How does deprivation vary between areas of London?

London is divided into 33 boroughs. These are administrative areas that make it easier to run such a large city, and they are also a useful way to show variations within the city. Figure 15.30 shows the percentage of people on state benefits in each borough, as a measure of deprivation. People receive benefits in the UK when they are unemployed to help them financially.

Another measure of deprivation is **life expectancy**. The more deprived a person is, the lower their life expectancy is likely to be. The variation in life expectancy in London is clearly demonstrated by travelling on the Underground from Knightsbridge, in the borough of Kensington & Chelsea, to West Ham in the borough of Newham. Life expectancy for those living in each area falls on average by one year for every station along the route (see Figure 15.31).

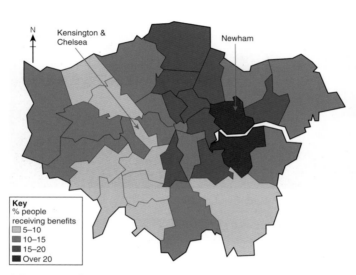

Key
% people receiving benefits
☐ 5–10
☐ 10–15
☐ 15–20
■ Over 20

▲ Figure 15.30 People on benefits in each London borough

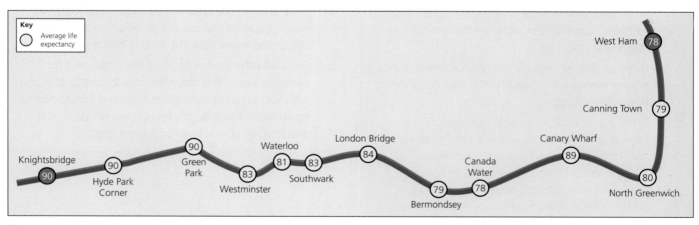

▲ Figure 15.31 Life expectancy falls on a journey from Knightsbridge to West Ham

Why is inequality still a challenge in London?

Despite years of economic success, **inequalities** are still a challenge in London. Differences in life expectancy still exist. Low life expectancy in the most deprived parts of the city is closely linked to poor diet, housing and education, as well as lack of employment.

Kensington & Chelsea, one of London's richest boroughs, does better than Newham, one of the poorest boroughs, for all the measures of deprivation (see Figure 15.32).

▼ Figure 15.32 Inequality between Kensington & Chelsea and Newham

Measure of deprivation	Kensington & Chelsea	Newham
Male life expectancy	83.7	75.7
Female life expectancy	87.8	79.8
Unemployment	3.9%	9.4%
Pupils achieving five + good GCSE grades	80%	62%
Households with joint income < £15,000	9%	26%
Households with joint income > £60,000	26%	7%

It's lovely here. Everyone is very friendly. It feels like a village. You see people running, see the mums power-walking on the school run. There's an organic food shop and good shops like Waitrose and M&S.

Jessica Kelly, 22, student in Kensington & Chelsea

I have lived here for two years and I have a good standard of life. There is still a positive vibe since the Olympics. The best thing we can do is help each other whenever we can. People may be quite modest here and not realise their potential.

Emily Leslie, 30, teacher in Newham

▲ Figure 15.33 Views from the street in Kensington & Chelsea and Newham

→ Activities

1. Study Figure 15.30. Describe the pattern of social deprivation in London. In which parts of London are the most and least deprived boroughs?

2. Study Figure 15.31.
   a) Describe the changes in life expectancy on an Underground journey from Knightsbridge to West Ham.
   b) How can you explain these changes?

Geographical skills

1. Study Figures 15.32 and 15.33.
   a) Compare life in Kensington & Chelsea and Newham using:
      ■ the data in Figure 15.32
      ■ the photos in Figure 15.33
      ■ peoples' experiences in Figure 15.33.
   b) Which of these three sources do you think is the most reliable way to compare the two boroughs? Give reasons.

Fieldwork: Get out there!

'Levels of deprivation vary between areas of a city.'

■ Devise at least one fieldwork method you could use to test this hypothesis. For example, you could take photos or conduct interviews with people.

■ Suggest two areas of a city you know where you could carry out your fieldwork.

■ Suggest any other data you could use to compare the two areas, apart from what you will find out by doing fieldwork.

⊛ KEY LEARNING

➤ Why there is a shortage of homes in London

➤ The reasons for building on brownfield or greenfield sites

➤ Whether homes should be built on the greenbelt or not

New homes needed

Why is there a shortage of homes in London?

London's population is growing (see Section 15.3) by about 100,000 people every year, yet only about 20,000 new homes a year are being built. This has led to a severe housing shortage in London and the rest of South East England. The result is that house prices are rising faster in London than the rest of the country (Figure 15.34). They are also rising faster in inner London than outer London.

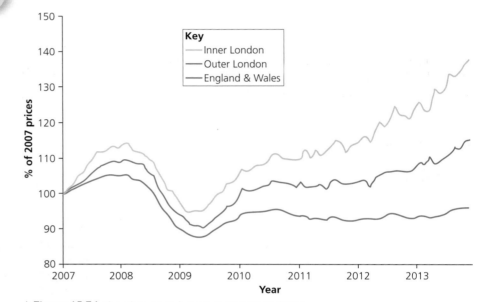

▲ Figure 15.34 The change in house prices since 2007

What are the reasons for building on brownfield or greenfield sites?

One possible solution to the shortage of homes in London is to build outside London on greenfield sites. These are areas of land that have not previously been built on – usually farmland on the **rural–urban fringe**. However, building on these sites can lead to urban sprawl and is not very popular with those people already living in the countryside.

The alternative is to build on **brownfield sites** in the city. These are areas of previously developed land which are often **derelict** now and have potential for redevelopment. Often, this is land that was previously used for industry where the ground may be contaminated by chemicals. There are many sites like this in London due to the decline of the manufacturing industry.

▼ Figure 15.35 Issues about building on brownfield or greenfield sites

Brownfield sites	Greenfield sites
• Sites are available since industry declined. • Reduces the need for urban sprawl. • Public transport is better in urban areas, so less need for cars. • Old buildings may need to be demolished first. • Ground may need to be decontaminated. • New development can improve the urban environment. • Land is more expensive in urban areas.	• Public transport is worse in rural areas, so more need for cars. • Increases urban sprawl. • Once land is built on, it is unlikely to be turned back to countryside. • Land is cheaper in rural areas. • No demolition or decontamination is needed. • Valuable farmland or land for recreation may be lost. • Natural habitats may be destroyed.

Should new homes be built on the green belt?

Around many cities in the UK, including London, is a **green belt**. This is land on which there are strict planning controls. It was established in 1947 to prevent further urban sprawl. Since then, it has helped to preserve farmland, woodland and parkland around London (Figure 15.36).

Now, with the pressure for more housing in London, people are questioning whether we can afford to keep the green belt. They suggest that less valuable areas of green belt land could be used for building new homes on greenfield sites.

As the population of London grows and house prices rise, more people move to commuter settlements around London. This forces population and house prices in the rest of South East England to rise too. The problem of urban sprawl has shifted to commuter settlements outside the green belt. Cities like Reading and Chelmsford, within a half-hour train journey to London, are growing rapidly. Urban development, in the form of new housing estates and business parks, encroaches into the surrounding countryside.

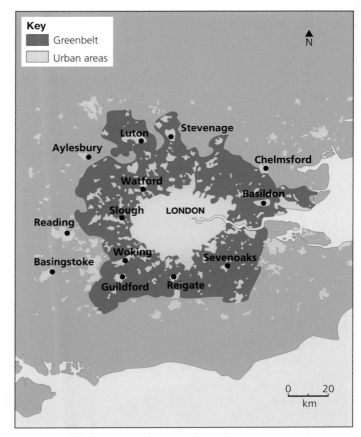

Key
- Greenbelt
- Urban areas

▲ Figure 15.36 The green belt around London

→ Activities

1 Study Figure 15.34.

  a) Compare the change in house prices in inner London, outer London and England and Wales.

  b) Explain why this has happened.

2 Study Figure 15.35.

  a) Classify the issues about building on brownfield and greenfield sites into advantages and disadvantages.

  b) Redraw the table like this and list the impacts in the correct boxes.

	Brownfield sites	Greenfield sites
Advantages		
Disadvantages		

3 Study Figure 15.36. Think about each of these people's interests in the green belt. Would they be for or against protecting it? In each case, explain why.

- A resident of inner London
- A resident of outer London (on the rural–urban fringe)
- A resident of Woking
- A farmer in the green belt

4 Prepare for a class debate about the green belt. You can base your ideas on the green belt around London or another city you know. Write a short speech for or against keeping the green belt.

⊗ KEY LEARNING

➤ How serious the air pollution in London is
➤ How new cycle superhighways will help
➤ What happens to London's waste

London's pollution problem

How serious is the problem of air pollution in London?

Compared to years gone by, pollution in London is less of a problem than it used to be. In the mid-twentieth century, when coal was burnt to power factories and provide domestic heating, the city used to experience smog, or a dense mixture of smoke and fog.

However, **air pollution** is still a problem (see Figure 15.37). The main problem now is emissions from road vehicles and modern heating systems. It is made worse by the dense road network in London and the tall buildings that trap air between them.

London has a worse pollution record than most other European cities, though not so bad as many cities in Asia. One of the worst modern pollutants is nitrogen dioxide (NO_2) that comes primarily from road vehicles, especially diesel engines. London regularly breaks European Union (EU) regulations on air quality. Most of central London is above the EU limit of 40 mg/m³ for NO_2 (see Figure 15.37). There are over 4,000 premature deaths a year in London due to long-term exposure to air pollution.

▲ Figure 15.37 Air pollution over London

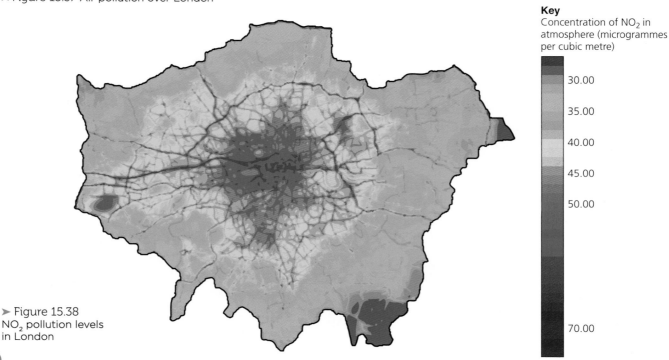

Key
Concentration of NO_2 in atmosphere (microgrammes per cubic metre)

30.00
35.00
40.00
45.00
50.00
70.00

➤ Figure 15.38 NO_2 pollution levels in London

How will new cycle superhighways help?

New cycle superhighways planned for London should encourage more people to cycle and reduce traffic and harmful emissions from vehicles (Figure 15.39). Cyclists have increased from one per cent to fifteen per cent of road users in London over the past 50 years. The percentage should increase further with the new cycle superhighways.

What happens to London's waste?

Almost a quarter of London's waste still goes to landfill sites outside London (Figure 15.40). In the past this was acceptable because the waste was out of sight, out of mind. Now, we realise that landfill waste contributes to wider environmental problems, such as the production of methane that adds to the greenhouse gases in the atmosphere (see Section 4.4). And, of course, 'waste' is just that – a waste of potentially valuable resources.

More of London's waste is now recycled or incinerated (burnt to generate electricity). The target is for zero waste to go to landfill by 2030, by focussing on waste reduction and by managing resources more efficiently.

▲ Figure 15.39 Plans for one of London's new cycle superhighways. Notice the separate lane for cyclists and the reduced width of the road for traffic

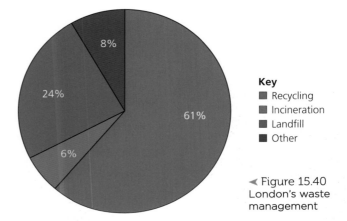

Key
■ Recycling
■ Incineration
■ Landfill
■ Other

◄ Figure 15.40 London's waste management

➜ Activities

1 Study Figure 15.37. Explain why air pollution in London:

   a) was a problem in the past.
   b) is still a problem today.

2 Study Figure 15.38.

   a) Draw a sketch map of Greater London to show the most and least polluted areas.
   b) Identify these places and label them on your map – central London, Heathrow Airport, North Circular Road (around central London).
   c) Explain the pattern on the map.

3 Study Figure 15.40.

   a) What happens to most of London's waste?
   b) Why are we trying to send less waste to landfill?
   c) What types of waste can be recycled, and what can't be recycled?
   d) What other types of waste management can you think of? (Clue: think about food waste.)

 Fieldwork: Get out there!

Cyclists in London represent about fifteen per cent of road users. What percentage is it in your area? Would the number increase with safer cycle routes?

■ Suggest how you could find out the percentage of cyclists on the road.

■ Suggest how you could find out if more people would cycle if there were safer cycle routes.

■ Predict what results you expect to get from your fieldwork in your area.

⊛ KEY LEARNING

➤ Why the Lower Lea Valley was in need of regeneration

➤ What obstacles had to be overcome to regenerate the site

➤ Why the London 2012 Olympic bid was successful

Urban regeneration: the Olympic plan

Why was the Lower Lea Valley in need of regeneration?

The Lower Lea Valley in East London was the site for the 2012 Olympics (Figure 15.40). The River Lea is a tributary of the Thames and the Lea Valley was once one of the main industrial areas in London. You will also remember that Newham, along with the other boroughs around the site, is in one of the most deprived parts of London (Figure 15.42).

By 2007, when work began to create the Olympic Park, many of the industries had already gone and some of the site was derelict and overgrown. But the land around the River Lea was far from being empty (Figure 15.42).

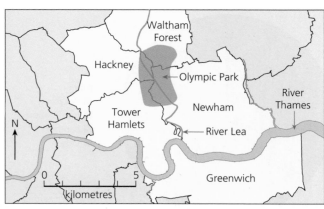

◀ Figure 15.41 The site of the Olympic Park in East London

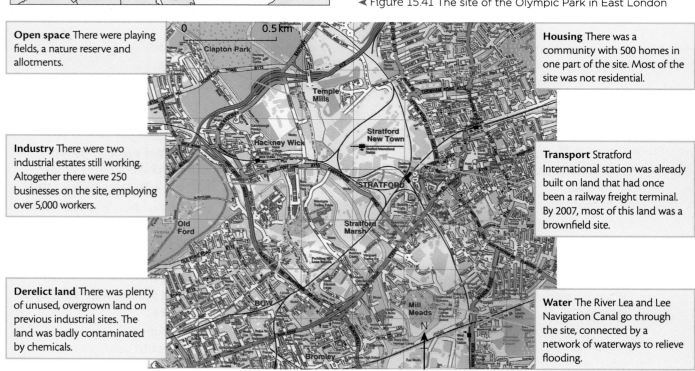

Open space There were playing fields, a nature reserve and allotments.

Housing There was a community with 500 homes in one part of the site. Most of the site was not residential.

Industry There were two industrial estates still working. Altogether there were 250 businesses on the site, employing over 5,000 workers.

Transport Stratford International station was already built on land that had once been a railway freight terminal. By 2007, most of this land was a brownfield site.

Derelict land There was plenty of unused, overgrown land on previous industrial sites. The land was badly contaminated by chemicals.

Water The River Lea and Lee Navigation Canal go through the site, connected by a network of waterways to relieve flooding.

▲ Figure 15.42 The Olympic Park site before work began in 2007

What obstacles had to be overcome to regenerate the site?

The construction of the Olympic Park in just five years, from 2007 to 2012, was an impressive achievement. Eventually, it is expected to lead to the regeneration of this part of east London, though the whole process will take much longer (see Section 15.12). Before construction or regeneration could begin, obstacles had to be overcome:

■ The land had to be brought together under one new owner, the Olympic Delivery Authority (ODA), which was set up by the government.

■ Existing landowners and users had to leave the site by 2007. Some of them protested (Figure 15.44). The land was bought from them by the ODA.

■ Land that was previously polluted by industry had to be decontaminated before building could begin.

■ Electricity pylons had to be removed and overhead cables buried below ground to improve the appearance of the landscape (Figure 15.43).

■ Waterways and railways crisscrossed the site, so bridges were built to link the area together.

▲ Figure 15.43 The Olympic Park site before 2007

Why was the London 2012 Olympic bid successful?

As you know, London won the bid to host the 2012 Olympics. There were a number of reasons why London beat the other rival cities around the world:

■ There was a large area of available land, even though there were also businesses and homes on the site.

■ East London has very good transport connections, particularly Stratford station, where most spectators arrived in 2012.

■ London's diverse population made it the natural city to host guests from around the world. Newham is the most diverse borough in London.

■ The Olympic bid promised to leave a lasting legacy that would help to regenerate east London.

▲ Figure 15.44 Local protests against the Olympics

→ Activities

1. Study Figure 15.41. Describe the location of the Olympic Park in London. Mention the boroughs that surround the site.

2. Study Figures 15.42 and 15.43. Explain why the Lower Lea Valley was in need of regeneration.

3. a What does the slogan on the banner in Figure 15.44 mean?

   b How true was it? (Hint: read more in Section 15.12 to find out what happened next.)

4. East London was an area in need of regeneration. Was that an advantage or disadvantage for London's bid? Explain your answer.

5. Would each of the following have supported the Olympic bid, or not? In each case, give reasons.

   ■ A local resident living on the site

   ■ A local resident living outside the site

   ■ A business owner on the site

   ■ The Mayor of London

   ■ The British government

⭐ KEY LEARNING

➤ How the environment of the Lower Lea Valley has changed
➤ What social and economic changes there have been

Urban regeneration: the Olympic legacy

How has the environment of the Lower Lea Valley changed?

The Queen Elizabeth Olympic Park has completely transformed the environment of the Lower Lea Valley.

Gone are the:

- old factories, industrial estates and homes
- derelict and overgrown sites
- electricity pylons and overhead cables
- contaminated soil and polluted waterways
- the Olympic Delivery Authority.

In their place have appeared:

- stunning new sports venues, including the Aquatics Centre (see Figure 15.45), stadium and velodrome
- a landscaped park with tourist attractions and natural habitats
- the Athletes' Village, now converted into a residential community
- clean soil and waterways
- the London Legacy Development Corporation (LLDC) (who have taken over from ODA).

▲ Figure 15.45 The Aquatics Centre beside the River Lea

Here East The new name for the Media Centre is now a hub for creative and media industries with 5,000 jobs.

Queen Elizabeth Olympic Park With over 100 hectares of open space, this is the largest new park in London for over a century.

Olympic stadium The new home of West Ham United FC but still an athletics stadium in the summer.

LLDC This organisation controls an area including the park and surrounding neighbourhoods to make the legacy a reality.

East Village The new name for the Athletes' Village, it now provides 2,800 homes for local people and newcomers.

Westfield Stratford City Not really part of the Olympic legacy, but next door to the park and employs 10,000 people.

The International Quarter A new commercial development of high-rise offices which will employ 25,000 people.

The Aquatics Centre and Velopark Two new sports venues open to the public and used by schools.

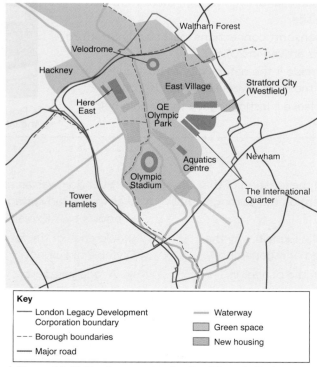

Key

— London Legacy Development Corporation boundary
--- Borough boundaries
— Major road

— Waterway
▨ Green space
▨ New housing

▲ Figure 15.46 The legacy plan for the Olympic Park

What social and economic changes have taken place?

The big promise of the 2012 Olympic Games was a lasting legacy to help regenerate one of the most deprived parts of London. London spent £9.3 billion of public money on the Games so people expected to see long-term social and economic benefits.

In 2012 the LLDC was set up to plan regeneration after the Games. It is likely to take until 2030 for the process to be completed, so it is still too early to judge how successful the changes will be.

However, the Athletes' Village (now East Village) has already been converted into new homes (see Figure 15.47). By 2030, another five new residential communities are planned, with a further 8,000 new homes, turning Queen Elizabeth Olympic Park into a new part of London.

East Village – a new community in east London with a new postcode, E20.

- 2,800 new homes, half for private rent and half for affordable rent.
- A range of homes from one-bedroom apartments to four-bedroom town houses.
- The site occupies 27 hectares, including 10 hectares of park and public open space.
- 35 small independent shops, cafes, bars and restaurants, a supermarket and a gym.
- A new school for 1,800 students aged from 3 to 18.
- Close to bus routes, a new local station and Stratford International Station.

▲ Figure 15.47 East Village

→ Activities

1. Study Figure 15.45. Compare the photo with Figure 15.42 in 15.11. Describe how the environment of the Lower Lea Valley has changed.

2. Study Figure 15.46.
   a Classify the changes in the Olympic Park into social, economic and environmental changes. You can include other ideas from these two pages.
   b Draw a table to list the three types of changes under the correct heading.

3. Study Figure 15.47.
   a East Village was built on a brownfield site. What are the advantages of this?
   b What are the benefits of East Village for the residents who live there?

→ Going Further

Compare Olympic regeneration with Docklands regeneration in another part of east London (see 16.5). The Olympics is an example of partnership regeneration, involving both public and private investment. The Docklands regeneration was an example of market-led regeneration, involving mainly private investment.

1. In your view which project has brought, or is likely to bring, more benefits to:
   a local people?
   b large companies?
   c people in the rest of London and the UK?
   d the environment?
      In each case, give reasons for your view.

2. What lessons do you think could be learnt for future regeneration projects in the UK?

16 Sustainable development of urban areas

Urban sustainability

What impact do cities have on the environment?

A sustainable city is one that can meet its needs without making it more difficult for future generations to meet their needs. Cities put pressure on the natural environment by using inputs, like food, water and energy and, at the same time, by producing outputs, like waste and pollution (see Figure 16.1).

Despite this, living in a city can be more sustainable than living in the countryside. In cities:

- people need to make fewer road journeys because everything they need is closer
- careful planning of things like public transport helps to save resources
- people work together to generate ideas or produce goods and services that benefit the economy.

Food – Most of it is grown outside the city on farms, or is imported.

Water – It is taken from rivers or from below ground and stored in reservoirs.

Energy – Most energy comes from burning fuels that are drilled or mined.

Other resources – building materials, like timber and concrete, plus other resources we consume.

Waste – A lot of it ends up in landfill sites outside the city, or is burnt.

Sewage – It is treated in sewage works before it is returned to a river.

Pollution – It can spread beyond the city in the air or water.

▲ Figure 16.1 A city has inputs and outputs

How large is our urban ecological footprint?

One way to think about the impact of cities on the environment is their **ecological footprint**. This is the area of land or sea that is needed to produce all the inputs a city uses and to dispose of its outputs.

A city's ecological footprint is always much larger than the city itself. In the case of London, it is estimated that each person uses six global hectares (gha). That means London's total footprint is about twice the size of the UK! In reality, London's footprint spreads globally to all the places where its inputs come from and where its outputs end up.

How could cities become more sustainable?

Many cities in the UK and around the world are taking initiatives to be more sustainable. These initiatives can include:

- recycling more waste
- improved public transport
- more green spaces
- local energy schemes (see Section 16.2)
- better cycling routes (see Section 16.3)

In 2004 the UK government devised a framework for thinking about sustainable communities (Figure 16.2). It included social, economic, political and environmental aspects of sustainability.

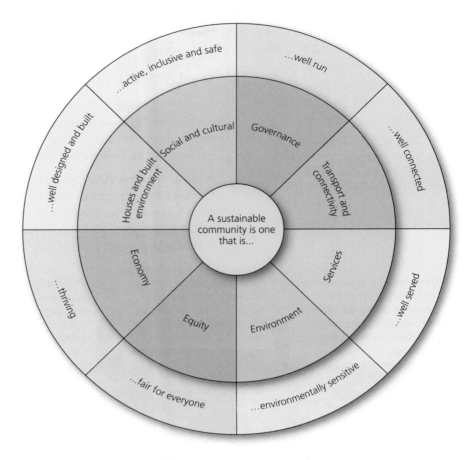

▲ Figure 16.2 A framework for sustainable communities

→ Activities

1 Do you think that cities are sustainable places to live? Give at least three reasons to support your opinion.

2 Look at Figure 16.1.

  a) Draw a simple diagram of a city like Figure 16.1. List the inputs and outputs.

  b) Now, draw a model of a more sustainable city that recycles, with reduced inputs and outputs. List the things that could be recycled.

3 a) Identify at least five things that contribute to a city's ecological footprint, (e.g. dumping waste in landfill sites). Suggest why London's footprint is so large.

  b) Suggest at least five ways in which cities could be more sustainable (e.g. recycling more waste).

👣 Fieldwork: Get out there!

How sustainable is my community?

- Look at Figure 16.2. Make a list of the features you would expect to find in a sustainable community. List them under these headings – Transport, Services, Environment, Economy, Buildings and Social. Think of at least two features under each heading.

- Design a sheet you could use to assess sustainability in your community. List sustainable and unsustainable features on each side, with spaces to score between +2 and –2, like this:

Sustainable features	+2	+1	0	–1	–2	Unsustainable features
Transport Close to a train station						Transport Far from a train station

- Use your sheet to assess your community. Give a score for each pair of features. Work out a total score.

⭐ KEY LEARNING

➤ What makes East Village a sustainable community

➤ How East Village minimises the use of water and energy

➤ The green spaces that have been created in East Village

Sustainable urban living

What makes East Village a sustainable community?

One of the more **sustainable urban communities** in the UK is East Village, part of the Olympic legacy in London (see Section 15.12). You may remember that it was built as the Athletes Village for the 2012 Olympics and then converted into new homes after the Games.

A key aim of the 2012 Olympics was for London to be 'the most sustainable Games ever'. Structures built for the Olympics were planned to have a long-term function after the Games. In the case of East Village, it was to provide 2,800 new homes for both newcomers and local residents. It was built to high standards of sustainability (see Figure 16.3).

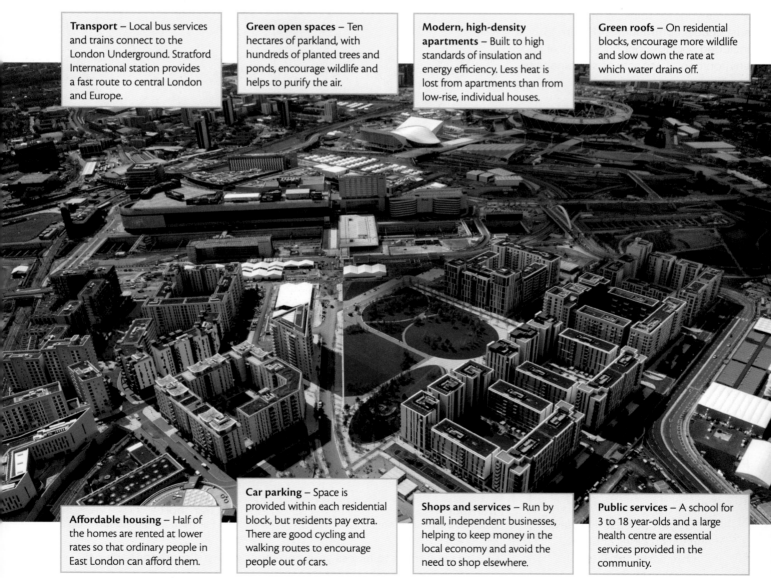

Transport – Local bus services and trains connect to the London Underground. Stratford International station provides a fast route to central London and Europe.

Green open spaces – Ten hectares of parkland, with hundreds of planted trees and ponds, encourage wildlife and helps to purify the air.

Modern, high-density apartments – Built to high standards of insulation and energy efficiency. Less heat is lost from apartments than from low-rise, individual houses.

Green roofs – On residential blocks, encourage more wildlife and slow down the rate at which water drains off.

Affordable housing – Half of the homes are rented at lower rates so that ordinary people in East London can afford them.

Car parking – Space is provided within each residential block, but residents pay extra. There are good cycling and walking routes to encourage people out of cars.

Shops and services – Run by small, independent businesses, helping to keep money in the local economy and avoid the need to shop elsewhere.

Public services – A school for 3 to 18 year-olds and a large health centre are essential services provided in the community.

▲ Figure 16.3 East Village: a sustainable urban community

What green spaces have been created in East Village?

East Village is a high-density urban area, yet there are ten hectares of green open space within a total area of 27 hectares. This is equivalent to the proportion of green space in London as a whole (see Section 15.7). It has:

- a wetland area with ponds where water is recycled, surrounded by parkland
- a large central park and an adventure play area for children
- green roofs on top of apartment blocks
- shared private green space within each apartment block
- an orchard with fruit trees and a children's play area.

How efficient is East Village?

East Village uses less water and energy than most urban areas.

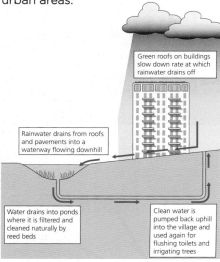

▲ Figure 16.4 Water recycling in East Village

Water

Water use is 50 per cent less than an average urban area. This is achieved by recycling water within the area (Figure 16.4). Rainwater is filtered and cleaned naturally in ponds before being recycled for toilet flushing and irrigating plants. Drinking water is part of a separate system.

Energy

Energy use is at least 30 per cent less than an average urban area. This is achieved by using a **combined heat and power (CHP)** system (Figure 16.5). CHP is more efficient because it generates electricity and produces heat from the same source of energy – in this case,

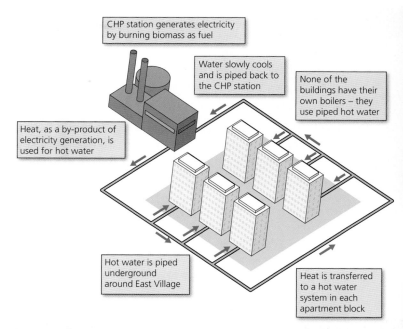

▲ Figure 16.5 A combined heat and power system in East Village

burning biomass. CHP systems only work on a local scale because hot water can only be piped a few kilometres underground before it loses heat.

→ Activities

1 Look at Figure 16.3.
   a) Identify the sustainable features of East Village.
   b) List them under the same headings as in the framework for a sustainable community in Figure 16.2 on page 243.
   c) In what ways is East Village a more sustainable community than where you live?

2 Look at Figure 16.4.
   a) Explain how water recycling works in East Village.
   b) How does this reduce water use by 50 per cent?

3 Look at Figure 16.5.
   a) Explain how a combined heat and power system works.
   b) How does this reduce energy use by 30 per cent in East Village?

4 Draw a sketch of one section of East Village from Figure 16.3. Annotate your sketch to show the types of green space in the village.

Sustainable urban transport

Why does Bristol need a sustainable urban transport strategy?

Like many cities in the UK, Bristol is developing a more sustainable urban transport strategy. As a growing city with a densely populated historic centre, transport is a key issue. Thousands of daily journeys are still made by car. Reliance on cars is leading to traffic congestion, poor air quality and ill health (see Section 15.10), as well as making the streets less friendly to people.

Cycling is easy, cheap and pollution-free (Figure 16.6). The number of people cycling has doubled in Bristol over ten years, but the city still has a long way to go to achieve the levels of cycling in some European cities, like Copenhagen or Amsterdam. To help with this, there is now a Bristol Cycling Strategy.

▲ Figure 16.6 An advert for cycling in Bristol

What are the benefits of cycling?

Cycling has many social, economic and environmental benefits, each contributing to a more sustainable city (Figure 16.7).

How could Bristol's cycling strategy be applied to other cities?

People cycle in cities for different reasons:

- To get into the city centre for work or shopping
- To go sight-seeing or shopping around the city centre
- To travel to work, school or local shops within a zone in the city
- To come in or out of the city from surrounding places

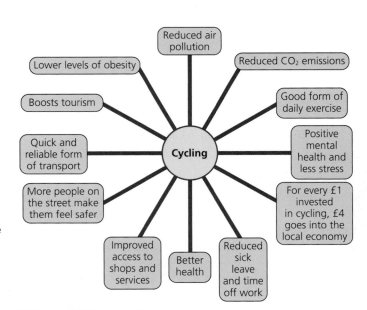

▲ Figure 16.7 The benefits of cycling

The Bristol Cycle Strategy includes all of these types of journey (Figure 16.9). It provides a network of cycle routes, both on-road and off-road, throughout the city. Wherever possible, cyclists are directed along quiet routes that avoid heavy traffic. When this is not possible, for example in the city centre, street space is split to keep traffic and cyclists apart (see Section 15.10).

Other urban transport strategies in Bristol

In addition to the cycling strategy, several other strategies exist in Bristol:

- a metro-style rail service linking Bristol with other nearby towns by reopening old railway lines (**MetroWest**);
- a new generation of rapid transit buses to improve journey times to Bristol (**MetroBus**);
- a network of charging points for electric vehicles at car parks (**Source Bristol**);
- **20 mph limits** in neighbourhoods across Bristol, to make streets safer for cyclists and pedestrians;
- and three **Park and Rides** around the city where visitors can park their cars and travel into the city centre by bus.

▼ Figure 16.8 Cycling in UK cities

City	Residents who cycle at least once a month (%)
Bristol	19
Newcastle	16
London	15
Manchester	15
Leeds	13
Sheffield	13
Birmingham	12
Liverpool	12
Nottingham	12

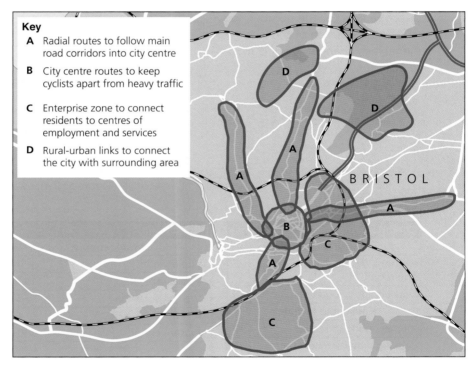

Key

A Radial routes to follow main road corridors into city centre

B City centre routes to keep cyclists apart from heavy traffic

C Enterprise zone to connect residents to centres of employment and services

D Rural-urban links to connect the city with surrounding area

▲ Figure 16.9 The Bristol Cycle Strategy

→ Activities

1. Look at Figure 16.6. Do you think cycling should be part of a sustainable transport strategy? Give reasons.

2. Look at Figure 16.7. Classify the benefits from cycling into social, economic and environmental benefits. List them in a table.

3. Look at Figure 16.8. Compare levels of cycling in UK cities. Can you think of any reasons they should differ? Make a list.

→ Going further

Plan a cycling strategy for your town or city. If you live in a large city, you could select one part of the city.

- Think of the types of journeys people make in your town or city. These could be similar to Bristol or different.
- Draw cycle zones onto a map of the city (see Figure 16.9) to show where people make these types of journeys.
- Explain how you would encourage more people to cycle on routes in these zones.

Question Practice

Unit 2 Section A

1 Study Figure 13.1.

    a) Define the term 'megacity'.   [1 mark]

    b) Name one megacity in Africa.   [1 mark]

    c) Name one megacity in Asia.   [1 mark]

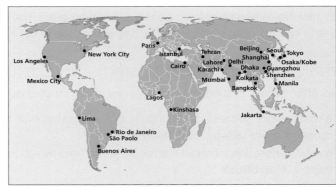

▲ Figure 13.1 The world's megacities in 2015

2 Compare Figure 13.1 with Figure 13.2.

  Describe how (a) the number and (b) the distribution of the world's megacities changed from 1975 to 2015.   [2 marks]

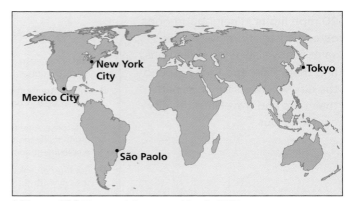

▲ Figure 13.2 The world's megacities in 1975

3 Study Figure 14.14

Condition	%
Housing density	
Households living in more than one room	25
Households living in one room	75
Household facilities	
Kitchen, bath and toilet	10
Lacking either kitchen, bath or toilet	52
No kitchen, bath or toilet	38

Condition	%
Water supply	
Piped water	11
Public tap	14
Well or borehole	55
River	4
Water vendor	16

Condition	%
Toilets	
Septic tank (underground tank in which sewage collects)	10
Pit latrine (sewage soaks straight into the ground)	55
Pail latrine (sewage is poured into a drain or river)	33
Bush (i.e. no toilet!)	2

▲ Figure 14.4 Population growth in Lagos

Outline **two** challenges of living in a squatter settlement from the data.   [2 marks]

4 To what extent do squatter settlements in urban areas of lower income countries (LICs) or newly emerging economies (NEEs) provide opportunities, as well as challenges, for people? Refer to evidence from a city you have studied.   [6 marks]

> The phrase, 'to what extent' is asking you how much you agree or disagree with an idea. If possible, you should include points for and against the idea in your answer.

5   Study Figure 15.8

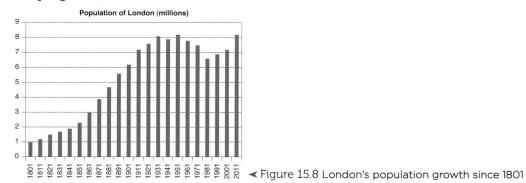

Population of London (millions)

◄ Figure 15.8 London's population growth since 1801

Which **two** of the following statements about London's population are
true? Select two.

    A London's population has grown continuously since 1801

    B London's population declined from 1939 to 1991

    C London's population was five times greater in 2015 than it was in 1801

    D London's population in 2015 was 8.6 million

    E London reached its highest population in 1939 and has declined since   [2 marks]

6   Explain how natural increase has led to population growth in London.   [2 marks]

7   Describe what Figure 15.10 (page 223) reveals about London's ethnic diversity.   [2 marks]

8   Study Figure 15.46.

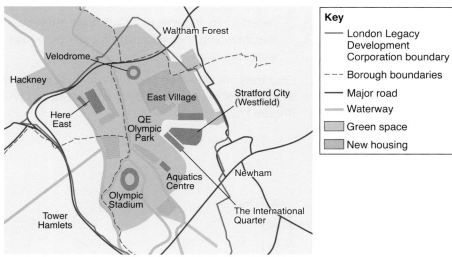

Key

— London Legacy Development Corporation boundary

--- Borough boundaries

— Major road

— Waterway

  Green space

  New housing

▲ Figure 15.46 The legacy plan for the Olympic Park

Suggest how **two** features shown in Figure 15.46 could help regenerate
East London.   [2 marks]

9   Evaluate the impacts of a regeneration project that you have studied.   [9 marks]
    [+3 SPaG marks]

When evaluating, it might help
to consider the social, economic
and environmental effects or
impacts of an event.

SPaG marks are extra marks given for spelling,
punctuation and grammar. However, it is worth writing
clearly, even if the extra SPaG marks are not offered.

17 Economic development and quality of life

⭐ KEY LEARNING

➤ How countries are classified

➤ The world map of development

World development

How are countries classified?

When we talk about a country's level of **development**, we are describing how far it has grown economically and technologically, and the typical **quality of life** (Figure 17.1).

Economic
Income
Job security
Standard of living
(housing, personal mobility)

Physical
Diet/nutrition
Water supply
Climate
Environmental quality/hazards

QUALITY OF LIFE

Social
Family/friends
Education
Health

Psychological
Happiness
Security
Freedom

▲ Figure 17.1 What quality of life means

Low-income countries (LICs)	This group of around 30 countries is classified by the World Bank as having low average incomes (GNI per capita) of US$1,045 or below (2015 values). Agriculture still plays an important role in their economies.
Newly emerging economies (NEEs)	These are around 80 countries that have begun to experience higher rates of economic growth, usually due to rapid factory expansion and industrialisation. Transnational corporations (TNCs) invest in these NEEs, which have subgroups like BRICS and MINTS (see Chapter 19, page 261). The NEEs roughly correspond with the World Bank's 'middle-income' group of countries. The number of NEEs has increased rapidly in recent decades: this is linked to the spread of globalisation.
High-income countries (HICs)	This group of around 80 countries is classified by the World Bank as having high average incomes (GNI per capita) of US$12,736 or above (2015 values). Around half are sometimes called 'developed' countries. These are states where office work has overtaken factory employment, creating a post industrial economy (Figure 17.3). There are also around 40 smaller high-income countries (of roughly 1 million people or less), including Bahrain, Qatar, Liechtenstein and the Cayman Islands.

▲ Figure 17.2 The three main global groups

A country's level of development is shown firstly by the average wealth of its citizens. One way of finding this out is to use a measurement called **gross national income (GNI)**. This is calculated by adding together:

- the total value of all the goods and services produced by its population
- the income earned from investments that its people and businesses have made overseas.

To compare the level of economic development for different countries, the GNI is:

1. divided by the population of the country to produce a per capita (per person) figure
2. this is then converted into US dollars to help make the comparison clearer
3. finally, each figure can be adjusted for each country based on its income. In LICs, goods often cost less, meaning that wages go further than might be expected in an HIC.

Countries are classified into three main groups according to their level of economic development based on GNI (Figure 17.2).

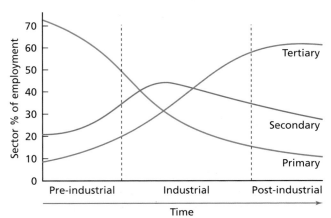

▲ Figure 17.3 The sector model of industrial change over time. Note: LIC, NEE and HIC can be mapped across the three sectors

What does the world map of development look like?

Figure 17.4 shows the world in 2015. The majority of the HICs lie in the northern hemisphere, with the exception of Australia and New Zealand. There are clusters of HICs in Western Europe, North America, the Middle East (including Saudi Arabia, Qatar and UAE) and East Asia (including Japan, South Korea and Singapore).

The distribution pattern for NEEs and LICs is complicated and is changing constantly. Key features are that:

- South American countries are NEEs
- Asia now has more NEEs than LICs
- Africa still has more LICs than NEEs
- Eastern European countries, including some EU members, are mainly NEEs.

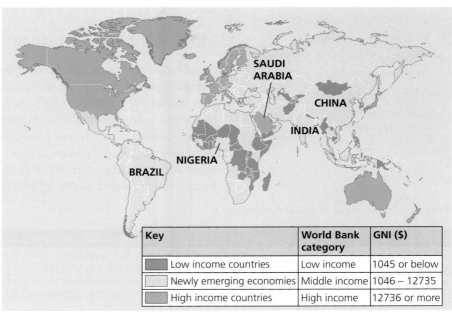

Key	World Bank category	GNI ($)
Low income countries	Low income	1045 or below
Newly emerging economies	Middle income	1046 – 12735
High income countries	High income	12736 or more

▲ Figure 17.4 The world map of development

The global pattern of economic development has changed radically over time. In the 1980s, there was still a clear divide between the rich 'global north' and the poor 'global south'. This division was marked on maps by the Brandt Line (Figure 17.5). This crude division is increasingly of historical interest only, especially when you consider that:

- China is now the world's largest economy
- Several of the world's highest-income countries lie south of the Brandt line (including Qatar, Kuwait and Singapore)
- The GNI per capita of some EU members, including Hungary and Bulgaria, is lower than that of Brazil and Malaysia
- Large numbers of millionaires and billionaires can be found in every populated continent, including Africa (see Chapter 19).

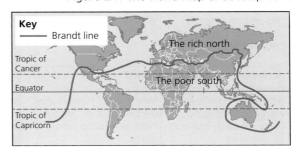

▲ Figure 17.5 How we viewed the world in the 1980s

→ Activities

1 a) How does an LIC differ from an NEE? Create a table and include as many points of difference as you can.

   b) Identify the steps that need to be taken before GNI data can be used to compare the economic development of different countries.

2 Look at Figure 17.4. Describe the pattern of income across the world. Remember to make reference to lines of latitude and the names of the major continents.

3 Look at Figure 17.3. Describe how the importance of different sectors changes over time as a country develops.

4 Look at Figure 17.5. Explain how the geography of world development has changed since the Brandt line was drawn. You could address the following points:

   a) What has happened south of the line in each continent?

   b) Where are today's HICs located?

   c) What variations exist among countries north of the line?

✪ KEY LEARNING

- The reliability of economic development data
- How social development is measured
- Why the human development index is important

Different measures of development

How reliable are economic development data?

Gross national income data sometimes provide a misleading picture of what the typical level of economic development is for a society, especially in LICs. First, the mathematical mean is a very crude way of generating a 'typical' figure. If one millionaire shares a street with 99 people who own nothing, the mean wealth of each person is counted as £10,000. Secondly, people in LICs and NEEs often work very hard, but the value of their efforts is not included in the GNI data. This is because their work consists of either subsistence farming (see Chapter 11) or informal sector work, neither of which are officially recognised.

Additionally:

- Data may not always be accurate: some people may lie about their earnings.
- Data may be hard to collect due to conflict or a disaster.
- The rapid migration of people into cities makes it hard to know how many people live in a place and what they earn.
- All GNI data are converted into US dollars, but the value of currencies changes every day.
- Errors and omissions can creep into the calculations. Some African countries like Nigeria did not include earnings from entertainment and the internet in their official calculations until very recently, meaning that they had under-estimated the value of their economy in some previous years.

Given the World Bank's categorisation of LICs, it is possible that some LICs might really be NEEs, or vice versa.

▼ Figure 17.6 Measures of social development

Measure	Global variations	Limitations of this measure
Literacy rate (the percentage of people with basic reading and writing skills.)	Most EU countries have a literacy rate of 99 per cent. In some LICs, the figure is as low as 40–50 per cent.	Carrying out surveys to determine literacy in rural populations, especially in conflict zones, or squatter settlements in LICs is difficult.
People per doctor (the number of people who depend on a single doctor for their health care needs).	The UK doctor-to-patient ratio is 1:350, whereas in Afghanistan it is 1:5,000.	In India and other NEEs, people in rural areas are now using their mobile phones to get healthcare advice. This is not taken into account by the 'people per doctor' measurement.
Access to safe water (the percentage of people who have access to water that does not carry a health risk such as cholera).	All EU citizens must have access to safe water by law. In rural Angola, just 34 per cent of people have access to safe water.	Water quality can decline due to flooding or poor maintenance of pipes. Rising cost of water in cities sometimes forces poor people to start using unsafe sources. Official data may underestimate these problems.
Infant mortality rate (IMR) (the number of deaths of children under one year of age per 1,000 live births).	The UK figure is just four deaths per year but Somalia has over 100.	In the world's poorest countries, not all infant deaths are recorded. Sadly, many children are buried in unmarked graves. Again, official data may be underestimated.
Life expectancy (the average number of years a person can be expected to live).	Most NEEs now have a high life expectancy of 65–75 years or more. In LICs, a figure of 55 (as in Chad) is more typical. In HICs, this is usually 75+.	In countries where infant mortality is high, the life expectancy of those who survive childhood is actually far higher than the mean life expectancy suggests. Can you see why this is the case?

How is social development measured?

The concept of development is linked with the idea of progress and civilisation: there is more to this than just money! Figure 17.6 shows some of the most important social measures. There is always a strong correlation between social development measures and economic measures like GNI per capita (Figure 17.7).

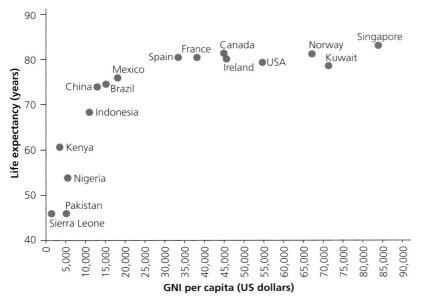

▲ Figure 17.7 Investigating the relationship between economic and social development

Income
The HDI uses a measure of wealth derived from the estimated per capita gross national income of a country

HDI

Life expectancy
This is the numbers of years that a person can be expected to live from birth. Women generally live longest in any society

Education
The HDI uses an education index based on the average number of years of schooling for people in a country

▲ Figure 17.8 The Human Development Index (HDI). In its current form, it has been used since 2010

Why is the Human Development Index important?

All of the measures shown in Figure 17.6 ought to be factored in to calculate a country's overall level of development. There are additional criteria not covered here that some people believe to be very important, such as human rights and even happiness. Some studies rank the world's countries according to their level of political corruption and gender inequality. With so many lists to think about, we use composite measures instead. These combine several development measures into one easy-to-use formula. The most widely used and reliable of these is the **Human Development Index** or HDI (Figure 17.8). The three 'ingredients' are processed to produce a number between 0 and 1. The world's highest-and lowest-scoring countries in 2015 are shown in Figure 17.9.

▼ Figure 17.9 The highest and lowest HDI scores in 2014

HDI rank and score	Country	HDI rank and score	Country
1 (0.944)	Norway	183 (0.374)	Sierra Leone
2 (0.933)	Australia	184 (0.372)	Chad
3 (0.917)	Switzerland	185 (0.341)	Central African Republic
4 (0.915)	Netherlands	186 (0.338)	DR Congo
5 (0.914)	United States	187 (0.337)	Niger

→ Activities

1 What are the advantages of using HDI rather than GNI per capita to measure development?

2 Describe the limitations of four indicators of development.

3 Why might it be difficult or impossible to collect development data for some countries or regions?

4 Look at Figure 17.7. (a) Describe the relationship between the two development indicators. (b) Suggest reasons for the relationship shown in the graph. (c) Why might it be misleading to draw a straight best-fit line on the scattergraph? (To help you with the last part of this question, you could think about what would happen to your own life expectancy if the UK's GNI were to double or triple next year.)

5 Average figures can sometimes give a misleading picture of how developed a place is. Explain why this is the case. Can you suggest an alternative way of showing what is 'typical' rather than calculating the mean average?

Development, population change and the demographic transition model

What are the population characteristics of countries with different levels of development?

Birth and death rates are sometimes used as social development measures. Over time, all of the world's HICs have progressed from having high rates to low rates of both. Based on the historical record of these changes, the population dynamics of LICs and NEEs can be studied in relation to a timeline called the **demographic transition model** or DTM (Figure 17.10).

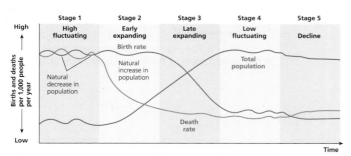

▲ Figure 17.10 The demographic transition model

- LICs are in stage 2 of the DTM. Even the world's poorest countries have experienced a fall in death rate due to global efforts to tackle hunger and diseases like smallpox. In Sierra Leone the death rate fell from 33 people out of every 1,000 per year in 1960s to 17 people out of every 1,000 per year in 2013. Like other stage 2 countries, Sierra Leone's birth rate is still very high at about 37 births per 1,000 people per year

- NEEs are mostly in stage 3 of the DTM. Compared with LICs, far fewer families in NEEs still live by subsistence means. This means that parents do not need to have large numbers of children to help farm the land. Improved health care means that contraception may be widely available. Independent working women in NEEs are choosing to have fewer children. In Bangladesh, the fertility rate is now just 2.2 children per woman on average. In 1970, it was seven per woman! In India, women's lives have changed similarly (Figure 17.11). This is a staggering developmental change in such a short period of time.

These rates do not always correspond with the level of development as we might expect. Occasionally, the death rate rises temporarily due to conflict or a natural disaster. It is also dependent on age. Some HICs are experiencing a rising death rate due to the growing proportion of people aged 80 and over who live there. When it comes to the birth rate, some cultures may be far more resistant to cultural change than others as they develop economically. In general, countries that promote education for women have seen a steep fall in the birth rate. Chile is an example of this.

▲ Figure 17.11 Young independent Indian women are working in call centres in Bangalore, India.

What causes rapid population growth in developing countries?

In an LIC where the birth rate remains high while the death rate has fallen, there is a high rate of natural increase. This is shown in Figure 17.11 by the gap that opens up between the birth rate and death rate: it is widest at the end of stage 2, corresponding with rapid population growth. This steep gradient is the 'population explosion' that every country on the planet has experienced or is experiencing. The UK's population explosion took place in the 1800s. India experienced it between 1950 and 2000.

Global population growth is slowing as more countries gain higher levels of development. It is expected to level off by 2050, at around 9 billion people. Only a minority of the world's countries are still experiencing rapid growth – mostly sub-Saharan countries like Niger and Chad, where the fertility rate remains high, at seven children per woman.

How does rapid population growth impact on development?

▲ Figure 17.13 The UK's growing population built the London Underground

Rapid population growth can lead to overpopulation, where there are too many people for the available land and resources. Symptoms of overpopulation are shown in Figure 17.12.

▼ Figure 17.12 Symptoms of overpopulation and their impact on the development process

Overpopulation symptom	Impact on economic and social development
Falling incomes	High unemployment, out-migration and low wages (because too many people are chasing too few jobs so employers can pay less).
Environmental degradation	Over-grazing of the land and water supplies can lead to soil erosion in areas on the fringes of hot deserts.
Reduced health and happiness	Malnourishment due to insufficient food and the spread of disease (people's immune systems weaken when they are hungry).

Population growth is rarely the sole cause of overpopulation and its problems. When famine occurred in Ethiopia in the 1980s, population growth was a contributing factor. However, the physical cause of drought played a role. So did the ongoing civil war and lack of government assistance. To say that the cause of the famine was 'too many people' is a dangerous over-simplification. Most European countries did not suffer from overpopulation, despite rapid population growth in the 1800s. This was partly because they had extensive overseas empires. As their own populations grew, they took the resources they needed from other countries.

Population growth should not be seen solely as a risk or cost for society. In fact, every HIC and many NEEs have benefited over time from population growth. This is because people are the **human resources** that industries need.

➔ Activities

1 State what natural increase means. Based on information provided on these pages, what is the rate of natural increase in Sierra Leone?

2 Look at Figure 17.10. Describe how the birth rate and death rate change over time as a country develops. Explain what the impact of these changes is on population growth.

3 Briefly discuss the strengths and weaknesses of using the birth rate as a measure of a country's development. Suggest reasons why life expectancy increases when a country begins to develop.

4 Explain how rapid population growth can both help and hinder the development process. To help with your answer, think about:

   • what happens if a place becomes overpopulated

   • possible reasons why population growth is beneficial for some places but not others.

Factors influencing development

What are the historical reasons for varying levels of national development?

Historically, colonialism harmed many countries, and sometimes created conflict which continues today. During the 1700s and 1800s (and earlier in some cases), most of the 'global south' was colonised by European nations such as Britain, France and Spain. Their aims were simple: to build global influence in order to better compete against rival European states and to access raw materials and labour. South American, Asian and African cultures were badly affected, especially those that became part of the transatlantic slave trade.

▲ Figure 17.14 Zatari refugee camp in Jordan

Colonialism ended, for the most part, in the twentieth century. For example, India and Nigeria gained independence from the UK in 1947 and 1960. But independence sometimes creates new problems. When the Democratic Republic of Congo (DR Congo) gained freedom from Belgium in 1960s, there were reputedly just fourteen university graduates amongst its population, so badly had the Belgians neglected the education system. Power struggles often took place in newly independent countries, especially if rich natural resources like diamonds were at stake.

Conflict remains a major development obstacle for some LICs. Persistent political problems stem from the way that ancient African, Asian and Middle Eastern countries were divided and re-assembled in ways by competing European countries. The modern borders of many Middle Eastern and central African countries fit badly with the distribution of different ethnic groups across these regions. Five million deaths have been linked with ethnic conflicts in DR Congo, Uganda and Rwanda in the 1990s. Since the conflict in Syria began in 2012, six million people have lost their homes. Many now live in refugee camps (Figure 17.14), more than half are aged under seventeen and 90 per cent of them no longer receive an education.

How is development affected by economic factors?

In the 1800s, European powers just took the raw materials they needed from other countries. Today, their TNCs buy materials and food from LICs, but at low prices that jeopardise economic development. There are several reasons for these low prices:

■ International organisations like the World Trade Organization (WTO) have been criticised for not doing enough to establish fair terms of global trade for food and raw materials.
■ Sometimes corrupt leaders of LICs, have profited personally from selling resources cheaply to TNCs.

■ Food prices fluctuate wildly depending on competition and the quality of the crops (Figure 17.15). The price of cocoa beans halved in the 1990s due to overproduction, slowing down economic development for Ghana and Ivory Coast.

In contrast, NEEs have benefited from global trade. China's leaders in particular have focused on developing manufacturing industries, resulting in rapid economic growth in recent decades.

What role do physical factors play in the development process?

Physical factors undoubtedly play a role in hindering development. However, it is wrong to ever place the blame entirely on physical factors. Can you name a country with an extreme climate that suffers from frequent earthquakes and hurricanes? One answer could be the USA, an HIC. Physical factors alone do not explain a country's development. All the factors in Figure 17.16 can be overcome with human ingenuity and money.

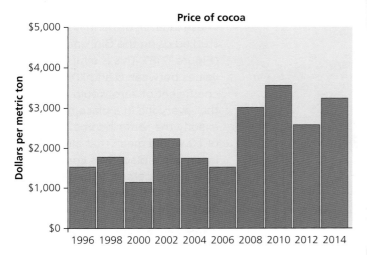

▲ Figure 17.15 World cocoa prices 1996–2014

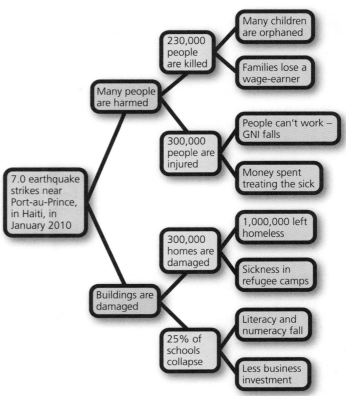

▲ Figure 17.17 How the earthquake in Haiti affected the country's rate of development

▼ Figure 17.16 Physical factors that can impact on development

Factor	How it influences the rate and level of development
Coastlines	There is quite a strong link between the lack of a coast and lower levels of development. With a few exceptions, the world's 45 landlocked countries are LICs or NEEs. Of the fifteen lowest-ranking HDI countries, eight have no coastline. The greatest development challenge is not being able to trade goods easily without ports.
Natural hazards	In 2010, a devastating earthquake struck Port-au-Prince, capital city of the Caribbean island of Haiti. 230,000 people died. Since then, the country has struggled to develop (Figure 17.17). But many countries with a high level of development suffer from hazards too. Japan, Italy and Iceland are all HICs located on plate boundaries (see Chapter 2).
Climate	The influence of climate on development is not very clear. For every poor, hot desert country like Chad there is a rich one like Saudi Arabia. The same is true of tropical rainforest countries: while central Africa's equatorial climate could be viewed as a development challenge, Brazil is the world's eighth largest economy.

→ Activities

1. a) State what is meant by colonialism.
   b) Suggest how colonialism may have contributed to the different level of development of LICs and HICs today.
2. a) Identify two LICs that have suffered from conflict.
   b) Explain two reasons why conflict has occurred in these places.
   c) Suggest why conflict can slow down a country's rate of economic development.
3. Study Figure 17.16.
   a) Rank the three physical factors in order according to how important you think their influence is on the development process.
   b) Give reasons for the order you have chosen.
4. Look at Figure 17.17. Describe the impact of the earthquake on Haiti's economy.
5. Using Figure 17.17 and your own ideas, suggest how the earthquake has affected the long-term social development of Haiti.

The consequences of uneven development

How does uneven development affect the wealth and health of people in LICs and NEEs?

For some groups of people in LICs and NEEs, quality of life is not improving. In some cases it may even have worsened. Newfound trading wealth has helped Nigeria get 'promoted' from LIC to NEE status, but this wealth has not been distributed fairly. Nigeria is now one of the most uneven societies on Earth (see Chapter 19).

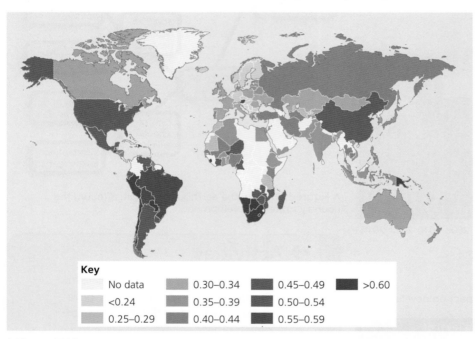

Key

No data	0.30–0.34	0.45–0.49	>0.60
<0.24	0.35–0.39	0.50–0.54	
0.25–0.29	0.40–0.44	0.55–0.59	

▲ Figure 17.18 World map showing variations in the Gini coefficient, 2009

These internal disparities can be studied using the Gini coefficient (Figure 17.18). This is a ratio with values between 0 and 1.0. A Gini coefficient of zero would mean that everyone in a place had exactly the same income. A score of 1.0 would mean that all the income in a place was controlled by a single person. In general, LICs and NEEs have a high Gini coefficient.

Another global-scale consequence of uneven development is how LICs have become dependent on HICs and some NEEs for aid. Many LICs have had to borrow money from the World Bank to pay for hospitals and health care and are now heavily in debt. The shortcomings of health care in some West African countries was shown by the Ebola outbreak of 2014–15, which resulted in over 11,000 people dying from the virus (see Section 19.2). Most deaths were in Sierra Leone, Guinea and Liberia. Sierra Leone's hospitals have deteriorated over time due to the country's low income, indebtedness and its civil war. This reminds us that many of the problems experienced by LICs are connected with one another.

◄ Figure 17.19 Pakistani woman Malala Yousafzai survived being shot. She now lives in the UK and campaigns for social development for women in LICs and NEEs

KEY LEARNING

➤ How uneven development affects people in LICs and NEEs
➤ How uneven development leads to international migration

How does uneven development lead to international migration?

At a global scale, uneven development leads to unequal flows of people between places. Some migrants move voluntarily in search of a better life. These are economic migrants. Others are forced to flee persecution or disasters and are called refugees.

- International migration from poor countries reached an all-time high in 2015. A combination of poverty and conflict in places like Syria and North Africa resulted in a record fourteen million people being forced from their homes.
- Also, people in LICs have become more aware of the **development gap** that exists between themselves and the NEEs and HICs. Despite their low incomes, many people in LICs are finding out about the 'bright lights' of richer places as technology spreads. In 2015, there were seven mobile phones in Africa for every ten people. Increasingly, even poor migrants have mobile phones and can share information.

One highly visible manifestation of people from LICs on the move is the African migrants who have been trying to reach Europe by boat. In the first half of 2015, 50,000 migrants landed in Italy. Many were refugees fleeing conflict and persecution. Thousands more have drowned in the attempt. Later in 2015, hundreds of thousands of migrants began crossing into Europe by land.

It is not only poor and desperate people from LICs who cross borders. The UK receives computer engineers from India and doctors from Poland. These are highly skilled people: the countries they are leaving had invested time and money in training them. This so-called 'brain drain' of skilled human resources could mean that the rate of development of an NEE like Poland begins to slow.

However, there is a positive consequence that stems from migration caused by unequal development. Migrants from poorer countries send home **remittances**. For Nepal, remittances contribute 25 per cent of GNI (Figure 17.20).

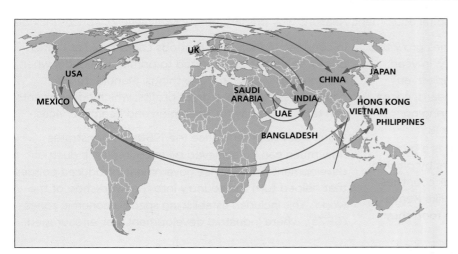

▲ Figure 17.20 Some important global remittance flows

→ Activities

1 (a) Define 'development gap'.
  (b) Suggest why there is more than one global development gap (refer to HICs, NEEs and LICs in your answer).

2 Look at Figure 17.20.
  (a) Define 'remittances'.
  (b) Describe how the flow of remittances connects HICs with other places.
  (c) Explain why these flows are a consequence of uneven global development.

3 Look at Figure 17.18.
  (a) What does a high Gini coefficient indicate?
  (b) Which countries have the highest Gini scores?
  (c) Describe how the Gini coefficient of HICs varies.

4 The 'brain drain', the Ebola outbreak and international migration are three issues raised on these pages. Choose one and write a letter to a newspaper explaining why it is a worrying world issue. Suggest one or more ways in which HICs could take action to help tackle the issue.

18 Reducing the global development gap

▲ Figure 18.1 Banana farmers

Industrial development and investment

Why is industrial development important for poorer countries?

For thousands of years, nations have traded with one another to generate wealth. Very few countries can meet all of their own needs. Not every country has the climate for bananas, cocoa or tomatoes, for instance (see Chapter 23). Some countries lack the raw materials they need for industry, like iron ore and copper, or fossil fuels. In theory, a country's strengths provide it with the opportunity to trade with other countries, thereby generating the income needed for economic development to take place. In reality, this does not always happen.

LICs which only trade in **primary products** (raw materials and agricultural produce) do not always receive a good price. This means they have insufficient money to import important manufactured products from HICs. Development goals are harder to achieve without computers for schools or specialised hospital equipment. Reasons why primary products achieve low prices include:

■ overproduction – when too many countries grow the same crop it pushes down prices globally. In years when crops are good, the problem is made even worse. In some years, prices for coffee beans, cocoa beans or bananas have fallen very low, bringing misery to producer communities (Figure 18.1).

■ import taxes – the European Union (EU) is a group of countries that protects its own farmers by placing import tariffs on food imports from other countries. As a result, farmers in non-EU countries like Kenya find it harder to get a good price for the food they sell to European supermarkets.

In contrast, manufactured goods can be sold at higher prices. Value has been added to primary products when they are processed to make things. Manufacturing companies often make great profits, which governments can tax to help pay for education and health services.

China's development since the 1980s demonstrates how much can be achieved by encouraging industrial development. The Chinese government introduced policies that helped turn the country into the 'workshop of the world'. This included establishing special economic zones (SEZs), where industrial development was encouraged.

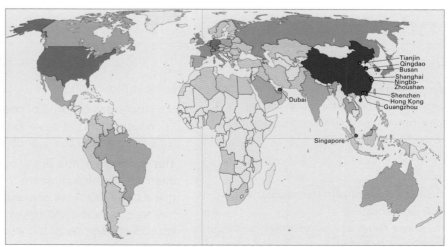

WORLD TRADE
Percentage share of total world exports by value (2013)

Over 10%
5 – 10%
2.5 – 4%
1.0 – 2.4%
0.1 – 0.9%
Under 0.1%
No data

Top ten container ports ●

International trade is dominated by a handful of powerful maritime nations; the members of 'G8' (Canada, France, Germany, Italy, Japan, Russia, UK and USA) and the 'BRICS' nations (Brazil, Russia, India, China and South Africa).

Tianjin
Qingdao
Busan
Shanghai
Ningbo-Zhoushan
Shenzhen
Hong Kong
Guangzhou
Dubai
Singapore

▲ Figure 18.2 Map of world trade

How does investment by TNCs help NEEs and LICs to develop?

In the early 1900s, many of the world's largest companies exported their products but still produced everything in a single place. Over time, companies have grown in size to become **transnational corporations (TNCs)**. They rely less on exporting and prefer instead to produce goods and services inside the borders of many different countries.

The cash injected into other countries by TNCs is called **foreign direct investment (FDI)**. It helps the development process to take place in different ways. Local people are employed to build factories or offices. Other people will work in them. A **multiplier effect** can also develop: investment by a TNC can help other local businesses begin to thrive, creating work for even more people. Increasingly, NEEs have their own TNCs that invest globally too. Investment from Chinese and Indian companies is helping African LICs to develop (Figure 18.4).

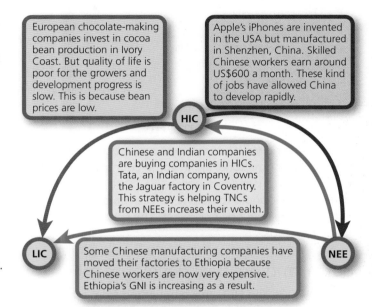

European chocolate-making companies invest in cocoa bean production in Ivory Coast. But quality of life is poor for the growers and development progress is slow. This is because bean prices are low.

Apple's iPhones are invented in the USA but manufactured in Shenzhen, China. Skilled Chinese workers earn around US$600 a month. These kind of jobs have allowed China to develop rapidly.

Chinese and Indian companies are buying companies in HICs. Tata, an Indian company, owns the Jaguar factory in Coventry. This strategy is helping TNCs from NEEs increase their wealth.

Some Chinese manufacturing companies have moved their factories to Ethiopia because Chinese workers are now very expensive. Ethiopia's GNI is increasing as a result.

HIC
LIC
NEE

▲ Figure 18.4 How global patterns of investment can affect development and quality of life

▼ Figure 18.3 Reasons why TNCs invest in other countries and the development this brings

Reducing transport and import costs	Looking for new markets	Looking for cheap labour
By assembling their products close to the people they will be selling them to, companies can reduce transport costs and avoid import tariffs. Guinness brews beer in Nigeria. It is much cheaper to do this than to export barrels of beer all the way from Ireland. This investment brings employment to Nigeria.	Worldwide, over one billion people in NEEs now have a 'middle-class' income and lifestyle. A range of TNCs, including McDonald's, Apple and Ikea, have invested in NEEs by building retail stores there. Globally, McDonald's has invested in 130 countries.	Because the cost of labour is high in HICs, many companies have relocated their operations to other countries to save money. TNCs have invested in the creation of new farms, factories and offices in many LICs and NEEs. While this can help development, workers are sometimes exploited and may have a low quality of life.

→ Activities

1   Using examples, explain how manufacturing industries can help a country to develop.

2   a) Make a list of items in your fridge that may have come from other countries.
    b) Do the same for the clothes in your wardrobe (look at the labels).
    c) Why are so many of the things sold in UK shops imported from other countries (and not made or grown in the UK)?

3   a) What is a transnational corporation (TNC)?
    b) Explain how investment by a TNC helps a county to develop economically and also socially. As part of your answer, think about (i) what workers will spend their incomes on that helps their families and (ii) what a 'multiplier effect' is, and how it works.

4   Look at Figure 18.4. Describe how investment from TNCs is helping to 'bridge' different development gaps.

⊕ KEY LEARNING

➤ How international aid helps development

➤ The role of intermediate technology

➤ The importance of the work of the Fairtrade Foundation

Aid, fair trade and development

How does international aid help development?

International aid is a gift of money, goods or services to a developing country. Unlike a loan, the gift does not need to be repaid. The donor may be a country, or a group of countries such as the European Union. Individuals in HICs give aid to poorer countries by making donations to charities like Oxfam. Most international aid is targeted at specific long-term development goals for people in LICs and some NEEs.

Economic development

Large-scale power and transport projects have been funded by international aid in some countries. In 2014, DR Congo was provided with a US$73 million grant by international donors to help it build a major new dam and **hydroelectric** power project. The scheme involves many new jobs.

Social and political development

■ Many LICs and NEEs have benefited from gifts of money and equipment to help with education. The One Laptop per Child project is part funded by Google. It has helped distribute free laptop computers to hundreds of thousands of children and teachers in South America and Africa.

■ The health of pregnant women in LICs is a spending priority for the United Nations. This issue is linked with a wider aim to improve the political rights of women across the world. A large part of Finland's international aid budget is targeted at helping women.

We can see particular geographical patterns in the way that development aid is distributed internationally:

■ Flows of aid from the UK are directed towards **Commonwealth** countries (see Figure 20.36 on page 300). This is partly explained by the history the UK shares with its former colonies. For example, although India is an NEE, half a billion of its people are still very poor and need help. The UK government also hopes to strengthen its ongoing friendship with nations such as India, now a very powerful country.

■ India and China provide aid to LICs across Africa. India has spent US$6 billion on education projects there. The Tazara railway that links Tanzania and Zambia was funded with international aid from China. The flow of aid from NEEs to LICs is an important new feature of the geography of development.

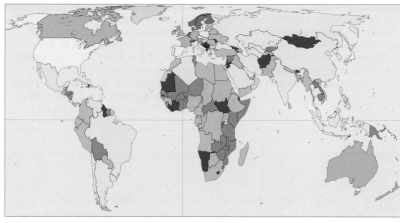

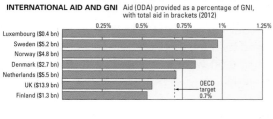

INTERNATIONAL AID AND GNI   Aid (ODA) provided as a percentage of GNI, with total aid in brackets (2012)

Luxembourg ($0.4 bn)
Sweden ($5.2 bn)
Norway ($4.8 bn)
Denmark ($2.7 bn)
Netherlands ($5.5 bn)
UK ($13.9 bn)
Finland ($1.3 bn)

0.25%   0.5%   0.75%   1%   1.25%

OECD target 0.7%

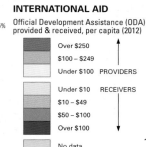

INTERNATIONAL AID

Official Development Assistance (ODA) provided & received, per capita (2012)

Over $250
$100 – $249
Under $100   PROVIDERS

Under $10   RECEIVERS
$10 – $49
$50 – $100
Over $100

No data

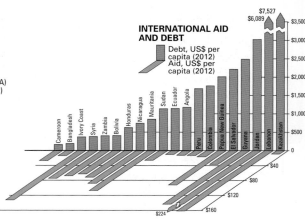

INTERNATIONAL AID AND DEBT

Debt, US$ per capita (2012)
Aid, US$ per capita (2012)

▲ Figure 18.5 International aid and GNI

What role does intermediate technology play in the development process?

Charities and NGOs use **intermediate technology** in LICs and NEEs to help them reach their development goals, such as **sustainable energy** or more efficient ways of cooking. It is technology that the local community can take ownership of and learn to maintain. For example, WaterAid does vital work providing aid for improved water supplies in poor countries, using intermediate technology such as the Afridev hand pump to help provide clean water. Diarrhoea from polluted water accounts for at least 20 per cent of infant deaths in Tanzania. To help tackle this development challenge, WaterAid helped provide the community of Chessa village with a well 24 metres deep, fitted with an Afridev hand pump (Figure 18.6). The villagers can now drink safe underground water.

The Afridev hand pump is not very sophisticated, but when more advanced machines break, a specialist engineer is needed. This can leave local people without water while they wait for repairs. In contrast, an Afridev hand pump will break down fairly often but is repaired quickly by the community. This is seen as the best strategy to help safeguard people's quality of life. It also contributes to the long-term development of Tanzania in two important ways:

- Life expectancy has increased due to fewer deaths from disease.
- Education has improved now children are missing fewer days of schooling due to illness.

How important is fair trade?

The work of the **Fairtrade** Foundation is very important for people in LICs and some NEEs. The aim is to give producers a better price for the goods they produce, and a price guarantee. If the global price for a particular crop like coffee collapses, Fairtrade farmers will still receive their regular income. This protects their quality of life.

Consumers in HICS have increased what they spend on Fairtrade food and goods over time. Examples of Fairtrade produce include chocolate, bananas, wine, footballs and even higher-value clothing items such as jeans. Some people are happy to pay a little more, knowing that a higher proportion than usual of the bill will find its way directly into the pay packets of some of the world's poorest people.

The benefits of the Fairtrade system are clearly shown by the experience of 18-year-old Sameena Nyaz. She lives in a village called Chagelen in the northeast of Punjab province in Pakistan. Sameena stitches Fairtrade footballs for a living. She did not attend school, but now that the family has more money, her younger sisters go. When Sameena needed a serious operation, she was able to afford it. The villagers belong to a health care scheme that is paid for by the Fairtrade scheme.

However, the higher price of Fairtrade products means that many shoppers in HICs avoid buying them, especially during times of economic hardship. This means there is a limit to the number of farms or villages that can be part of the scheme.

▲ Figure 18.6 An Afridev handpump

→ Activities

1. Describe how aid differs from a loan of money.
2. Give two reasons why HICs give aid to LICs.
3. a) What is intermediate technology?
   b) How can it improve people's quality of life?
   c) How does its use in a country lead to an increased rate of development?
4. How can the Fairtrade system help people in LICs and NEEs? As part of your answer, refer to social development measures (see pages 252-3).
5. Do you think all trade should be part of the Fairtrade system? Explain your answer.

KEY LEARNING

- Why many developing countries have suffered a debt crisis
- How microfinance loans are helping the world's poorest people

Borrowing, debt relief and development

Why have many developing countries suffered a debt crisis?

During the period between 1960 and 1980, some HICs loaned many LICs and NEEs staggering amounts of money to develop their countries. The debt crisis began in 1982, when Mexico admitted it had no way of paying back the US$80 billion it had borrowed. The lenders were:

- the World Bank and the International Monetary Fund (IMF) – organisations established after the Second World War to help re-stabilise the world economy. Both are based in Washington DC, USA. They lend money on a global scale to countries that apply.
- large commercial banks – during the 1970s, US and UK banks lent large amounts of money to countries in the developing world. Levels of interest on bank loans were very high at the time.

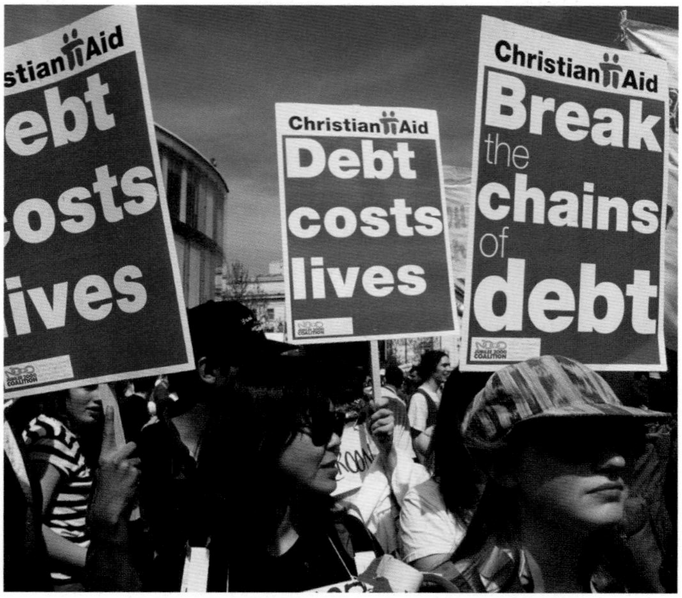

▲ Figure 18.7 The dam on the Nam Theun River

Is it a bad thing for countries to borrow large sums of money? Not necessarily. If the money is invested wisely, it can generate enough wealth to pay back the loan and help the borrowing country to develop too. The World Bank lent Laos US$1 billion to build a dam on the Nam Theun River (Figure 18.7). The dam generates hydroelectric power (HEP). Laos can now earn US$2 billion by selling electricity to its neighbour Thailand over the next 25 years. This will be enough money to repay the loan and increase the GNI of Laos too.

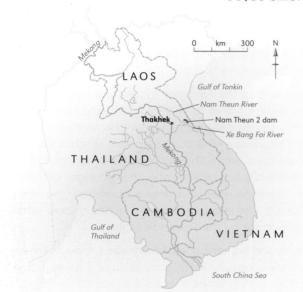

▲ Figure 18.8 A protest march supporting debt relief

The need for debt relief

In some cases, borrowing led to serious problems. For instance, the World Bank funded the speedy modernisation of Indonesia with large loans during the 1970s. Roads, power stations and ports were all built in order to attract investment from TNCs. In the process, a lot of money went missing: it had been siphoned off by Indonesia's ruling family!

In DR Congo, the outcome was even more disastrous. Previous Prime Minister Joseph Mobutu pocketed US$4 billion that had been lent to his country. That money has never been recovered. After Mobuto's death, the World Bank decided it would be unfair to expect the desperately poor people of DR Congo to pay back the stolen money. DR Congo's debt was written off. As a result of pressure from charities and protests, other LICs have been offered **debt relief** too (Figure 18.8).

Debt relief can also be achieved through 'conservation swaps'. A richer country may agree to write off part of a poorer country's debt if that poorer country agrees to protect its physical environment. The USA agreed to let Indonesia keep US$30 million of borrowed money in exchange for increased protection of Sumatran forests, home to endangered rhinos and tigers.

How are microloans helping the world's poorest people?

At a very different scale, poor people in LICs and NEEs borrow small sums of money called **microfinance loans** (Figure 18.9). The most well-known provider of microloans is the Grameen Bank in Bangladesh. It has lent money to nine million people, 97 per cent of whom are women. In contrast to the billions of dollars lent to countries, microfinance loans involve just a few hundred dollars, but can play a crucial role in kick-starting development at a local level. The theory is that if enough villages are helped then, in time, an entire country can develop.

Microloans are needed because subsistence farmers find it hard to escape poverty (Figure 18.10). They can only grow enough food for their own needs, rather than to sell the food. The seeds they use do not always yield good enough crops. The soil they plant them in may not be fertile. This is where microfinance loans come in.

Microfinance loans provide farmers with the vital cash their families need to escape a cycle of poverty. A microloan is not a 'free hand-out'; it must be paid back. One advantage of a small commercial loan like this, when compared with charitable aid, is that poor people feel they can stand on their own two feet instead of being dependent on others.

> A small loan of money is all that is needed to buy better seeds and some fertiliser.

> Within a year, crops are growing so well that the farmers have a surplus that can be sold at a market.

> The profit is then divided between the farmers and the Grameen Bank. Over time the entire loan is repaid.

> The farmers can use their share of the profit to pay for their children to be educated.

> Family healthcare needs can also be met once there is money to pay for medicine.

⋏ Figure 18.9 Microloans flowchart

→ Activities

1. a) What is debt relief?
   b) Why can some countries not pay off their debts?
   c) How can debt relief help countries to develop?

2. What are microfinance loans?

3. Describe how microfinance loans help improve the quality of life of people in LICs.

4. Explain how a 'conservation swap' can benefit people and the environment.

5. 'Microfinance loans are too small to make any difference to how a county develops.' What is your view on this statement? Do you agree or disagree? Write a response to this that looks at both sides of the argument before arriving at a final judgement.

⋏ Figure 18.10 A subsistence farmer in Ethiopia harvests his onion crop

⊛ KEY LEARNING

➤ Tunisia as a popular tourist destination

➤ The impact of tourism on Tunisia's development gap

➤ The sustainability of tourism

Tourism in Tunisia

Key facts

■ In 2013, tourism brought 6.2 million people to Tunisia.

■ This North African NEE had a GNI per capita of US$4,230 in 2014.

■ Since 1960, life expectancy in Tunisia has risen from 42 to 75, showing social development.

Why has Tunisia become a popular tourist destination?

Before the 1970s, Tunisia was classified as an LIC. Agriculture still made up a very large part of its economy. A series of government reforms helped the country's economy to diversify and grow. A development strategy was introduced that promoted tourism alongside manufacturing industries. Factors that favoured the growth of Tunisia's tourist industry included:

■ climate – the country's northerly coast enjoys a Mediterranean climate, with hot summers and mild, warm winters. Summer temperatures reach 40 °C in the coastal city of Sousse (Figure 18.11), making Tunisia a popular destination for sun-seeking Brits and other Europeans.

■ links with Europe – Tunisia's northern coastline is close to Europe, making it easily accessible. French colonial rule ended in Tunisia in 1956, so French is widely spoken and understood, which attracts French and French-speaking tourists.

■ history and culture – Tunisia contains seven UNESCO World Heritage Sites. These include the ancient remains of the city of Carthage and the El-Jem amphitheatre built by the Romans.

■ physical landscape – Tunisia's physical geography is diverse, ranging from Mediterranean beaches to the Dorsal Mountains and Saharan Desert. Many films,

▲ Figure 18.11 Tunisia

such as the Star Wars films, were shot there. Tourists can go and visit their favourite film locations.

■ cheap package holidays – the Tunisian government worked with private companies like Thomas Cook to develop the country into a tourist destination. Since affordable package holidays were introduced in the 1960s, Tunisia's tourist industry has grown from strength to strength. In 2009, the industry provided 370,000 jobs.

The economic benefits of tourism have spread well beyond the most popular resorts due to the linked multiplier effects. In addition to money spent on hotels, tourism boosts the income of many Tunisian businesses. Visitors buy rugs and other goods in the local *souk* (market), which helps formal and informal businesses to thrive. Benefits have also spread to the agricultural sector, which supplies hotels with food, and taxi companies.

What impact has tourism had on Tunisia's development gap?

Tunisia is now one of the wealthiest countries in Africa. The income of Tunisians quadrupled in the 1970s. Higher incomes quickly translated into longer life expectancy as diet and health improved. Tunisia's government invests almost four per cent of its annual GDP in the health system.

Literacy rates have increased markedly over time, rising from 66 per cent to 79 per cent since 1995. More families can now afford to send their children to school and even university. Tourism has contributed to all of this through job creation, and by helping to connect Tunisia to other places and their cultures. This may help explain changing attitudes towards girls' education. Schooling is now compulsory for girls, and women are entering higher education in increased numbers. Greater equality for women is an important international development goal for any country, and Tunisia has made good progress.

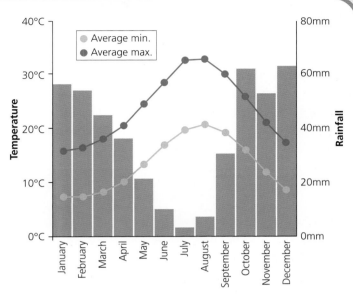

▲ Figure 18.12 The hot climate of Carthage attracts visitors from northern Europe

Is tourism a sustainable development strategy for Tunisia?

There may be limits to how much more the **development** gap can be narrowed by tourism. Areas of concern include:

- pollution of the environment – some of Tunisia's Mediterranean beaches have been polluted with untreated sewage from hotels.
- 'leakage' of profits – foreign companies like Thomas Cook send holidaymakers to Tunisia but keep a large percentage of the profits. This limits how much money becomes reinvested locally and slows down the rate of economic development.
- terrorism – in 2015, there were two terrorist attacks aimed specifically at tourists. One took place in the Museum Bardo in Tunis, and the second at a beach resort in the holiday destination of Sousse. As a result, European governments say Tunisia is no longer a safe destination for their citizens. This sadly means less foreign investment in Tunisia's economy. The terrorists have attacked the tourist industry because they object to some of the ways in which Tunisia is developing, such as greater equality for women.

→ Activities

1 What evidence is there that tourism has helped Tunisia to develop? Refer to both economic development and social development as part of your answer.

2 Look at Figure 18.12.
   a) Using evidence from the climate graph, explain why tourist arrivals might vary from season to season in Carthage.
   b) Suggest why this could hinder the economic development of some groups of people living in Tunisia.

3 In your view, what is the single most important reason for the take-off of tourism in Tunisia? Give reasons for your answer.

4 a) What is meant by 'the development gap'?
   b) Other than tourism, suggest an alternative economic strategy for a country like Tunisia that could help to close the development gap.
   c) Explain how your suggested strategy would help to close the development gap.

19 Economic development in Nigeria

⭐ KEY LEARNING

➤ Nigeria's location and importance in Africa
➤ How Nigeria's population is growing
➤ How Nigeria's economy is growing

Nigeria's place in the world

What is Nigeria's location and importance in Africa?

Nigeria is a country in West Africa that is over three times larger than the UK. It lies just north of the Equator, with its south coast on the Gulf of Guinea, which is part of the Atlantic Ocean.

Nigeria is sometimes known as the 'Giant of Africa'. With a population of 184 million people (2015), its population is much larger than any other African country (Figure 19.1), and almost three times the size of the UK's population.

During the twenty-first century, Nigeria has graduated from being a low-income country (LIC) to becoming a newly-emerging economy (NEE). In 2014, it overtook South Africa as the largest economy in Africa.

How is Nigeria's population growing?

Like many other LICs and NEEs, Nigeria's population is growing fast. It has a high proportion of young people and a high birth rate, and therefore, a high rate of natural increase. In 2015 Nigeria ranked seventh in the world by population, but by 2050 it is predicted to rank fourth, behind India, China and the USA (Figure 19.2).

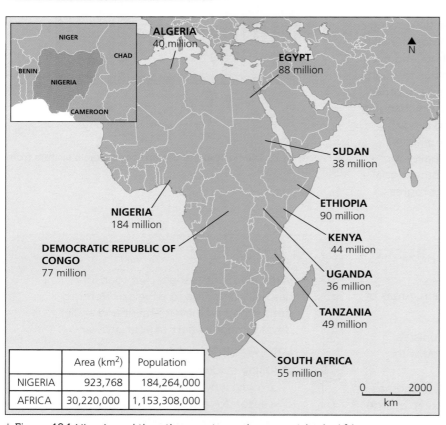

	Area (km²)	Population
NIGERIA	923,768	184,264,000
AFRICA	30,220,000	1,153,308,000

▲ Figure 19.1 Nigeria and the other most populous countries in Africa

▼ Figure 19.2 The world's top ten countries by population in 2015 and 2050

Country	Population 2015 (millions)	Country	Predicted population 2050
China	1,356	India	1,692
India	1,236	China	1,296
USA	319	USA	403
Indonesia	254	**Nigeria**	390
Brazil	203	Indonesia	293
Pakistan	196	Pakistan	275
Nigeria	184	Brazil	223
Bangladesh	166	Bangladesh	194
Russia	142	Philippines	155
Japan	127	DR of Congo	149

How is Nigeria's economy growing?

In 2001, four countries with the world's fastest-growing economies were identified. They are the **BRIC** economies – Brazil, Russia, India and China. In 2014, four more countries, following in the footsteps of the BRICs, were also identified. These are the **MINT** economies – Mexico, Indonesia, Nigeria and Turkey.

Nigeria is now one of the fastest-growing economies in the world (Figure 9.3). By 2020, it is predicted to become one of the world's top twenty economies, and by 2050, it could be ahead of economies like France and Canada (see Figure 19.4).

One reason for Nigeria's predicted economic growth is its youthful population. It has a high proportion of educated young people due to start working in the next twenty years. They will provide the country with a plentiful supply of skilled labour to work in manufacturing and services.

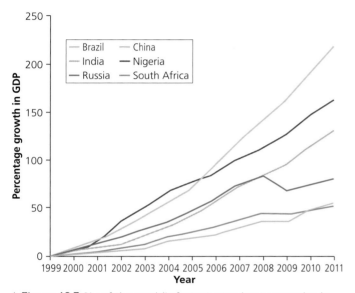

▲ Figure 19.3 Six of the world's fastest-growing economies in the twenty-first century

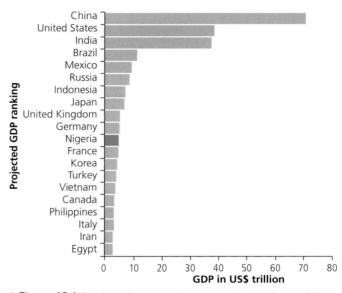

▲ Figure 19.4 Predicted top twenty economies in the world by 2050

→ Activities

1 Study Figure 19.3.

   a) Name the four BRICs on the graph.

   b) How does economic growth in Nigeria compare with that of the BRICs?

2 Study Figure 19.4.

   a) Where is Nigeria predicted to rank among world economies in 2050?

   b) Where will the other BRICs and MINTs rank?

Geographical skills

1 Study Figure 19.1.

   a) Rank the countries labelled on the map by their populations.

   b) Show this information in the form of a suitable graph.

2 a) Work out Nigeria's population density in people/km². Divide its population by the area (a calculator will help).

   b) Do the same for Africa. How does Nigeria's population density compare with population density for the whole of Africa?

3 Study Figure 19.2.

   a) Calculate Nigeria's predicted percentage population growth from 2015 to 2050.

   b) Compare Nigeria's percentage growth with other countries in the table. Will it be faster or slower?

✪ KEY LEARNING

➤ The social and cultural context of Nigeria

➤ The environmental context

➤ How the political context is changing

Nigeria: the geographical context

What is the social and cultural context?

Modern-day Nigeria was formed in the twentieth century under British rule. Until then, the country was comprised of many smaller tribal kingdoms. Nigeria gained independence from Britain in 1960.

Nigeria has more than 500 different ethnic groups, each with its own language. However, three ethnic groups dominate – the Igbo, the Yoruba and the Hausa. The south of Nigeria, where the Igbo and Yoruba live, is predominantly Christian, while the north, where the Hausa live, is mainly Muslim (Figure 19.5).

Rapid urbanisation in recent years has led to a shift of population (see Chapter 13). Rural–urban migration of people from the countryside into cities has broken down some of the traditional boundaries. However, ethnic identities still exist, even within modern cities.

What is the environmental context?

Nigeria is located 5–12° north of the Equator in tropical Africa. Moving north from the Equator, the climate becomes drier and this determines the type of vegetation in each area. Tropical rainforest grows in the hot, humid climate in the south of Nigeria, and savanna grassland in the hot, dry climate further north (Figures 19.6 and 19.7). Much of the natural vegetation in Nigeria has been replaced by agriculture. Cocoa and oil palm are grown in the south, and peanuts are grown in the north.

How is the political context in Nigeria changing?

Since independence in 1960, Nigeria has progressed from civil war (1967–70), through several military dictatorships when the army ruled the country (until 1998), to a stable **democracy** today. The country now holds regular elections, when people vote to choose their government.

However, there is still conflict in Nigeria. In the north of the country, Boko Haram, an extremist organisation, wants to abolish democracy and set up its own government under its version of Sharia (Islamic) law. At least 17,000 people have been killed in the conflict since 2002, and over half a million people have fled the region.

One of the most shocking incidents happened in 2014, when Boko Haram kidnapped 276 schoolgirls as it opposes education, especially for girls. Up to now, the girls have still not been found. The rise of Boko Haram has been blamed on inequality, with extremists able to exploit the growing gap between rich cities and poor rural areas in Nigeria.

▲ Figure 19.5 The main ethnic groups and major cities in Nigeria

Key
- Hausa and Fulani
- Yoruba
- Igbo

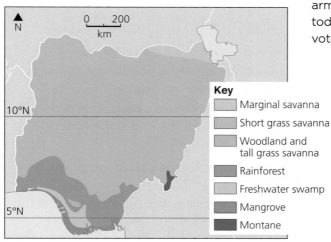

Key
- Marginal savanna
- Short grass savanna
- Woodland and tall grass savanna
- Rainforest
- Freshwater swamp
- Mangrove
- Montane

▲ Figure 19.6 Natural vegetation in Nigeria

Mangrove – grows on tropical coasts	Freshwater swamp – grows inland in flooded areas	Tropical rainforest – grows in hot, wet tropical climate	Woodland and long grass savanna – grows in hot, tropical climate with a short dry season	Short grass savanna – grows in hot, dry tropical climate with a short wet season

◄ Figure 19.7 Changing natural vegetation, south to north in Nigeria

On the other hand, a sign of progress in Nigeria was the way the country dealt with the Ebola outbreak in 2014. This was the world's worst ever outbreak of the disease, killing over 11,000 people (Figure 19.8). Only eight people died in Nigeria, which managed to contain the disease with good healthcare and planning. Every person in Nigeria who came into contact with an Ebola victim was traced and screened for the disease. Other countries later copied Nigeria's method to prevent the disease from spreading.

→ Going further

■ Find out what has happened in Nigeria since 2014.

■ Has Boko Haram killed more people or has the government stopped them? What happened?

■ Has Ebola been eradicated or have any more outbreaks happened? What has been done to prevent any future outbreaks?

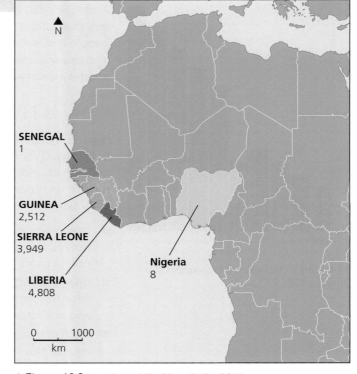

SENEGAL
1

GUINEA
2,512

SIERRA LEONE
3,949

LIBERIA
4,808

Nigeria
8

0   1000
km

▲ Figure 19.8 Numbers killed by ebola, 2014

→ Activities

1 Study Figure 19.5.

a) Describe the distribution of the three main ethnic groups in Nigeria.

b) How might this change with urbanisation?

2 Study Figures 19.6 and 19.7.

a) Describe how natural vegetation changes from south to north in Nigeria.

b) Explain how this change is related to climate.

3 Look at Figure 19.8 and consider the kidnapping of the 276 schoolgirls by Boko Haram.

a) Suggest how both events in 2014 could have been affected by politics.

b) Should the Nigerian government take any of the blame or credit for either of these events, do you think? Explain your ideas.

4 To what extent do you think there has been political progress in Nigeria?

a) Outline the main political events since 1960.

b) Do each of these events demonstrate progress or not? Give reasons for your opinions.

> ## ✪ KEY LEARNING
>
> ➤ How Nigeria's economy is changing
> ➤ The industrial structure of Nigeria
> ➤ The importance of oil and manufacturing to Nigeria's economy

Nigeria's economy

How is Nigeria's economy changing?

As you know, Nigeria has the largest economy in Africa and is among the world's fastest-growing economies (see Section 19.1). Figure 19.9 shows the recent increase in Nigeria's gross domestic product (GDP) – the total value of goods and services the country produces.

On the graph, you may notice a dramatic change in 2011, when GDP more than doubled. Until 2011, many of Nigeria's new industries had not been included in the figures. With the new, more accurate way of measuring the economy, the contribution of manufacturing and service industries increased.

Although GDP has grown, most people in Nigeria are still poor, living on less than US$1.25 a day. There is growing inequality, with a few very wealthy people and a minority of people working in cities in well-paid jobs. There are also regional inequalities, with most wealth in the south around Lagos, and greater poverty in the north and south-east.

How is Nigeria's industrial structure changing?

Nigeria is changing from a mainly agricultural economy into an industrial economy. Over half of the country's GDP now comes from manufacturing and service industries (Figure 19.10). This reflects the change from a mainly rural to an urban population, brought about by urbanisation (see Section 13.2). Some of the fastest-growing industries in Nigeria include:

- telecommunications – in 1990 there were less than a million landline telephone customers in Nigeria. Now, there are over 115 million mobile phone users.
- retail and wholesale – many small businesses that used to be in the informal sector (see Section 14.3) are now part of the formal economy and are included in the calculation of GDP.
- the film industry in Nigeria – known as Nollywood, it is now the third largest film industry in the world, after Hollywood (USA) and Bollywood (India).

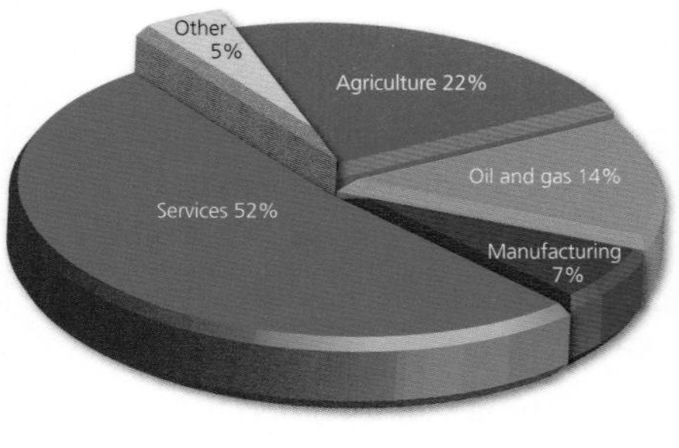

▲ Figure 19.9 Growth of Nigeria's GDP since 2006

▼ Figure 19.11 Employment in different sectors of Nigeria's economy

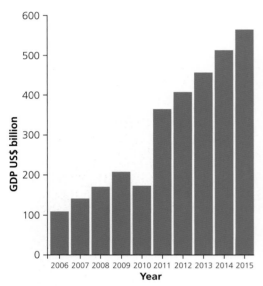

▲ Figure 19.10 Nigeria's GDP by economic sector

Economic sector	% employment
Agriculture	30.5
Manufacturing (including oil and gas)	14.3
Services	
Accommodation, food, transport	12.2
Education, health, science and technology	6.3
Retail, repair and maintenance	24.9
Finance and insurance	4.2
Telecommunication, arts and entertainment	1.8
Other services	5.8

What is the importance of oil and manufacturing in the Nigerian economy?

Oil was discovered in Nigeria in the 1950s and is a vital part of the country's economy (Figure 19.12). Oil and gas account for about fourteen per cent of Nigeria's GDP and 95 per cent of its export earnings. Income from oil has helped Nigeria to make the transition from low-income country to newly emerging economy. However, the country's dependence on oil makes it vulnerable to changes in the world oil price. When oil prices fall, as they did in 2015, it damages the Nigerian economy.

Aliko Dangote is Africa's richest person. He is a billionaire many times over and the founder of Dangote Cement, one of Nigeria's largest companies. The company has three giant cement plants in Nigeria. In a continent that is urbanising rapidly, cement is in high demand and Dangote Cement has expanded into thirteen other African countries. Since the major earthquake in Nepal in 2015, it also has plans to open a plant in Nepal.

New manufacturing industries, like Dangote, are increasing the pace of economic development in Nigeria in several ways:

- Improving the standard of living by the products of industries such as cement.
- Producing manufactured goods in the country reduces the need to import goods and can be cheaper.
- New industries create jobs, give people an income and contribute to the country's wealth through taxes.
- The expansion of Nigerian companies into other countries increases Nigeria's influence in the region.

▲ Figure 19.12 Drilling for oil in Nigeria

→ Activities

1 Study Figure 19.9.

   a) Describe the changes in Nigeria's GDP since 2006.

   b) Explain why an increase in GDP does not mean everybody is wealthier.

2 Study Figure 19.10.

   a) Describe the importance of four sectors in Nigeria's economy.

   b) Suggest how the importance of each sector of the economy is changing.

3 Compare Figures 19.10 and 19.11.

   a) Draw a pie chart for the data in Figure 19.11, showing employment in each sector of Nigeria's economy. Compare it with Figure 19.10.

   b) Try to explain the differences between the two pie charts for agriculture, manufacturing and services. (Hint: think about how technology is used and how much people might be paid.)

4 Look at Figure 19.12.

   a) How has the oil industry helped Nigeria's economic development?

   b) What is the problem with dependence on oil? (You will find out more on this in Section 19.4.)

5 Consider Aliko Dangote, Africa's richest person.

   a) Suggest why Aliko Dangote was in the right place at the right time.

   b) How would the company's expansion into Nepal help (a) Nepal? (b) Nigeria?

⭐ KEY LEARNING

➤ The role of transnational companies in Nigeria's oil industry

➤ The environmental impact of oil in Nigeria

➤ The advantages and disadvantages of transnational companies

The role of transnational companies

What is the role of transnational companies in Nigeria's oil industry?

The oil industry in Nigeria is located in the Niger Delta region, a vast area of wetlands on the delta of the Niger River, where it flows into the Gulf of Guinea (Figure 19.13). The oil boom in Nigeria took off here in the 1970s. It depended on the expertise and money of large transnational corporations (TNCs) based in Europe and the USA, including:

- Royal Dutch Shell (UK, Netherlands)
- Chevron (USA)
- Exxon-Mobil (USA)
- Agip (Italy)
- Total (France).

The companies erected drilling platforms on the oil and gas fields around the Niger Delta region, linked by pipelines to export terminals in the Gulf of Guinea, where the crude oil is piped onto tankers. The oil is shipped to Europe and the USA where it is refined into petrol and other oil products. The companies make most of their profit from refined oil.

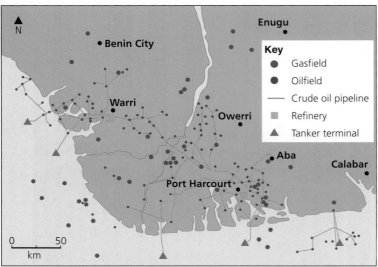

▲ Figure 19.13 The oil and gas industry in the Niger Delta region of Nigeria

When oil was discovered in Nigeria

Chief Sunday Inengite remembers the day the oilmen came hunting for oil in 1953. Members of his small community weren't sure what the men were after. Most thought they were on the lookout for palm oil, which had been exported from West Africa since 1832. Nigeria accounted for 75 per cent of this exported oil by the 1870s. But it became clear this was not the case: Inengite says, 'It wasn't until we saw what they called the oil – the black stuff – that we knew they were after something different.' (93)

While several attempts to find oil had taken place in the early twentieth century, throughout the mid-twentieth century companies discovered numerous oil wells, many along the Niger Delta. By 1959, Royal Dutch Shell-BP had 53 wells operating in the area.

While oil quickly became a lucrative export for Nigeria, it came at a grave environmental cost. 'You see fish floating on the surface of the water, something we didn't know before,' Inengite says. 'It may be difficult to make a catch that will be enough to feed your family for one day.'

▲ Figure 19.14 A news article about oil in Nigeria (adapted from numerous sources)

What is the environmental impact of oil in Nigeria?

Some people believe, rather than being a blessing, oil has become a curse for Nigeria. In order to keep some control over the oil industry, the Nigerian government set up the Nigerian National Petroleum Corporation (NNPC) to form joint ventures with TNCs. This ensures that part of the profit from oil stays in Nigeria.

The oil industry also causes environmental damage. The Delta region contains important wetland and coastal ecosystems. Most people depend on the natural environment for their livelihood, either through farming or fishing.

- Oil spills from leaking pipelines damage farmland so crops no longer grow.
- Gas flares are used to burn off gas from the oil. Apart from being wasteful, the fumes affect people's health and contribute to global warming.
- Oil heated by the Sun becomes highly flammable and can burn out of control.
- Oil pollution, which occurs offshore from tankers, kills fish in the sea.

▲ Figure 19.15 An oil spill on farmland in the Delta region

What are the advantages and disadvantages of TNCs?

▼ Figure 19.16 The advantages and disadvantages of TNCs

Advantages of TNCs	Disadvantages of TNCs
• Bring new investment into the country's economy. • Provide jobs, often at higher wage levels than average in the local economy. • Bring expertise and new skills that the country does not have. • Have international links that bring access to world markets. • Provide new technology that helps economic development.	• Take profits out of the country to pay shareholders or to invest elsewhere. • Wage levels in LICs and NEEs are usually lower than in HICs. • Can cause environmental damage and deplete natural resources. • TNCs can withdraw their investment from a country if they wish. • They are powerful organisations and can exert political influence over the government in a country

→ Activities

1 Study Figure 19.13. Describe the distribution of oil and gas fields in Nigeria.

2 Explain the role of transnational corporations in the oil industry in Nigeria.

3 Study Figures 19.14 and 19.15. Imagine two conversations between Chief Sunday Inengite and an oil company executive, (a) in 1953 and (b) today.

Write a short dialogue for each conversation. What questions would Chief Sunday Inengite ask? What answers would he get?

4 Identify the main advantages and/or disadvantages of TNCs for each of these people:

- an oil worker in Nigeria
- a farmer in the Niger Delta
- a Nigerian government minister.

On balance, would each person be for or against the role of TNCs in Nigeria? Give reasons.

Nigeria looks east

What relationship did Nigeria have with Britain?

Britain has had a trading relationship with West Africa for over 300 years. From 1650, the British traded enslaved African people and took them to America and the Caribbean. When slavery was made illegal in 1807, trade with West Africa turned to palm oil, used in Britain to make soap.

In the late nineteenth century, Nigeria, along with much of Africa, became part of the British Empire (Figure 19.17). The country was ruled by Britain until it gained independence in 1960. By then, a pattern of trade was established where Nigeria exported natural commodities to Britain and, in exchange, imported manufactured goods.

Exports

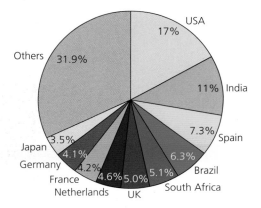

Imports

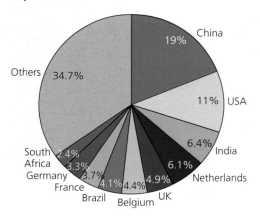

▲ Figure 19.18 Nigeria's main export and import partners

▲ Figure 19.17 Much of Africa was ruled by Britain as part of the British Empire. In this 1892 cartoon, the British imperialist Cecil Rhodes is pictured ruling over Africa, from Cape Town to Cairo.

How are Nigeria's trade relationships changing?

Nigeria still trades with the UK, but more of its trade is now with some of the world's largest economies, which include the USA, China, India and other countries in the European Union (Figure 19.18). Since independence, oil has replaced other natural commodities as Nigeria's main export, but the country still imports manufactured goods like machinery, chemicals and transport equipment.

What influence does China now have on Nigeria's economy?

Nigeria's main import partner for manufactured goods is now China. But China's influence on Nigeria's economy goes beyond the goods it sells. There is also growing Chinese investment in Nigeria and other African countries (Figure 19.19). Both China and Nigeria benefit from this relationship.

- Nigeria needs huge investment in infrastructure, particularly its transport network and power supply (see Chapter 14). China, with recent experience in building its own infrastructure, is now able to bring that expertise to other countries. In 2014, the China Railway Construction Corporation won a US$12 billion contract to build a new 1,400 kilometre railway along the coast of Nigeria.
- China's fast-growing economy needs more resources than the country can provide for itself. It can find these resources abroad, in countries like Nigeria. In 2014, another Chinese corporation agreed to invest $10 billion in exploration and drilling in a new oilfield in Nigeria.

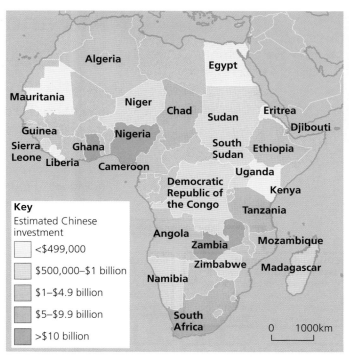

Key
Estimated Chinese investment
- <$499,000
- $500,000–$1 billion
- $1–$4.9 billion
- $5–$9.9 billion
- >$10 billion

0   1000km

▲ Figure 19.19 New Chinese investment in Africa, 2010–2014

→ Activities

1   Study Figure 19.17. What does the image suggest about Britain's relationship with Africa in the British Empire?

2   Study Figure 19.19. Describe the pattern of Chinese investment around Africa. Name the countries with the most investment.

3   Suggest how a new railway along the coast of Nigeria could support economic development. Mention Lagos and the Niger Delta.

Geographical skills

Study Figure 19.18. Complete a map to show Nigeria's trading partners. On an outline world map, draw arrows from and to Nigeria linked to its trading partners, showing (a) exports and (b) imports.

▲ Figure 19.20 CRC will be building bullet trains in Nigeria

⊕ KEY LEARNING

➤ The types of international aid Nigeria receives

➤ How Nigeria got in to (and out of) debt

➤ Whether Nigeria still needs aid or not

Aid and debt

What types of international aid does Nigeria receive?

International aid or 'aid' is help given by one country to another in the form of money, food, technology or advice. Usually the help is from high-income countries (HICs) like the UK to low-income countries (LICs) like those in Africa.

Aid can involve international organisations, governments, charities or individuals (Figure 19.21). As part of the Millennium Development Goals to eradicate extreme poverty by 2015, the UN set a target for high-income countries to commit 0.7 per cent of their GDP as aid. The UK achieved this target in 2013.

Nigeria has one of the highest death rates from malaria in the world. Malaria is a disease transmitted by a mosquito bite that can cause death or long-term health problems. It can easily be prevented by sleeping under a mosquito net at night (Figure 19.22). Each net costs as little as £2.

From 2009 to 2013, 60 million mosquito nets were distributed to households across Nigeria as part of an international aid project funded by the World Bank, IMF and USA government.

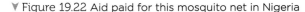

```
                    ┌──────────────────┐
                    │ International aid │
                    └──────────────────┘
          ┌────────────────┴────────────────┐
```

Official development assistance (ODA) is given by governments and paid for by taxes. For this reason, it is sometimes unpopular with taxpayers in those countries.	**Voluntary aid** is given by individuals or companies and distributed through charities and non-governmental organisations (NGOs), like OXFAM.

Multilateral aid is given by countries through international organisations, like the World Bank or International Monetary Fund (IMF).	**Bilateral aid** is given directly by one country to another. Sometimes, it is tied aid, with conditions attached. For example, the recipient may be required to buy goods from the donor country with the aid money.	**Short-term emergency relief** is to cope with immediate problems caused by disasters, like earthquakes and wars.	**Long-term development assistance** helps people to improve their lives through education, healthcare or agricultural development.

▲ Figure 19.21 Types of aid

▼ Figure 19.22 Aid paid for this mosquito net in Nigeria

How did Nigeria get into (and out of) debt?

During the 1980s and 1990s, many low-income countries like Nigeria faced a **debt crisis**. They were unable to repay their debts without cutting essential government spending (see Figure 19.23). While spending cuts in high-income countries, like the UK, can be inconvenient, in low-income countries they can be a matter of life or death.

It was impossible for the low-income countries to escape from the cycle of debt repayment. Although millions of US dollars were repaid, this did not keep up with the interest repayments. So, debt continued to increase. In 2005, leaders of the world's richest countries finally agreed to debt relief for 39 of the world's most highly indebted poor countries (HIPCs), which included Nigeria. Some, or all, of the low-income countries' debt was cancelled, so it no longer had to be repaid.

Does Nigeria still need aid or not?

Since 2005, the Nigerian economy, like the economies of many other African countries, has been growing. Nigeria has graduated from being a low-income country (LIC) to a newly emerging economy (NEE). Given this achievement, does Nigeria still need aid?

The UK gives £300 million a year in aid to Nigeria. Critics point out that Nigeria now funds its own space programme so, surely, it does not need aid. The UK government's response is that Nigeria's space programme is about investment in weather satellites that will help to improve food production. While Nigeria's economy is growing, it should not be forgotten that 60 per cent of the population still live below the poverty line on less than US$1.25 a day.

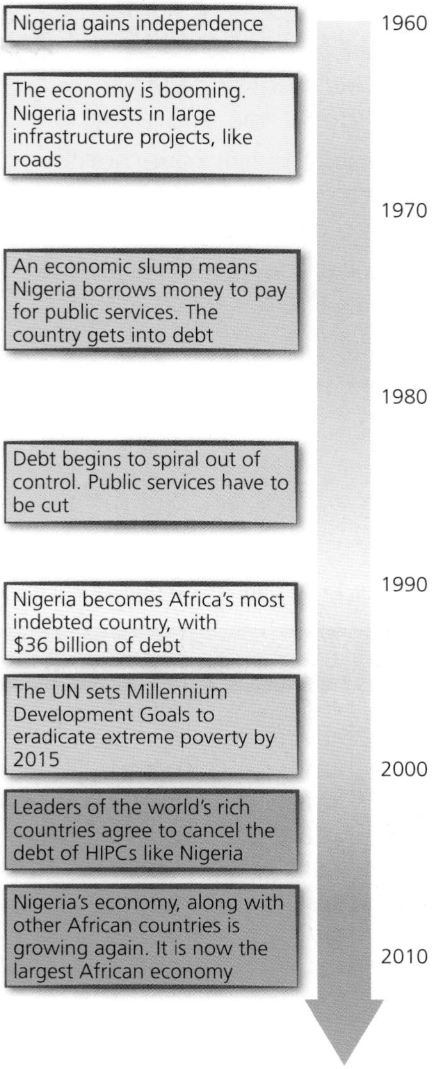

Nigeria gains independence — 1960

The economy is booming. Nigeria invests in large infrastructure projects, like roads

1970

An economic slump means Nigeria borrows money to pay for public services. The country gets into debt

1980

Debt begins to spiral out of control. Public services have to be cut

1990

Nigeria becomes Africa's most indebted country, with $36 billion of debt

The UN sets Millennium Development Goals to eradicate extreme poverty by 2015

2000

Leaders of the world's rich countries agree to cancel the debt of HIPCs like Nigeria

Nigeria's economy, along with other African countries is growing again. It is now the largest African economy

2010

▲ Figure 19.23 Nigeria's debt timeline

→ Activities

1 Study Figures 19.21 and 19.22.

   a) Into which category of aid does the mosquito net project fall? Give reasons.

   b) Apart from saving lives, what other benefits could the project have for Nigeria? Make a list.

2 Study Figure 19.23. Explain the terms (a) debt crisis and (b) debt relief.

3 Which do you think is the most effective way of helping the Nigerian economy to grow – aid or debt relief? Give reasons to justify your answer.

4 Does Nigeria still need aid? Argue the case, using evidence either for or against giving aid to Nigeria.

KEY LEARNING

➤ How quality of life in Nigeria has improved

➤ How the improvements are connected with economic development

➤ Why economic migrants risk their lives to leave Nigeria

Nigeria: development for all?

How has quality of life in Nigeria improved?

Quality of life can be measured using the UN's Human Development Index (HDI) (see Section 17.2). It is a combined measure of three important aspects of peoples' quality of life:

■ Life expectancy as a measure of health

■ Years of schooling as a measure of access to education

■ Gross national income (GNI) per capita as a measure of wealth (Figure 19.24).

Nigeria's is ranked 152 out of 187 countries by its HDI, which puts it in the low category of human development. However, over a longer period, since 1980, there have been significant improvements in life expectancy, years of schooling and GNI per capita in Nigeria.

▼ Figure 19.24 Changes in quality of life in Nigeria since 1980

Year	Life expectancy	Expected years of schooling	GNI per capita	HDI
1980	45.6	6.7	4,259	
1985	46.4	8.6	3,202	
1990	46.1	6.7	2,668	
1995	46.1	7.2	2,594	
2000	48.6	8.0	2,711	
2005	48.7	9.0	3,830	0.466
2010	51.3	9.0	4,716	0.492
2011	51.7	9.0	4,949	0.496
2012	52.1	9.0	5,176	0.500
2013	52.5	9.0	5,353	0.504

How are the improvements connected with economic development?

Nigeria's improved quality of life is connected with the country's economic development. With the new jobs that come with development, people are able to earn more money to pay for the things they need. The government also earns more money through taxes (Figure 19.25).

However, the benefits of economic development are not equally shared. The data in Figure 19.24 are averages. What they do not tell us is how many people are above or below the average. There are large differences between:

■ north and south of the country

■ urban and rural areas

■ educated and uneducated people.

▼ Figure 19.25 The benefits of economic development

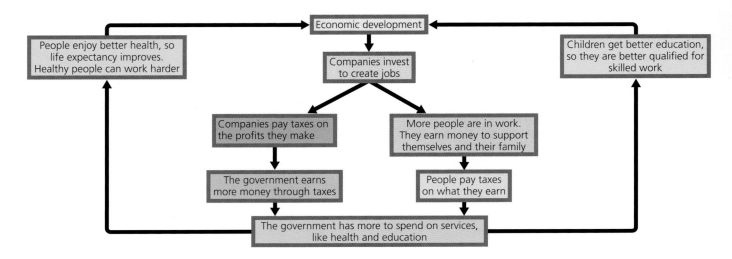

Why do economic migrants risk their lives to leave Nigeria?

In 2014, over 100,000 migrants from Africa and the Middle East crossed the Mediterranean Sea in makeshift boats to reach Europe; 9,000 of them came from Nigeria. Unlike some of the other countries from which the migrants came, Nigeria was not at war. Most of the Nigerian migrants were not refugees, but **economic migrants**, escaping from a life of poverty to find jobs in Europe.

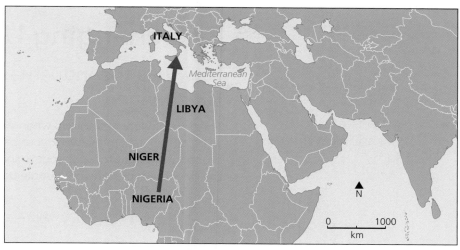

▲ Figure 19.26 The route for migrants from Nigeria to Europe

The journey from Nigeria to Europe is long and hazardous (Figure 19.26). It involves a journey across the Sahara Desert to Libya, a country that was involved in a civil war in 2014. In Libya, migrants pay hundreds of pounds to sail across the Mediterranean in boats that are often unfit for the voyage (Figure 19.27). Over 3,000 migrants died making the crossing in 2014.

The aim for most Nigerian migrants is to earn enough money in Europe to send home to support their families and, perhaps, one day to return to Nigeria to enjoy a better life with the money they have earned.

▲ Figure 19.27 Migrants on the Mediterranean

→ Activities

1 Study Figure 19.25.

   a) Explain how economic development can lead to improvements in quality of life.

   b) Which groups are most likely to benefit from economic development in Nigeria:
      ■ north or south?
      ■ urban or rural?
      ■ educated or uneducated?

      In each case, give reasons.

2 Study Figures 19.26 and 19.27.

   a) What are the risks of a migrant's journey from Nigeria to Europe?

   b) Suggest why many Nigerian migrants believe it is worth the risk.

   c) What does this tell you about economic development in Nigeria?

Geographical skills

1 Study Figure 19.24. Draw suitable graphs to show changes in quality of life in Nigeria, using the data in the table. Draw graphs to show:

   a) life expectancy   c) GNI/capita

   b) years of schooling   d) HDI.

2 a) Write a sentence about each graph to describe what it shows about changes in quality of life in Nigeria.

   b) Explain why HDI in Nigeria increased from 2005.

→ Going further

While this book was being written in 2015, the issue of African migration to Europe was never far from the news. Find out what has happened since then.

■ Do people still migrate across the Mediterranean?

■ Where are they from? How many are Nigerian?

■ Why are they migrating?

■ What measures have been taken to reduce migration and/or make it safer?

20 Economic change in the UK

⭐ KEY LEARNING

➤ How the industrial structure of the UK has changed
➤ The effects of globalisation on the UK economy

The changing UK economy

How has the industrial structure of the UK changed?

The type of work you will do when you leave school or university is likely to be different from the work your parents do. And the jobs they do are probably different to what your grandparents did. The **industrial structure** of the UK – the types of work people do – is always changing (Figure 20.1).

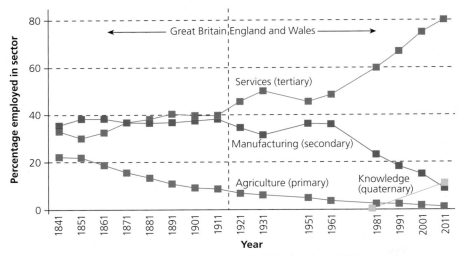

▲ Figure 20.1 The changing industrial structure of England and Wales

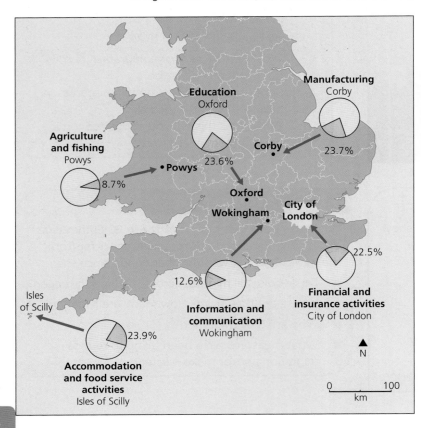

Back in 1841 at the height of the Industrial Revolution in the UK, more people worked in manufacturing than services. Almost a quarter of the workforce still worked in agriculture. By 2011, 80 per cent worked in services, nine per cent in manufacturing and just one per cent in agriculture. However, industrial structure varies around the country, with areas specialising in different industries (Figure 20.2).

◄ Figure 20.2 Places around the UK specialise in different industries

What impacts does globalisation have on the UK economy?

The UK economy is changing due to **globalisation** (see Figure 20.3). This is the way business, ideas and lifestyles spread rapidly around the world. For example, more businesses in the UK are now owned by foreign companies, while, in the same way, more British companies own businesses in other countries.

It would be almost impossible for the UK to be isolated from the global economy. We have been trading with the rest of the world for centuries. For the UK economy to thrive, we need to be part of the global economy.

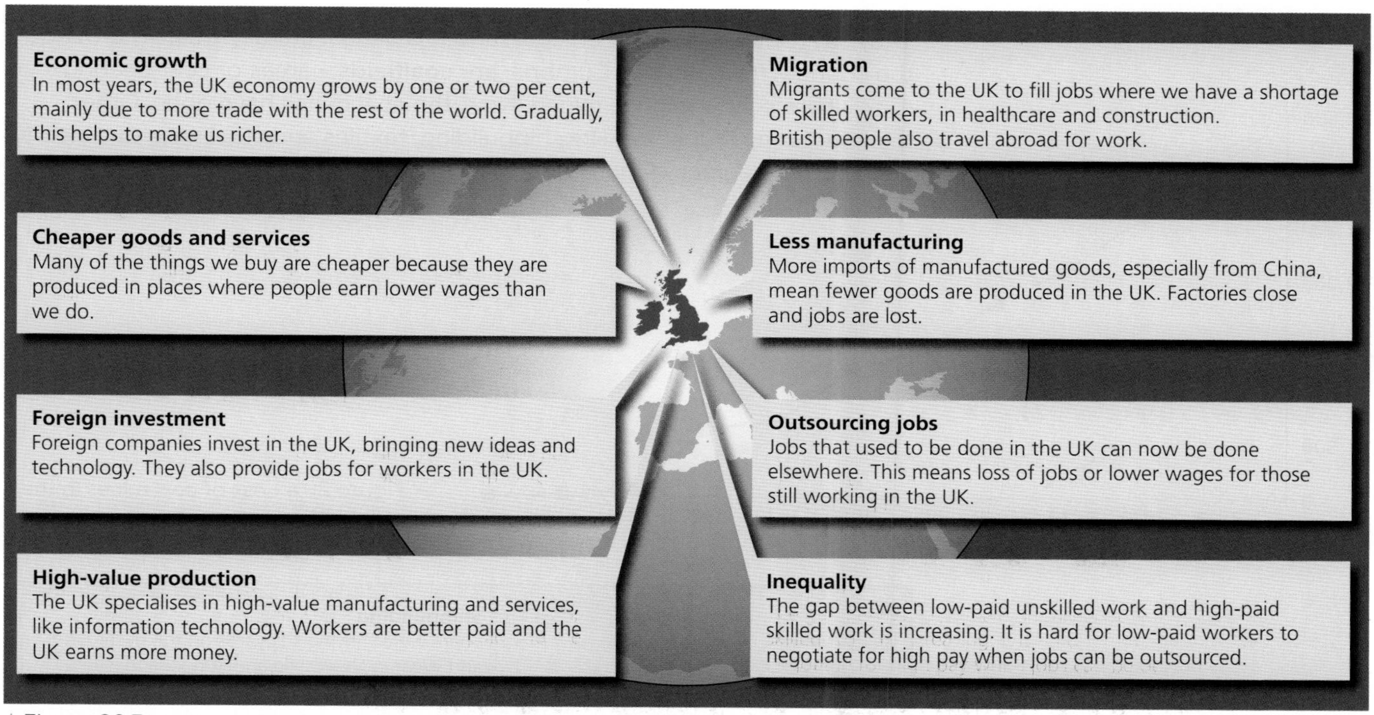

Economic growth
In most years, the UK economy grows by one or two per cent, mainly due to more trade with the rest of the world. Gradually, this helps to make us richer.

Migration
Migrants come to the UK to fill jobs where we have a shortage of skilled workers, in healthcare and construction.
British people also travel abroad for work.

Cheaper goods and services
Many of the things we buy are cheaper because they are produced in places where people earn lower wages than we do.

Less manufacturing
More imports of manufactured goods, especially from China, mean fewer goods are produced in the UK. Factories close and jobs are lost.

Foreign investment
Foreign companies invest in the UK, bringing new ideas and technology. They also provide jobs for workers in the UK.

Outsourcing jobs
Jobs that used to be done in the UK can now be done elsewhere. This means loss of jobs or lower wages for those still working in the UK.

High-value production
The UK specialises in high-value manufacturing and services, like information technology. Workers are better paid and the UK earns more money.

Inequality
The gap between low-paid unskilled work and high-paid skilled work is increasing. It is hard for low-paid workers to negotiate for high pay when jobs can be outsourced.

▲ Figure 20.3 The impacts of globalisation in the UK

→ Activities

1 Think about jobs in your family. Classify them into primary, secondary or tertiary jobs.

- What job do you hope to do?
- What jobs do the adults in your family do?
- What jobs did your grandparents do?

From the three generations in your family, is there any evidence that jobs in the UK are changing? You could carry out a class survey to get a bigger sample.

2 Study Figure 20.1.

a) What percentage of people worked in agriculture, manufacturing and services in 1841?

b) Describe the change in industrial structure from 1841 to 2011.

3 Study Figure 20.2. Most people in the UK now work in services.

a) Make a list of the service jobs people do in different parts of the UK.

b) Think of at least five more service jobs to add to your list. You could include jobs in your area or look at London in Figure 15.22 on page 227 for more ideas.

4 Study Figure 20.3.

a) Classify the impacts of globalisation into benefits and problems. Draw a table to list them.

b) Overall, do you think globalisation is good or bad for the UK? Give reasons for your opinion.

De-industrialisation

How have traditional industries declined in the UK?

Nothing illustrates the story of **de-industrialisation** in the UK better than coal mining. During the twentieth century (Figure 20.4) the number of coal mines in the UK declined from over 3,000 to just 30. The last working coal mines in the UK closed down in 2015.

They say 'Britain was built on coal', and it is almost literally true. All around the country, from Kent to Cumbria, are **coalfields** – areas of coal-bearing rock where coal was once mined. Now, the coal mines have gone. Other manufacturing industries, like shipbuilding, textiles and steel have declined as well in some regions (Figure 20.5). This process is called de-industrialisation.

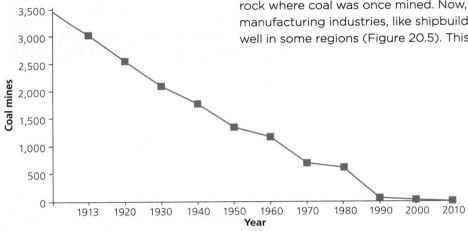

◄ Figure 20.4 Loss of coal mines in the UK since 1913

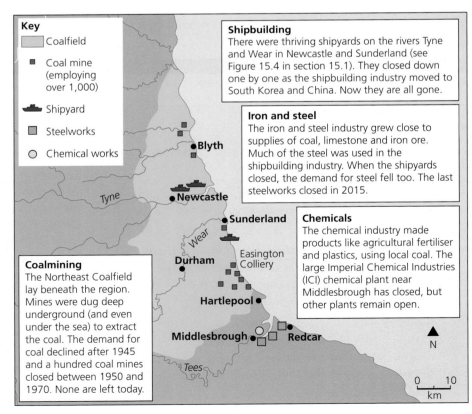

Shipbuilding
There were thriving shipyards on the rivers Tyne and Wear in Newcastle and Sunderland (see Figure 15.4 in section 15.1). They closed down one by one as the shipbuilding industry moved to South Korea and China. Now they are all gone.

Iron and steel
The iron and steel industry grew close to supplies of coal, limestone and iron ore. Much of the steel was used in the shipbuilding industry. When the shipyards closed, the demand for steel fell too. The last steelworks closed in 2015.

Chemicals
The chemical industry made products like agricultural fertiliser and plastics, using local coal. The large Imperial Chemical Industries (ICI) chemical plant near Middlesbrough has closed, but other plants remain open.

Coalmining
The Northeast Coalfield lay beneath the region. Mines were dug deep underground (and even under the sea) to extract the coal. The demand for coal declined after 1945 and a hundred coal mines closed between 1950 and 1970. None are left today.

Key
Coalfield
Coal mine (employing over 1,000)
Shipyard
Steelworks
Chemical works

➤ Figure 20.5 Industry in North East England in the 1960s

What impact has de-industrialisation had on North East England?

North East England was one of the first industrial regions in the UK at the start of the Industrial Revolution. It also became one of the first to experience de-industrialisation with the closure of the coal mines and shipyards.

The impact on towns like Easington Colliery ('colliery' is another name for a mine) has been devastating. The town of Easington grew around its coal mine (Figure 20.6), and when the mine closed in 1993, over a thousand men were left unemployed. Over twenty years later, the town has not recovered. Unemployment is still high and people are on low incomes, so businesses in the town struggle to survive (Figure 20.7).

How has the government responded to de-industrialisation?

Successive governments have tried different strategies to revitalise north east England, such as:

- investment in new infrastructure, including roads and industrial estates
- encouraging foreign investment from large, transnational companies. Nissan, the Japanese car manufacturer opened a new car plant near Sunderland in the northeast in 1986. It now employs 7,000 people
- setting up a regional development agency in 1999, which was replaced in 2012 by a local enterprise partnership which supports businesses, improves skills and plans for economic growth.

▲ Figure 20.6 Easington Colliery when its mine was still open

▲ Figure 20.7 Easington Colliery today

→ Activities

1   Study Figure 20.4. Describe how the number of coal mines in the UK changed from 1913 to 2013. Include numbers and dates from the graph.

2   Study Figure 20.5.

a)   Outline the main industries in the North East in the 1960s. Explain how the industries were linked with each other.

b)   To what extent has the North East become de-industrialised? Which industries have gone and which remain?

3   Look at Figures 20.6 and 20.7. Create a flow diagram to show the impact of de-industrialisation on Easington.

a)   Write a list of the problems the town faces, arranged on a page in your book, like this:

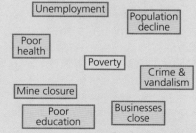

b)   Draw lines to link problems that are related to each other, e.g. mine closure and unemployment. Try to link all the problems to at least one other.

c)   Explain the links by annotating on the lines you have drawn (e.g. people were made unemployed by the mine closure).

Towards a post-industrial economy

Which types of industry are growing in the UK?

A **post-industrial economy** is one where manufacturing industry has been replaced by the **service industry** or tertiary jobs. A new sector of the UK economy that is growing rapidly in the twenty-first century is the **quaternary industry**. The quaternary sector is sometimes described as the 'knowledge economy' because it involves providing information and the development of new ideas. This includes **information technology**, biotechnology and new creative industries.

It is estimated that ten to fifteen per cent of the UK workforce now works in the quaternary sector, though it is often hard to distinguish quaternary from tertiary jobs.

▼ Figure 20.8 Ten UK cities with the most potential for growth

Rank	Cities (outside London)	Potential growth score
1	Cambridge	175
2	Reading	146
3	Manchester	131
4	Bristol	129
5	Oxford	128
6	Brighton & Hove	127
7	Milton Keynes	123
8	Leeds	114
9	Warrington	111
10	Nottingham	107

Where is most economic growth found?

Economists have identified a list of cities outside London that have experienced recent growth and have the greatest potential for future growth (Figure 20.8). Cities are given a score based on:

■ the number of quaternary industries with potential for growth

■ a highly skilled workforce, educated to degree level or above

■ new, start-up businesses with the potential to grow larger

■ good transport connections, including road, rail and air.

These cities are often the focus of **growth corridors**, following major transport routes, where the fastest economic growth is happening (Figure 20.9). While this map shows England, similar corridors could be identified in other parts of the UK, around Glasgow, Edinburgh, Cardiff and Belfast.

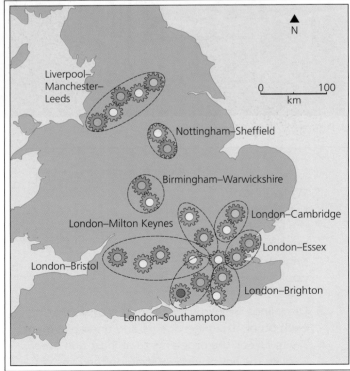

▲ Figure 20.9 Growth corridors in England

How does the M4 corridor contribute to the economy?

The M4 corridor, from London to Bristol, has become home to **hi-tech industry** over the past thirty years (Figure 20.10). Many well-known companies, like Microsoft, Sony and Vodaphone, are based there, usually in modern, out-of-town **business parks** (Figure 20.11). It is estimated that the M4 corridor produces eight per cent of the UK's economic output, as much as Manchester and Birmingham combined.

Recently, businesses in the M4 corridor have begun to be sucked into London. Vodaphone moved its global headquarters to London in 2009, and Google has opted to move to London rather than the M4 corridor. There are a number of factors drawing companies to London:

- the attraction of urban living for a young workforce
- the proximity of similar companies to swap ideas and workers
- new businesses require less space than the first generation of hi-tech industry.

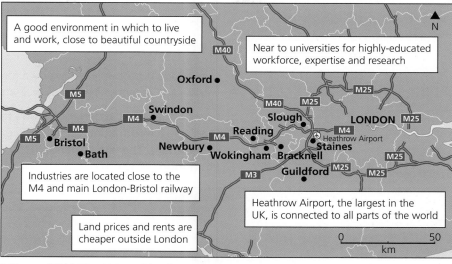

A good environment in which to live and work, close to beautiful countryside

Near to universities for highly-educated workforce, expertise and research

Industries are located close to the M4 and main London-Bristol railway

Land prices and rents are cheaper outside London

Heathrow Airport, the largest in the UK, is connected to all parts of the world

▲ Figure 20.10 The M4 corridor from London to Bristol

▲ Figure 20.11 A business park near Reading

→ Activities

1 Study Figure 20.8.
   a) Which cities have universities that you have heard of? Write a list.
   b) Explain why quaternary industries are often linked with university cities.

2 Study Figure 20.9.
   a) Describe the distribution of growth corridors in England.
   b) Suggest why London is so dominant in a post-industrial economy. (Hint: look back at Sections 15.4 and 15.5.)

3 Study Figures 20.10 and 20.11. Design an advert for a business park in the M4 corridor to attract new business. Think particularly about what your location offers that London cannot provide.

Fieldwork: Get out there!

How are the economic activities in this area changing?

How would you investigate changing jobs in your area? You will need:

- primary data (that you collect yourself through fieldwork) about the workplaces that are there now
- secondary data (that you obtain from another source) about workplaces that were there in the past.

You could classify jobs, both past and present, into primary, secondary, tertiary and quaternary sectors. How has the proportion changed?

➤ Why Cambridge is growing as a hub for hi-tech industry

➤ Cambridge as a location for industry

Cambridge: a hi-tech hub

Why is Cambridge growing as a hub for hi-tech industry?

Cambridge is probably best known for having one of the top universities in the world. Until recently, it was less well-known for its industry, but this is changing.

Cambridge is fast emerging as one the UK's main hubs for hi-tech industry. Over 1,500 information technology and biotechnology companies are now based there. The city lies about 80 kilometres north of London, close to the M11 (Figure 20.12), in one of the UK's growth corridors (see Section 20.3).

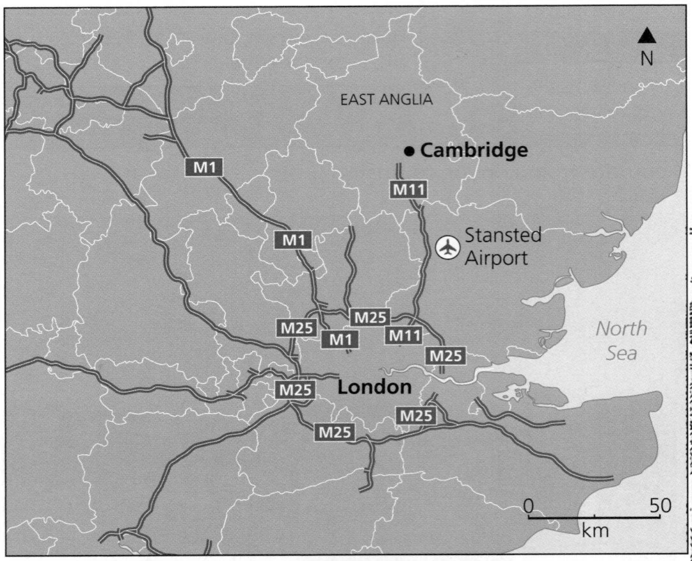

Hi-tech industry in Cambridge began with the Cambridge **Science Park**, in the north of the city (Figure 20.13), which was opened in 1970 by Trinity College. Trinity was the first of several Cambridge University colleges to make links with industry.

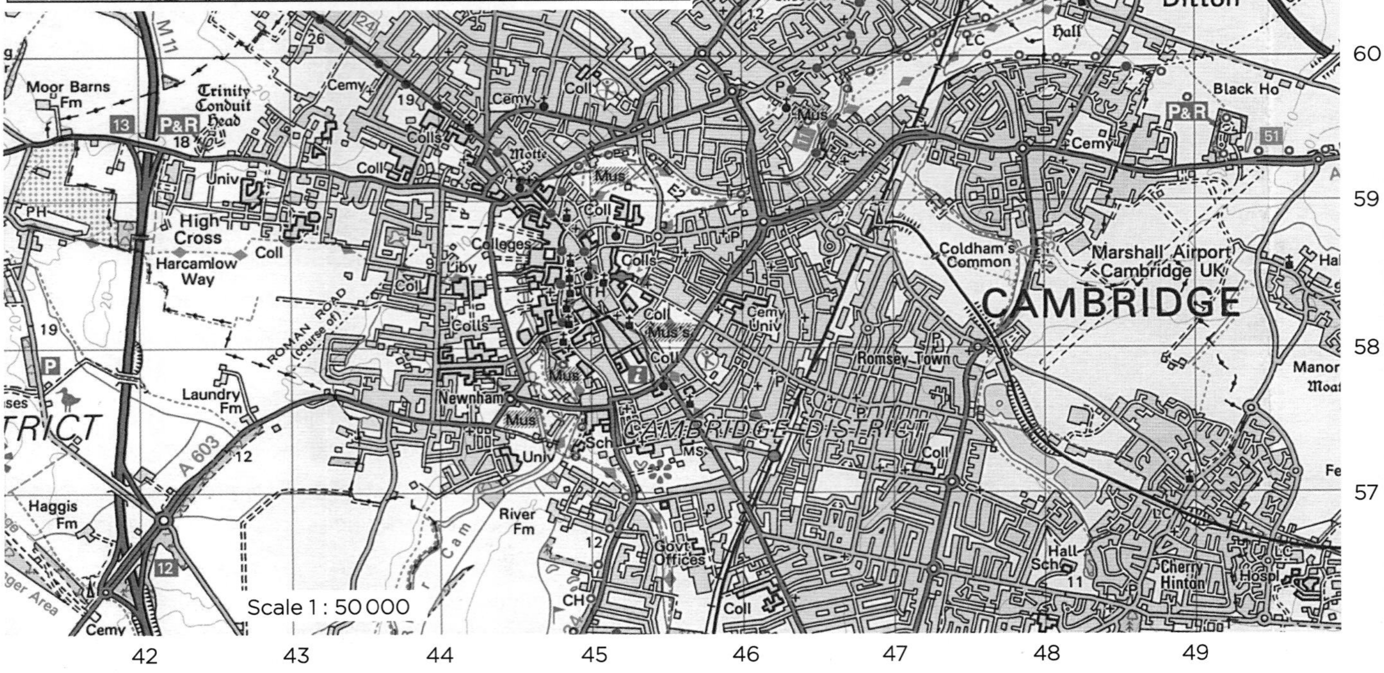

▼ Figure 20.12 The London–Cambridge growth corridor on the M11

▲ Figure 20.13 Ordnance Survey map extract of Cambridge, 1:50,000

What are the advantages and disadvantages of Cambridge as a location for industry?

Many of the hi-tech companies in Cambridge began as small start-up businesses, formed by university graduates who wanted to stay in the city when they finished their degrees. Some of these businesses, like the biotech company Abcam, have grown into successful companies.

Abcam (named for 'antibodies Cambridge') is based at Cambridge Science Park. It produces antibodies that are used in the treatment of diseases. The company is now worth £1 billion and employs 200 staff with PhD degrees – more than some universities!

▼ Figure 20.14 Advantages and disadvantages of Cambridge as a location for industry

Advantages	Disadvantages
• Good transport links, including the M11 motorway to London and Stansted Airport • Graduates from the university provide a highly educated workforce • There are few traditional industries to compete for space, so rents are lower • The city offers a good quality of life, with plenty of shops and open spaces • There are good links between colleges and industry, helping to develop new business ideas	• The city is overcrowded and congested, making it difficult to drive or park • House prices are high and still rising, making it expensive to live there • Road and rail routes need to be improved to speed up connections to other cities apart from London

Geographical skills

1. Study Figure 20.13. Find Cambridge Science Park in square 46 61. Draw a sketch map to show the location of the science park. Include the following features and label them on your map: Cambridge (urban area), M11, A14, city centre, Cambridge Science Park.

2. Find these grid references on the map in Figure 20.13: 470621, 452589, 487585, 480623. For each location:

    a) State what you find there.

    b) Explain why it might be a factor for a company to choose Cambridge as a location for their business.

→ Activities

1. Study Figure 20.14. What would be the main advantages and disadvantages for:

    a) a university graduate looking for a job in Cambridge

    b) a large biotech company thinking of moving to Cambridge.

Fieldwork: Get out there!

Where is the best location for a new business in your area?

- Choose a type of business that might want to locate in your area. It does not need to be a hi-tech firm. For example, you could choose a café.

- Think of the main criteria your business would have for finding the best location. For example, the number of people walking past.

- Select two or three possible locations where the business could be located.

- Carry out fieldwork at each location to find the best one for your business. For example, you could count the number of people walking past at different times of day.

Rural changes

How are rural areas changing?

Most people in the UK live in urban areas, but 19 per cent still live in rural areas. Although they might not look crowded, the population of most rural areas in the UK is actually growing (Figure 20.15) as a result of **counter-urbanisation**. People leave cities to live in the countryside for a better quality of life. The population of urban areas is growing faster, but this is mainly the result of natural increase and immigration.

Around major cities in the UK is the greenbelt – green open space in which further building development is not allowed (Figure 20.16). Within the greenbelt, and just beyond, are towns and villages that are desirable places to live and **commute** to work in the city. Without the greenbelt these areas might have experienced even greater population growth.

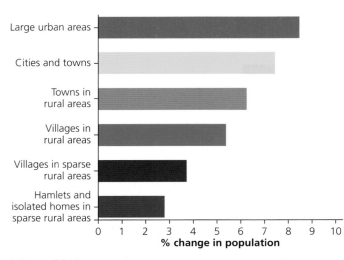

▲ Figure 20.15 Population change in urban and rural areas, 2001–11

While the greenbelt has been successful in preserving rural areas, it has also limited the amount of land available for building new homes. As there is a housing shortage in the UK there is pressure on the government to allow more building in the greenbelt.

Even **sparsely populated** rural areas furthest from cities have experienced population growth. Some of these areas are popular with tourists and second home owners, particularly in **national parks** like the Lake District.

There is high demand to live in both the greenbelt and national park areas. This has pushed up house prices, making homes for local people unaffordable. People are forced to rent locally or to move away to find affordable homes elsewhere.

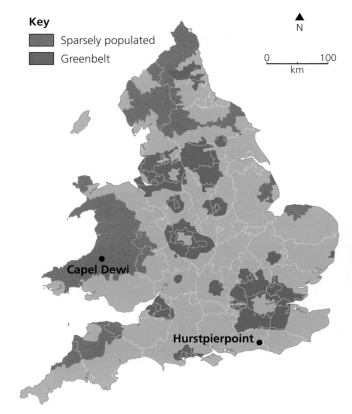

Key
- Sparsely populated
- Greenbelt

▲ Figure 20.16 Sparsely populated and greenbelt areas in England and Wales

What happens in an area of population growth?

The greatest population pressure on rural areas is in South East England, where many people want to live in rural surroundings but work in London (Figure 20.18). Population growth in these areas brings benefits and problems.

▼ Figure 20.17 Benefits and problems of population growth

Benefits of population growth	Problems of population growth
• It maintains the population of small towns and villages as a balance to rural–urban migration. • It brings new energy to rural areas. There is evidence that newcomers are more likely to start their own local business. • It helps to maintain the demand for rural services, like shops and schools, that might otherwise close.	• Older people retire to live in rural areas and this increases the average age. • Newcomers are often wealthy and this helps to push house prices up even further. • The arrival of newcomers and out-migration of local people, is changing rural culture.

The large village of Hurstpierpoint has a high street full of shops and a good pub. Trains from the nearest station, which is two miles away, take 50 minutes to journey to London. An annual season ticket costs £3,504. Properties range from two-bedroom cottages valued at £230,000, to modern four-bedroom houses selling for over £500,000 and period family houses from £650,000.

▲ Figure 20.18 An advert for property in the village of Hurstpierpoint, Sussex

What happens in an area of population decline?

The village of Capel Dewi is in a sparsely-populated area of mid-Wales. During the twentieth century its population declined as young people moved away to find work, leaving older people to continue farming. One by one, shops in the village closed as the number of customers fell until, eventually, none were left.

In 2012 the local community got together to open a new shop. It is a convenience store for the village run by volunteers. It saves people making long car journeys to supermarkets in town and is most useful for older people who do not drive.

▲ Figure 20.19 A house in the village of Hurstpierpoint, Sussex

→ Activities

1 Study Figure 20.15.

   a) Describe the different rates of population growth in urban and rural areas.

   b) Most rural areas of the UK have growing populations. Explain why.

2 Study Figure 20.16. Describe the distribution of, (a) greenbelt, and (b) sparsely populated areas.

3 Locate Capel Dewi and Hurstpierpoint on the map in Figure 20.16.

   a) Compare the locations of the two villages.

   b) How do their locations help to explain the characteristics of the two villages?

4 Study Figure 20.18. What would be, (a) the advantages, and (b) the disadvantages for someone moving to the village and commuting to London. Think of at least three of each.

5 Write a leaflet for people in Capel Dewi telling them why they should support the new community shop.

The UK's North–South divide

Is there a North–South divide in the UK?

In people's minds there has long been a **North–South divide** in the UK. Depending on where you live, you will often hear people talk about 'up north' or 'down south'. But, is there a real North–South divide in the UK, and, if so, does it matter?

Geographers have drawn a line on the UK map to show the North–South divide (Figure 20.20). North of the line:

■ are the hills and mountains of upland Britain

■ is where most manufacturing industry was located until de-industrialisation began

■ there are higher unemployment levels (Figure 20.21)

■ population is growing more slowly as people move south to find work

■ house prices are lower because there is less demand for housing.

South of the line:

■ is the flat, fertile farmland of lowland Britain

■ there was less manufacturing, so de-industrialisation has been less of an issue

■ higher employment levels are found

■ population is growing more quickly as people move here to find work

■ house prices are higher because there is more demand.

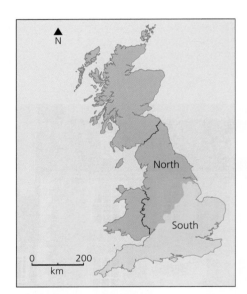

▲ Figure 20.20 The North–South divide in the UK

What are the exceptions to the North–South divide?

The North–South divide is not as simple as the map shows. There are exceptions. For example, although London has a booming economy (see Section 15.5), it still has higher levels of unemployment than other regions in the south (Figure 20.21). There are also inequalities within London (see Section 15.8).

Scotland is in the north but it has lower unemployment than other regions in the north of England. This is partly due to wealth from North Sea oil, but Scotland also has its own government, which has powers to raise and spend more money than the rest of the UK.

There are also individual cities that are exceptions to the North–South divide (Figure 20.22).

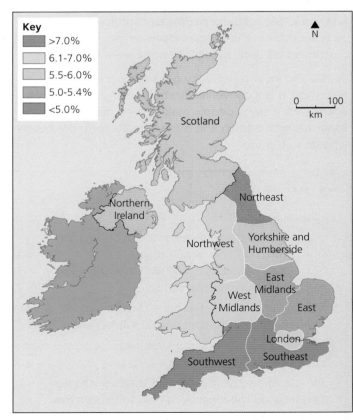

Key
▓	>7.0%
░	6.1-7.0%
░	5.5-6.0%
▒	5.0-5.4%
▓	<5.0%

▲ Figure 20.21 Unemployment in UK regions, 2015

▼ Figure 20.22 Cities with the highest and lowest employment levels in the UK (UK average 71.9)

Rank	City	Employed (%)	Rank	City	Employed (%)
1	Warrington	79.8	55	Bradford	66.4
2	Cambridge	78.9	56	Swansea	65.8
3	Swindon	78.0	57	Hull	64.8
4	Aldershot	78.0	58	Birmingham	64.2
5	Reading	77.2	59	Coventry	63.6
6	Aberdeen	77.1	60	Rochdale	62.8
7	Gloucester	76.8	61	Blackburn	62.6
8	Crawley	76.3	62	Liverpool	62.3
9	Brighton	75.5	63	Burnley	62.1
10	Ipswich	75.2	64	Dundee	61.9

What strategies can be used to reduce differences between the north and south?

Over the years, the UK government has attempted to reduce the North–South divide without much success. In the twenty-first century, the gap between the north and south has widened, with most economic growth happening in the south.

The strategies the government is trying, or hopes to try, include:

- identifying areas of the UK that need special help, called **assisted areas**, to provide money for new business (Figure 20.23)
- improving the transport infrastructure, linking cities in the North, including improvements to the M62 motorway and a proposed new high-speed rail link
- giving more power to individual cities to take decisions on how to raise and spend their own money (as Scotland does).

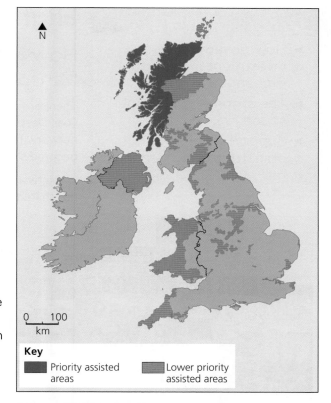

Key

■ Priority assisted areas   ■ Lower priority assisted areas

▲ Figure 20.23 Assisted areas in the UK

→ Activities

1. Look at Figure 20.20. For each of the following cities, say whether they are in the North or the South; Manchester, Sheffield, Bristol, Norwich, Leeds, Birmingham, Plymouth, Cardiff, Milton Keynes. You can check in an atlas.

2. Study Figure 20.21.
   a) Describe the regional pattern of unemployment in the UK. Mention differences between the North and the South and any exceptions to the pattern.
   b) Explain the pattern (Hint: look back at the previous Sections 20.1 to 20.3.)

3. Does the North–South divide matter, and does it matter that most wealth and economic activity are concentrated in the South? Give reasons to support your answer.

Geographical skills

1. Study Figure 20.22.
   a) Mark and label the cities in the correct location on an outline map of the UK. You can use an atlas to help you. Use two colours to mark the cities – one for the top ten and another for the bottom ten.
   b) What pattern do you notice on the map?

2. Study Figure 20.23.
   a) Describe the pattern of assisted areas on the map.
   b) To what extent does the map support or not support the idea of a North–South divide in the UK? Give evidence from the map.

High-speed rail

How is government investment in transport changing?

How government investment in transport is changing

There are over 35 million vehicles on the roads in the UK and this grows each year. Despite government investment over many years to improve the road network, traffic congestion remains a serious problem and journey times become slower rather than quicker (Figure 20.24).

In the twenty-first century, the focus for government investment on transport has shifted to an older form of transport – the railways. In particular, there are plans for a new **high-speed rail** network in the UK. Already, High Speed 1 runs from London to Kent on the same route as the Eurostar from London to Paris. Now, there are plans for High Speed 2 (HS2) from London to Birmingham, then on to Manchester and Leeds (Figure 20.25). The government's most recent road investment strategy, in 2014, is to increase road capacity with over 100 new road schemes by 2020 and over 100 miles of new lanes added to motorways. Additionally, there are plans to improve the M4 motorway by making it a 'smart motorway'. This involves helping reduce congestion, for example, by varying the speed limits to keep traffic moving smoothly.

⭐ **KEY LEARNING**

➤ How government investment in transport is changing

➤ The arguments for high-speed rail

➤ Supporters of and objectors to high-speed rail

▲ Figure 20.24 Congestion on the M25 motorway

What are the arguments for high-speed rail?

The main arguments in favour of high-speed rail, including HS2, are:

- It will take the pressure off the existing road and rail network, encouraging more people to travel by rail.
- It will reduce journey times between cities (Figure 20.26), so people spend less time travelling.
- It will bring economic benefits to the Midlands and northern England where de-industrialisation has led to a loss of jobs.

Even though HS2 is not planned to start running until 2026, there are already ideas for HS3, another high-speed rail route from Manchester to Leeds, to link cities in northern England.

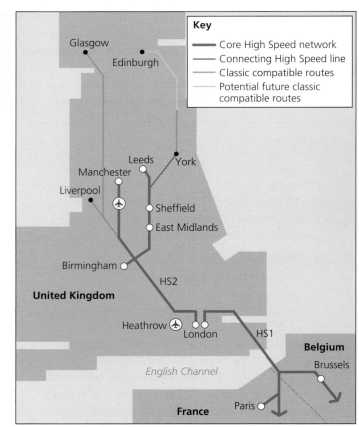

▲ Figure 20.25 The proposed route for High Speed 2

▼ Figure 20.26 Journey times for HS2 compared with current journey times

Journey	Current rail journey time	HS2 journey time
London – Birmingham	1hr 24 mins	49 mins
Birmingham – Manchester	1hr 8 mins	49 mins
Birmingham – Leeds	2 hrs	57 mins
London – Manchester	2hrs 8 mins	1hr 8 mins
London – Leeds	2hrs 12 mins	1hr 22 mins
London – Glasgow	4hrs 30 mins	3hrs 30 mins
London - Edinburgh	4hrs 30 mins	3hrs 30 mins

Who are the supporters of and objectors to high-speed rail?

The plan for HS2 has proved to be controversial. While there are supporters, there are also objectors (Figure 20.27).

Supporters of the plan include:

- the main UK political parties
- large cities, including Birmingham, Manchester and Leeds
- businesses in those cities
- the Scottish government.

Objectors to the plan include:

- county councils on the route, like Buckinghamshire and Oxfordshire
- residents living close to the route
- environmental organisations and the Green Party
- taxpayers' groups.

▼ Figure 20.27 What supporters of and objectors to HS2 say

What supporters say	What objectors say
• It will create thousands of jobs in the Midlands and northern England.	• It is more likely to create jobs in London and people will commute there instead.
• It is estimated HS2 will help to generate £40 billion for the UK economy.	• The cost of HS2 is estimated at £42 billion and it is difficult to predict how much money it will generate.
• It will increase the number of rail passengers and make transport more sustainable.	• Existing rail routes could be improved to increase the number of passengers.
• It will also reduce the number of people who fly between UK cities.	• The number of people flying within the UK is already falling.
• It will be a faster way to travel between cities.	• People do not want to travel any faster. Intercity routes are already fast.
• It will be carbon-neutral because it will reduce journeys that use other transport.	• It will increase carbon emissions because high-speed trains use more power.

→ Activities

1 Look at Figure 20.24. Should the government invest more money in roads or railways? Give reasons for your answer.

2 a) Rank the reasons for investing in high-speed rail in order of importance.
   b) Explain why you ranked them in this order.

3 Study Figures 20.25 and 20.26. Draw your own map of the HS2 route. Label HS2 journey times on the correct sections of the route.

4 Study Figure 20.27. Give reasons why each of the following either support or object to the plan for HS2.
   a) Birmingham City Council
   b) A business in Birmingham
   c) Oxfordshire County Council
   d) A resident living near the route

5 Either write a short newspaper article or present a short TV report about the HS2 controversy. Give arguments on both sides, for and against HS2.

❂ KEY LEARNING

➤ Where the main ports and airports in the UK are located

➤ Why a new port has opened on the Thames Estuary near London

➤ Whether Heathrow airport should expand or not

Ports and airports

Where are the main ports and airports in the UK located?

The locations of ports and airports in the UK are very different, as you might expect, because they serve different purposes (Figure 20.28).

Ports are found at coastal and estuary locations all around the UK. They are used mainly for the import and export of bulky raw materials as well as manufactured goods, usually in large metal containers. These are then transported by road or rail around the UK. Some ports are also for passengers travelling on ferries or cruise ships.

Airports are located close to major cities, especially London, and are used mainly by passengers travelling on international flights. By far the largest airport in the UK, London Heathrow also serves as a hub airport, used by passengers in transit from one country to another rather than staying in the UK. Airports are also used for the transport of less bulky high-value goods.

Why has a new port opened on the Thames Estuary near London?

The London Gateway on the Thames Estuary opened in 2013 (see Figure 20.29). It is the first port in London to open since the old docks closed in the 1970s (see Section 15.5) and there are plans for it to expand. Unlike the docks, which were too small for large container ships, London Gateway can accommodate the largest container ships in the world, up to 400 metres long and carrying up to 18,000 containers.

Map key

Key
- ● Port
- ✈ Airport

Map locations: Forth, Edinburgh, Glasgow, Newcastle, Tees and Hartlepool, Belfast, Manchester, Liverpool, Grimsby and Immingham, Birmingham, London Luton, Felixstowe, London Heathrow, London Stansted, London, Milford Haven, Bristol, Dover, Southampton, London Gatwick

0 200 km

▲ Figure 20.28 The main ports and airports in the UK

Once it is complete, the new port will employ 2,000 people, with another 6,000 employed at the new logistics park next to it. This is where many companies will base their distribution centres.

One advantage of London Gateway is that it brings the largest ships closer to London, the biggest market for consumer goods in the UK. It will reduce the distance lorries need to travel and help to cut carbon emissions.

▲ Figure 20.29 London Gateway, a new container port close to London

Should Heathrow Airport expand or not?

Heathrow is already by far the largest airport in the UK (Figure 20.30). By 2030 it could expand even further. A new runway would be built at an estimated cost of £18.6 billion. Heathrow currently operates at almost full capacity, with 480,000 flights a year. It would be impossible to increase the number of flights on its two existing runways.

There were alternatives to expanding Heathrow (Figure 20.31). Some people thought that Manchester Airport should be expanded to boost the economy in northern England and help reduce the North–South divide. However, Heathrow supporters pointed out that, unless Heathrow is allowed to expand, London would be in danger of losing its position as a leading world city.

▼ Figure 20.30 Passenger numbers at UK airports (2013)

Airport	No. of passenger (millions)
London Heathrow	73.4
London Gatwick	38.1
Manchester	21.9
London Stansted	19.9
London Luton	10.5
Edinburgh	10.2
Birmingham	9.7
Glasgow	7.7
Bristol	6.3
Newcastle	4.5

▼ Figure 20.31 Expansion of Heathrow or Manchester?

	Heathrow expansion	Manchester expansion
For	• It will help London to compete with rivals like New York and Paris. • The airport employs 76,000 people and supports a similar number of jobs in London. • Expansion would boost the UK economy by over £200 billion.	• The airport is further from the built-up area so fewer people will be affected by noise. • 22 million people live within a two-hour drive and HS2 will improve connections. • It would boost the economy of northern England.
Against	• It is already the largest emitter of CO_2 in the UK. This will increase when the airport expands. • Noise pollution will get worse for one million people who live below the flight path. • One village will be demolished and two others are threatened.	• The boost to the UK economy as a whole would not be as great as expanding Heathrow. • London would be less able to compete with rival cities. • CO_2 emissions would increase by 50 per cent if the runways double from one to two.

→ Activities

1 Study Figure 20.28.

   a) Describe the location of (i) ports and (ii) airports in the UK.

   b) Explain the different locations of port and airports.

2 Look at Figure 20.29. Suggest what benefits the new port has for, (a) London and (b) the environment.

3 Study Figure 20.31. Make the case for the expansion of either Heathrow or Manchester Airport. Give points in favour of your preferred airport and points against the alternative.

Geographical skills

1 Study Figure 20.30. Complete a map to show the number of passengers at UK airports.

   a) Mark and label each airport at the correct location on an outline map of the UK.

   b) Draw proportional bars at each airport on the map to show the number of passengers.

✪ KEY LEARNING

➤ The impacts of the car industry on the environment

➤ How the car industry can be more environmentally sustainable

Making industry more sustainable

What impacts does the car industry have on the environment?

The car industry is one of the few large-scale manufacturing industries left in the UK. More than 1.5 million new cars are made in the UK every year and most of them at just seven giant manufacturing plants. All of these are owned by foreign-owned TNCs such as Nissan, Honda and BMW.

The car industry does not enjoy the best reputation due to its impact on the environment, for example, most people know that in cities emissions from cars is one of the main causes of air pollution (see Section 15.10). Less well-known are the other environmental impacts cars have through their lifetime, from the resources used in their manufacture to their disposal at the end of the car's life (Figure 20.32).

Fuel consumption

Most cars run on petrol or diesel, which are both obtained from oil; the cause of many environmental problems:

- Drilling for oil uses energy and can endanger ecosystems.
- Shipping oil can cause oil spills.
- As oil is used up, new sources are harder to obtain and can cause more problems.

Resources

Cars are made from a range of resources including steel, rubber, glass, plastic, paint and fabric. Manufacturing and transporting these resources also uses energy.

Air pollution

Burning petrol or diesel in cars is a major cause of air pollution. The main pollutants are carbon dioxide (the main greenhouse gas), nitrogen dioxide (a cause of breathing problems) and particulates (tiny particles, which also cause breathing problems).

Manufacture

Cars consume a lot of energy even before they are driven. It is estimated that manufacturing a car uses as much energy as the car will consume in its lifetime on the road.

Disposal

Cars end up on the scrap heap at the end of their life. Some components like plastic are hard to recycle. Others, like the acid in batteries, can leak into the environment.

▲ Figure 20.32 Environmental impacts of the car industry

▲ Figure 20.33 The Nissan LEAF, an electric car produced in the UK

How can the car industry be more environmentally sustainable?

You may have noticed electric or hybrid (combined electric and petrol-powered) cars on the road (Figure 20.33). They are a sign that the car industry is becoming more environmentally aware. Car companies are responding to demand from consumers for more sustainable cars and to stricter government regulations.

By most measures, the car industry in the UK is becoming more sustainable (Figure 20.34). The amount of energy and water used in the production process has declined. There has been a dramatic fall in the amount of waste going to landfill sites at the end of a car's life. And, importantly, the average CO_2 emissions from new cars are falling.

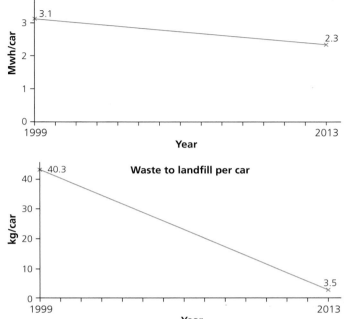

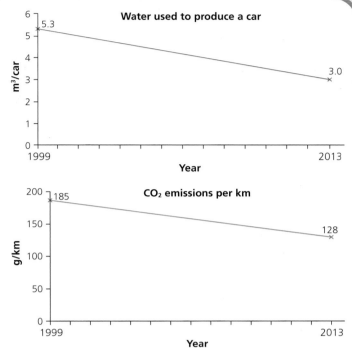

Figure 20.34 How the car industry has become more sustainable

Production begins at Nissan: 1986

Number of people employed: 7,000

Estimated number of jobs created in the UK by Nissan's car plant: 28,000

Number of cars produced in a year: 500,000 (one-third of all cars produced in the UK)

Models produced at the plant: Nissan Note, Nissan Qashqai, Nissan LEAF, Compact Infiniti

Amount of energy generated by wind turbines: 7%

Figure 20.35 The Nissan car manufacturing plant near Sunderland. Note the wind turbines

→ Activities

1   Study Figure 20.32 and the rest of the information on this spread. Which of the five environmental impacts of the car industry would be:

    a) easiest for the car manufacturer to reduce?
    b) easiest for the car owner to reduce?
    c) most difficult to reduce?

    In each case, give reasons for your answer.

2   Look at Figure 20.33. The Nissan LEAF is an example of a more sustainable car. In what ways:

    a) does it have a reduced environmental impact?
    b) does it still have a harmful environmental impact?

3   Study Figure 20.34. Describe the four ways in which the UK car industry has become more sustainable. Write four sentences, including data from each of the graphs.

4   Study Figure 20.35. Is it possible for a car plant both to contribute to the economic development of an area and to be more sustainable? Give evidence from Nissan's car plant to support your opinion.

> ➤ How the UK's place in the world has changed
> ➤ Where the UK's main international links are today

The UK's place in the world

How has the UK's place in the world changed?

The British Empire once covered about one-third of the world's land surface. It was described as 'the Empire on which the Sun never sets' because it was always daytime somewhere in the Empire. During the twentieth century, most countries in the Empire gained independence from the UK, so the UK is instead a member of the Commonwealth (Figure 20.36). All these countries share common values, including the promotion of democracy, human rights and trade. One consequence of Britain's historical role in these countries is that the English language is often used.

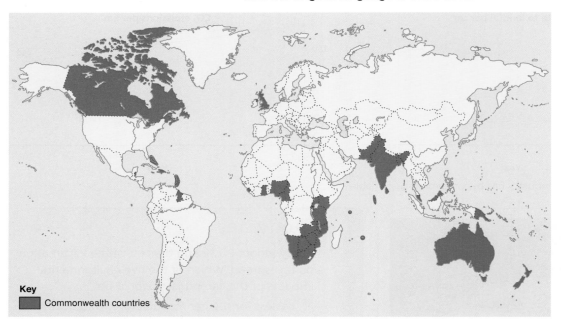

Key
▮ Commonwealth countries

▲ Figure 20.36 Map of Commonwealth countries

Commonwealth countries

Africa Botswana, Cameroon, Ghana, Kenya, Lesotho, Malawi, Mauritius, Mozambique, Namibia, Nigeria, Seychelles, Sierra Leone, South Africa, Swaziland, Uganda, Tanzania, Zambia

Asia Bangladesh, Brunei, India, Malaysia, Pakistan, Singapore, Sri Lanka

The Caribbean and America Antigua and Barbuda, Bahamas, Barbados, Belize, Canada, Dominica, Grenada, Guyana, Jamaica, St Lucia, St Kitts and Nevis, St Vincent and the Grenadines, Trinidad and Tobago

Europe Cyprus, Malta, United Kingdom

Pacific Australia, Fiji, Kiribati, Nauru, New Zealand, Papua New Guinea, Samoa, Solomon Islands, Tonga, Tuvalu, Vanuatu

Where are the UK's main international links today?

The UK maintains its links with the Commonwealth through trade, culture and also migration. Many people of British descent now live in Commonwealth countries like Australia, Canada and New Zealand.

And, of course, there are many people of Asian, African and Caribbean descent now living in the UK. Migration from these countries grew in the second half of the twentieth century and still continues, often filling gaps in the UK workforce as our population gets older.

In 1973, the UK joined the **European Union (EU)** (Figure 20.37). The EU allows the free movement of people, goods and services between the member countries.

There is also a single currency, which is shared by nineteen members, but not the UK.

Ten countries from Eastern Europe joined the EU early in the twenty-first century. Since then, migration from Europe to the UK has increased, as people from those countries move to the UK for better-paid work. Migration from the EU to the UK was approaching the same level as non-EU migration (Figure 20.38).

In 2016, people in the UK voted in a referendum to leave the EU. One of the reasons people gave for voting to leave was the high level of migration from other countries in the EU.

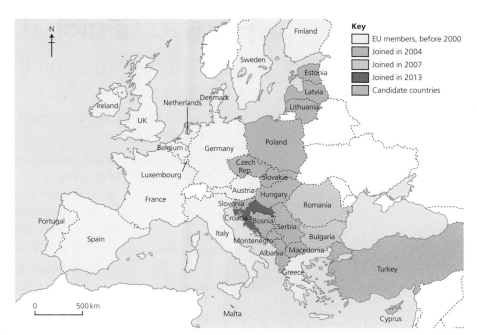

▲ Figure 20.37 Map of EU countries

UK citizens in other EU countries (2012)	
Spain	761,000
Ireland	291,000
France	200,000
Germany	115,000

EU citizens in the UK (2012)	
Poland	646,000
Ireland	403,000
Germany	304,000
France	136,000
Italy	133,000
Lithuania	130,000
Romania	101,000

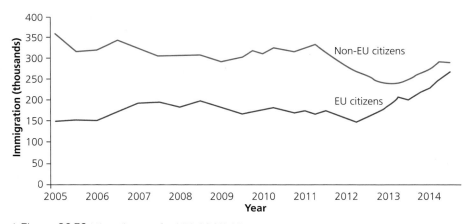

▲ Figure 20.38 Migration to the UK, 2005–14

Geographical skills

1 Study Figure 20.37.
   a) Describe how the EU has grown in the twenty-first century.
   b) Explain the impact this has had on migration to the UK.

2 Create a map of Europe to show the movement of people between the UK and other EU countries. Use proportional arrows to show the direction and numbers of people who moved to and from the UK.

→ Activities

1 Study Figure 20.36.
   a) Describe the distribution of Commonwealth countries around the world. Include the names of continents and oceans where they are located.
   b) 'The Empire on which the Sun never sets.' Why was this a good description of the British Empire?

2 a) Name as many Commonwealth countries on the map as you can. Check your ideas in an atlas.
   b) Complete a Commonwealth map on an outline map of the world. Label the countries.

3 Study Figure 20.38. Describe the changes in (a) non-EU migration and (b) EU migration to the UK from 2005 to 2014.

The UK's global links

Which countries are the UK's main trading partners?

Most of the UK's trade is with other countries in Europe (Figure 20.39). This is not surprising since:

- the UK is part of the European Union (EU), which encourages trade between member countries.
  - European countries are geographically close to the UK, so transport costs are cheaper.
  - European countries are among the world's wealthiest economies, so the volume of trade is greater.

However, the USA and China are also major trading partners for the UK.

Where do most international flights from the UK go to?

Heathrow is the largest UK airport, with the most international flights (Figure 20.40). The flight routes from Heathrow reflect the parts of the world with which the UK has the most links, as a result of business, visiting family and friends, and holidays.

They also reflect the parts of the world in which the UK has the most cultural and trade links. For example, festivals such as Diwali and Eid are known in the UK due to its strong relationship with India (see Sections 17.4 and 18.2).

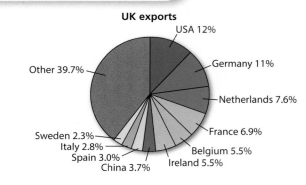

UK exports
- USA 12%
- Germany 11%
- Netherlands 7.6%
- France 6.9%
- Belgium 5.5%
- Ireland 5.5%
- China 3.7%
- Spain 3.0%
- Italy 2.8%
- Sweden 2.3%
- Other 39.7%

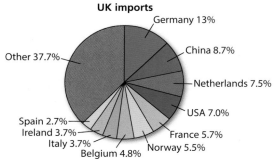

UK imports
- Germany 13%
- China 8.7%
- Netherlands 7.5%
- USA 7.0%
- France 5.7%
- Norway 5.5%
- Belgium 4.8%
- Italy 3.7%
- Ireland 3.7%
- Spain 2.7%
- Other 37.7%

▲ Figure 20.39 The UK's main trading partners

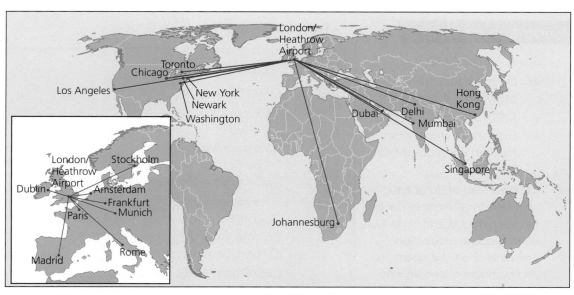

▲ Figure 20.40 Most popular international routes from Heathrow Airport

What is the impact of the internet on our global links?

The biggest change to our global links in the twenty-first century has come through the internet. Globally, the growth of the internet has become an almost unstoppable force. By 2014, almost three billion people had access to the internet, which is about 40 per cent of the world's population (Figure 20.41).

On average, 183 billion emails were sent and received each day in 2013 – that is 2.1 million emails every second! Of course, these figures will be out of date when you read them, because they increase all the time. Social media is one of the internet's biggest success stories. Facebook alone has over a billion users, meaning that if it were a country, it would be the third largest in the world.

The UK is one of the world's most connected countries. In 2014, 90 per cent of people in the UK used the internet, compared to just 27 per cent in 2000.

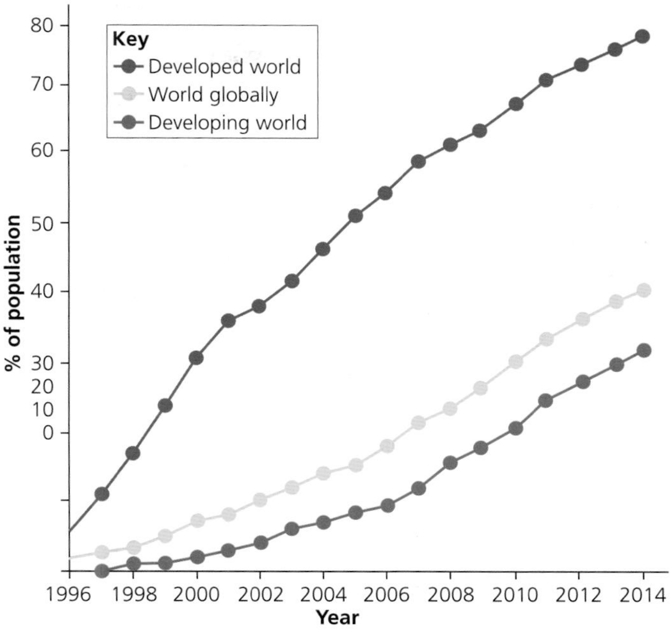

Key
- Developed world
- World globally
- Developing world

▲ Figure 20.41 The global growth of the internet

→ Activities

1   Map your personal global links on a world map. Include all the links you can think of, including;

- family and friends – where they live outside the UK
- clothing and other items – where they were made
- holidays – where you have been
- the internet – where you have linked with people by email or social media.

Mark and label each country you are linked to. Give the map a key to show different types of links.

2   Study Figure 20.39. Map the UK's exports and imports on a world map.

a)   Mark and label the countries.

b)   Use arrows in two colours to indicate exports and imports, from and to the UK.

3   Study Figure 20.40. Choose at least three cities around the world that are linked with the UK. From what you know of each city, suggest reasons for that link as a result of one or more of:

- business
- family and friends
- tourism.

→ Going further

There was a referendum in 2016 in which people in the UK voted to leave the EU. People under eighteen were not allowed to vote. But imagine they were. Would it make any difference to our decision to stay in the EU or not?

These are some of the main issues to consider:

- the influence of the UK within the EU
- the influence of the UK on the rest of the world
- the amount of trade the UK can do, inside and outside the EU
- the number of jobs created or lost in the UK
- how easy it is for UK residents to travel or live abroad
- how easy it is for EU residents to live and work in the UK.

1   Find out more about these issues. Do you think being a member of the EU makes them better or worse?

2   Write a short speech for or against the UK staying in the EU. You could give your speech as part of a class debate. Which way would your class vote? If people under 18 voted in the referendum, would it have made any difference to the result?

Question Practice

Unit 2 Section B

1   Using Figure 17.4, compare the GNI values of Australia and Asia.   [2 marks]

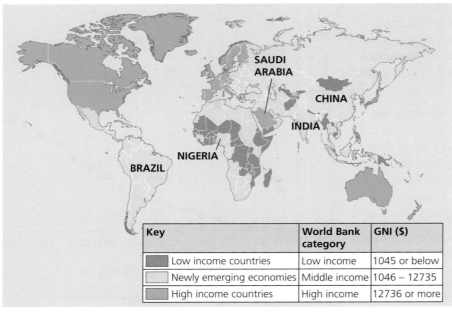

Key		World Bank category	GNI ($)
	Low income countries	Low income	1045 or below
	Newly emerging economies	Middle income	1046 – 12735
	High income countries	High income	12736 or more

▲ Figure 17.4 The world map of development

2   Look at Figure 17.7. Compare the GNI values in Indonesia and Singapore.   [2 marks]

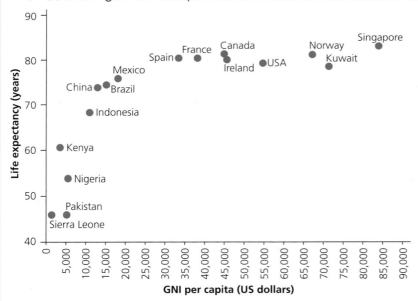

▲ Figure 17.7 Investigating the relationship between economic and social development

3   Outline one disadvantage of using a single measure of development such as life expectancy.   [2 marks]

4   Explain how the indicators of development in Figure 17.7 and Figure 17.9
    show the differences in the quality of life between Sierra Leone, Norway
    and the USA.

HDI rank and score	Country	HDI rank and score	Country
1 (0.944)	Norway	183 (0.374)	Sierra Leone
2 (0.933)	Australia	184 (0.372)	Chad
3 (0.917)	Switzerland	185 (0.341)	Central African Republic
4 (0.915)	Netherlands	186 (0.338)	DR Congo
5 (0.914)	USA	187 (0.337)	Niger

▲ Figure 17.9 The highest and lowest HDI scores in 2014   [4 marks]

5a Using Figure 20.13 (page 288), identify the feature found at grid reference
   470621.   [1 mark]

5b Explain why this feature might be considered a benefit for a company that
   is considering choosing Cambridge as the location for its headquarters.   [4 marks]

6   Study Figure 17.15.

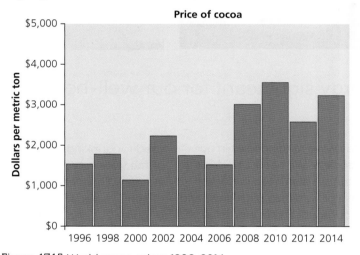

▲ Figure 17.15 World cocoa prices 1996–2014

Calculate the difference in the price of cocoa from 1995 to 2014.   [2 marks]

7   Suggest one reason why tourism can help social development.   [2 marks]

8   Outline one way that intermediate technology can help LICs and NEEs.   [2 marks]

9   'Microfinance loans are a better long-term aid to a country's development
    than international aid.' Do you agree with this statement? Justify your
    decision.   [9 marks]

21 Global resource management

Essential resources

What are the key resources needed for economic and social well-being?

The key **resources** of food, water and energy influence all elements of human well-being, including the basic material needs for a good life, health, good social relations, security, and freedom of choice and action. People are dependent on Earth's ecosystems and what they provide. When the supply of any of these resources is abundant, relative to the demand, there are benefits to human well-being, whether economically or socially. Where availability is relatively scarce, a small decrease can substantially reduce human well-being. If people have a good supply of food, water and energy, their quality of life, as well as their standard of living, improves.

▼ Figure 21.1 Key resources

How are food, water and energy significant for our well-being?

Food

The need for food is obvious. 'Calories in' (fuel for our bodies) are needed to work and enjoy ourselves, which equals 'calories out'. The calories needed per day depend on the type of job you perform, your age and gender. Average figures can be seen in Figure 21.2.

▼ Figure 21.2 Daily calories guidelines

Category	Calories
Women	2,000
Men	2,500
Child, 5–10	1,800
Girl, 11–14	1,850
Boy, 11–14	2,200

Water

Water has wide-ranging uses in our current society. We need it to drink to survive, but we also need it to wash, to dispose of waste, to both grow and process our food, and in industrial manufacturing processes. Figure 21.3 shows you the amount of water needed to produce your jeans, your cup of coffee and the paper for this book! The average person in the UK today uses 150 litres of water daily at home, of which only four per cent is used for drinking. Nearly 75 per cent of the water used in the UK is used in industry. This includes the production of items as varied as cakes and cars (Figures 21.3 and 21.4).

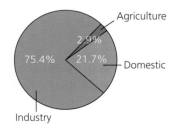

▲ Figure 21.3 Water use in the UK

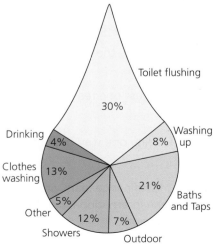

▲ Figure 21.4 Domestic water use in the UK

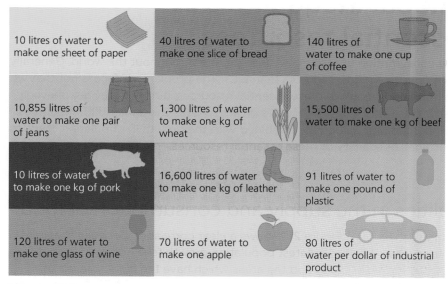

▲ Figure 21.5 Volume of water it takes to make common products

The following values appear in Figure 21.5:

- 10 litres of water to make one sheet of paper
- 40 litres of water to make one slice of bread
- 140 litres of water to make one cup of coffee
- 10,855 litres of water to make one pair of jeans
- 1,300 litres of water to make one kg of wheat
- 15,500 litres of water to make one kg of beef
- 10 litres of water to make one kg of pork
- 16,600 litres of water to make one kg of leather
- 91 litres of water to make one pound of plastic
- 120 litres of water to make one glass of wine
- 70 litres of water to make one apple
- 80 litres of water per dollar of industrial product

Energy

Energy is used to make the bricks for our houses, to heat our homes, transport us, power machinery and process food.

The amount and type of energy used depends on a variety of factors, including where people live and how wealthy they are. Traditionally, energy has come from burning naturally occurring fuels such as wood and coal. However, more and more natural resources are being harnessed to produce energy. Today, there is renewable energy from the **wind** and waves, as well as nuclear energy and solar power.

Conclusion

As the world's population grows, so the pressure on the supply of resources becomes greater. The rate of growth can cause huge problems, as the supply of resources struggles to keep up with demand. Technology cannot change or improve fast enough to provide the essential resources needed. One of the major problems facing us today is meeting the demand for these essential resources, and, in particular, solving the problems caused by their unequal distribution and consumption.

- Biomass heat — 11.44%
- Hydropower — 3.34%
- Solar hotwater — 1.85%
- Geothermal heat
- Ethanol
- Biodiesel
- Biomass electricity
- Wind power
- Geothermal electricity
- Solar PV power
- Solar CSP
- Ocean power

- Fossil fuels — 80.6%
- Renewables — 16.7%
- Nuclear — 2.7%

▲ Figure 21.6 World energy consumption by source (2010)

→ Going further

Choose one energy source and research the changes in consumption in the last ten years. Explain the changes you have found.

→ Activities

1. Keep a record of the food you eat in a day and use the calorie checker on the NHS choices website to look up the calories. Compare the calories you eat in a day with the average. Do you eat more or less? Why?

2. What are the three main uses of water in the UK?

3. Using Figure 21.6, describe the global consumption of energy by source.

Geographical skills

1. Use Figure 21.4 to calculate the percentage of water people use in washing and cleaning in the UK.

2. Using Figure 21.2 draw a bar graph to demonstrate the advised average calorie intake per person.

Global inequalities in essential resources

The consumption of resources varies greatly throughout the world. Generally, HICs consume more resources than LICs. The problem we face is not that we do not have enough of these essential resources, but that they are unevenly distributed. As the wealth of LIC s grows so does the demand for resources. This demand for resources, combined with the growth in population, leads to shortages or scarcity of these essential resources.

What are the global inequalities in the supply and consumption of food?

The average calorie consumption in a country such as the UK is 3,200 per person, while in a country such as Somalia it is 1,580 per person.

Figure 21.7 shows a clear correlation between the areas of greatest population growth (see Figure 23.3 on page 325) and the areas which have the highest levels of **undernourishment**.

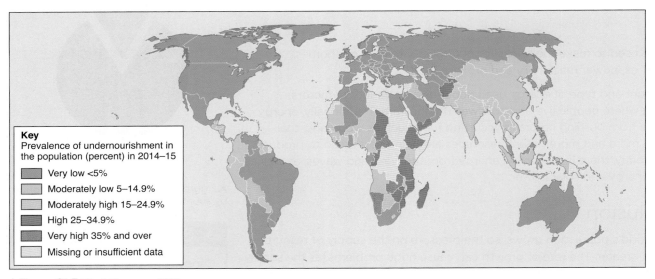

Key
Prevalence of undernourishment in the population (percent) in 2014–15

- Very low <5%
- Moderately low 5–14.9%
- Moderately high 15–24.9%
- High 25–34.9%
- Very high 35% and over
- Missing or insufficient data

▲ Figure 21.7 World hunger, 2015

What are the global inequalities in the supply and consumption of water?

The global supply of freshwater is limited and unequally distributed. The **water footprint** of countries can be calculated to compare consumption. This is the amount of water used throughout the day, for example, from a tap for drinking or showering. It also includes the water it takes to produce food, products, energy and even the water saved when products are recycled. This virtual water may not be seen, but it makes up the majority of a country's water footprint.

The global average water footprint is 1,240 litres per person. The water footprint of the USA is 2,483 litres per person. The water footprint of Bangladesh is just 896 litres per person.

Figure 21.8 shows the areas that suffer from water scarcity and those that have water to spare. Many countries may have water but may not have the money to access the water, such as Sudan. This is known as economic water scarcity. Others may not have as much water due to the physical conditions such as climate, for example, Saudi Arabia (physical water scarcity).

What are the global inequalities in the supply and consumption of energy?

Energy consumption varies considerably in different countries (Figure 21.9). The richest one billion people in the world actually consume 50 per cent of the world's energy, while the poorest one billion people consume only 4 per cent of the world's energy.

Why is the demand for essential resources growing?

The demand for essential resources has grown over time as we develop new processes, new products and change our way of life. As LICs and NEEs develop industrially and economically, their demand for resources has grown too. For example, as industry has grown in China, energy consumption has increased with it. Between 2003 and 2011, China saw an increase of 53 per cent in its consumption of energy.

→ Going further

Use the internet to research the impact of climate change on the global pattern of water supply.

→ Activities

1   Look back to your daily calorie consumption from Activity 1 on page 307. How does this compare to the average calorie consumption of a person in Somalia? Explain the difference.

2   Using the data from Figure 21.9, argue why some countries use more energy than others. Use figures to support your answers.

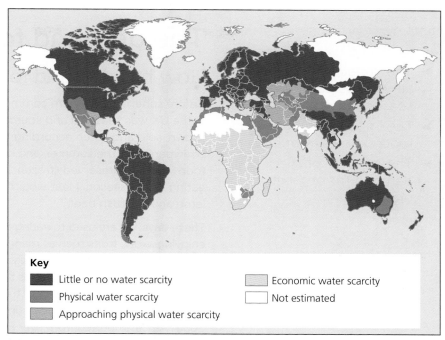

Key

■ Little or no water scarcity   ▨ Economic water scarcity

■ Physical water scarcity   □ Not estimated

▨ Approaching physical water scarcity

▲ Figure 21.8 Map to show global water scarcity

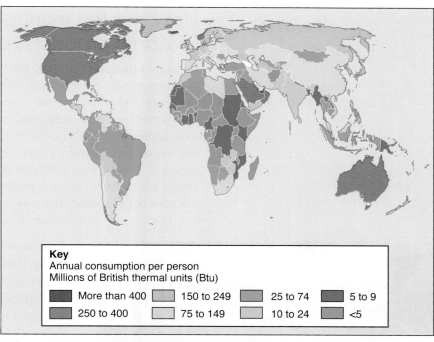

Key
Annual consumption per person
Millions of British thermal units (Btu)

| ■ More than 400 | ▨ 150 to 249 | ▨ 25 to 74 | ■ 5 to 9 |
| ▨ 250 to 400 | □ 75 to 149 | ▨ 10 to 24 | ▨ <5 |

▲ Figure 21.9 Energy consumption per person by country

Geographical skills

1   Using Figure 21.7, describe the global pattern of undernourishment.

2   Using Figure 21.8, describe how global water scarcity links with the global pattern of undernourishment.

22 Resources in the UK

The demand for food in the UK

How has demand for food in the UK changed?

Before supermarkets were commonplace, the majority of the food eaten in the UK was seasonal and sourced in the UK. Fruit and vegetables were grown, sold and eaten according to the seasons, for example, lettuce and strawberries in the summer, and cabbage and parsnips in the winter. More food was also preserved (frozen, bottled, or made into jam and pickles) for eating out of season. Meat would also have been produced, such as Welsh lamb and Scottish beef.

These days, we are used to eating fruit and vegetables all year round, and enjoying exotic fruits such as mangoes (which cannot be grown in the UK due to the climate). However, even seasonal fruit and vegetables in the UK are often imported from other countries as they can be grown more cheaply elsewhere. In September, you would expect to find seasonal fruits and vegetables such as apples and onions in your local supermarket, but you will also find apples from South Africa, imported from over 8,000 kilometres away, along with onions from Spain. In 2013, 47 per cent of the UK's food supply was imported.

How has the increase in demand for non-seasonal products had an impact on LICs?

Consumer demand in the UK affects what is imported from other countries. Consumers want out-of-season and exotic food available all year round. The UK imports food like this from places such as Kenya and the Caribbean. This means that land previously used to produce food for local people is now used to provide high-value food products for people in the UK. High-value foods and ingredients can fetch retail prices that are up to five times those of similar products. The high value may be due to the product itself, such as Madagascan vanilla, specialist honeys and gourmet coffees, or because they are luxury items available out of season that are in high demand. The cost of these products to the UK consumer is high, but there are also costs for the people in Kenya:

- Less land is available for locals to grow food to eat
- Often these crops need huge amounts of water in areas where the water supply is unreliable or poor
- Sometimes the people growing the crops are exposed to chemicals such as pesticides without protective clothing.

However, there are opportunities for Kenyans too:

- Jobs are created, for example in farming, packaging and transport.
- These jobs supply wages for local people.
- From the wages, taxes are paid to the government, which can then fund facilities for the country such as schools and hospitals.

▲ Figure 22.1 Green beans being packed for export in Kenya

How and why has the demand for organic produce changed?

Another change to the UK's eating habits has been the increasing demand for **organic produce**. Organic produce, including meat, fruit and vegetables, is produced by organic farming, a type of farming which does not include the use of chemicals such as pesticides and fertilisers.

- The aim is to protect the environment and wildlife by using natural predators to control pests, for example using ladybirds to eat blackfly.
- Farmers maintain the fertility of the soil by rotating crops and using a variety of natural fertilisers, including green manure and compost.
- Weeds are controlled by mechanical weeding rather than using chemical weed killers.
- Animals are farmed without the use of antibiotics and the regular use of drugs such as hormones to increase growth.

The demand for organic products has been rising steadily since the early 1990s, as people are increasingly concerned about the effect of what they eat on their health. Organic food is believed to be healthier than non-organic food. In a government survey of households in 2014, the main reasons for choosing organic products were:

- 'It contains fewer chemicals and pesticides'
- 'It's natural and unprocessed'
- 'It's healthier for me and my family'

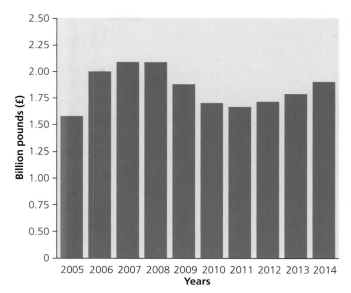

▲ Figure 22.2 Sales of organic food in the UK, 2005–14

A closer examination of the figures will show a decline in sales between 2009–11. This was due to the global recession and the reduction in incomes for many families, making expensive organic produce a luxury item. It is expensive because yields from organic farming tend to be lower, but many people claim that organic food tastes better. They are prepared to pay extra for this and the reduced impact on the environment.

Today, all the major supermarkets sell organic produce, providing about 75 per cent of all organic food sold. Other sources of organic food are local farmers' markets and vegetable box schemes, where households receive a box of seasonal organic fruit and vegetables weekly or monthly, usually delivered to their home. The highest-selling organic products are dairy produce including milk, cheese and yoghurt.

→ Activities

1 Why is organic produce more expensive than non-organic produce?

2 Make a list of the challenges and opportunities provided by:

   a) the increased demand for fruit and vegetables all year round

   b) organic products.

   Include the challenges and opportunities to:

   a) the consumers

   b) the producers.

4 Suggest what challenges importing seasonal food causes farmers in the UK.

Geographical skills

1 Using Figure 22.2:

   a) Describe the pattern of sales of organic products between 2005 and 2014.

   b) Calculate the change in organic sales between 2008 and 2014 in pounds.

KEY LEARNING

- Food miles and carbon footprints
- The impact of importing food on the UK's carbon footprint
- Alternatives to importing food
- The trend towards agribusiness

Food miles and carbon footprints

What are food miles and carbon footprints?

Food miles are the distance that food travels from producer to consumer. For example, green beans from Kenya travel 6,818 kilometres to reach the UK. This does not include the distance the beans travel from the airport to the supermarket, or from the supermarket to your home. In the UK, food travels over 30 billion kilometres every year. This includes transport by air, ship, train and road.

A **carbon footprint** is the measure of the impact that human activities have on the environment in terms of the amount of greenhouse gases they produce (see Section 4.3). It is possible to calculate an individual's carbon footprint as well as that for a country or a business.

How does importing food increase the UK's carbon footprint?

The transport used to import food into the UK adds over 19 million tonnes of carbon dioxide to the atmosphere every year which increases the UK's carbon footprint.

In theory, the further a product has travelled, the higher the food miles and the higher its emissions. However, there are other aspects of food production which add to the UK's carbon footprint. Carbon dioxide is also produced when food is grown and harvested, for example, when farm machines harvest the crops or when greenhouses are heated. Figure 22.3 shows how the production and transportation of food contributes to the UK's carbon footprint. Food contributes at least seventeen per cent of the total UK carbon dioxide emissions, but only eleven per cent of this is linked to transport.

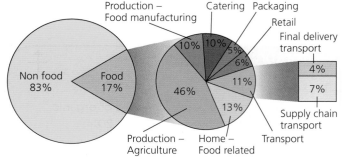

▲ Figure 22.3 The contribution of food to the UK carbon footprint

Emissions

The emissions created by producing a food product in the UK can sometimes create higher emissions than those imported from overseas. An example of this is tomatoes. Even including the transport emissions from the aircraft bringing them to the UK, the carbon footprint of Spanish tomatoes is smaller because growing tomatoes in the UK requires heated greenhouses whereas Spain's warmer climate means no additional heat is needed.

The actual emissions produced by different forms of transport used in transporting food also need to be taken into consideration when looking at food miles and carbon footprints. Food transported by plane generates around 100 times the amount of emissions of food transported by boat. This means that bananas from Ecuador which travel by ship are much more carbon friendly than avocados flown from Mexico, although both are a similar distance away from the UK. In general, food products which are perishable and have a high value relative to their weight are transported by plane, while others are sent by sea, which takes a lot longer.

▼ Figure 22.4 Transport emissions, kilogram CO_2 equivalents per tonne kilometres

Type of transport	Carbon emissions (kg CO_2/t.km)
Long-haul air, e.g. UK–Australia	1.762
Short-haul air, e.g. UK–Spain	0.733
Road	0.41
Rail	0.037
Water	0.038

What are the alternatives to importing food?

To reduce the amount of carbon emissions, we need to reduce the amount of food products that are flown into the UK. There are several ways to do this:

■ Eating seasonal produce grown in the UK.
■ Limiting imported foods to those we cannot grow in the UK, and limiting those that are transported by air. Many restaurants now label the origins of food and work in close collaboration with the food producers. This helps the farmers and producers and allows consumers to make informed choices. Some supermarkets such as the Co-operative have also started to stock only British meat, and all supermarkets now use the Red Tractor scheme (Figure 22.5).
■ Eating locally produced food, which reduces the amount of food miles that the food travels within the UK, and also supports local farmers and producers. There are a growing local farmers' markets or farm shops in the UK. To qualify to sell at a farmers' market the products must be produced within a specific local area.
■ Growing food at home or on an allotment. The move towards growing our own food has increased in the last five years, with over a third of people now growing their own fruit and vegetables.

◀ Figure 22.5 The Red Tractor label, which assures consumers that the source of food is British and has been inspected for safety, welfare and environmental impacts

Why is there a trend towards agribusiness?

Agribusiness refers to treating food production from farms like a large industrial business, making it a large scale, capital-intensive, commercial activity. This has meant increasing the size of farms by:

■ removing hedgerows
■ increasing field sizes and combining smaller family farms
■ using modern production methods (see Chapter 23)
■ increased mechanisation
■ using the latest technology, better seeds, and increased use of chemicals such as pesticides and fertilisers.

Today, the vertical integration of food production, commonly known as 'from farm to fork', is increasingly common. Large agribusinesses now often own not only the farms where food is grown, but also the processing factories, the transport and the retail outlets. Food-processing companies and supermarkets will often buy the crops before they are planted. Agribusiness has some huge advantages for the increased production of food, but can have considerable impacts on the environment and **local food** production (see Chapter 23) . This trend can be seen clearly in East Anglia, where farm sizes have increased dramatically over the last 40 years. This has led to a decline in agricultural employment, especially in more isolated areas.

➜ Activities

1 a) Choose your favourite meal and work out the basic ingredients needed to cook it.
b) Find out where these products come from by looking at the label or by using a supermarket website. The origin of every product should be on the label.
c) Use the following website to find out how far each part of your meal has travelled: www.foodmiles.com
d) Plot the locations on a world map and add up how far your 'dinner' has travelled.
e) Using the food miles website to help you, suggest an alternative meal that would have lower food miles

2 What are the advantages of buying local food products?

Geographical skills

1 a) Using Figure 22.4, draw a bar graph to show the different emissions produced by different methods of transport.
b) Explain why different food items are transported by different methods.

Water resources in the UK

How has the demand for water changed?

The UK does not really feel like a place that suffers from water shortage. We tend to complain that it rains a lot, but compared to some other parts of Europe, we have less rainfall. For example, London has a lower average annual rainfall then Rome. Added to the problem of water supply is that the demand for water has changed in recent years. The amount of water used by the average household in the UK has risen by 70 per cent since 1985.

The increase is due to:

■ the increase in wealth leading to more use of domestic appliances such as dishwashers and washing machines, which use a lot of water
■ changes in personal hygiene: most people now shower several times a week, whereas in the past, when houses did not have inside bathrooms, a weekly bath, shared between family members, was the norm.
■ the demand for out-of-season food, which requires additional watering, usually in greenhouses (see Chapter 24)

■ increased industrial production.
■ increased leisure use: for example, golf courses needing a great deal of watering to maintain the grass.
■ the increased UK population (see Section 15.1).

The average person in the UK uses 150 litres of water per day; compare that with the average person in Africa, who uses just 47 litres.

Where are the areas of water deficit and surplus in the UK?

The population of the UK is not evenly distributed throughout the country, and the areas of high population density do not correspond with the areas of high water supply. One-third of the UK's population live in the South East, the driest part of the UK. In addition, as the population of the UK grows the major cities such as London and Manchester are the areas that are growing most rapidly.

Compare Figures 22.6, 22.7 and 22.8, which show the population density, rainfall and areas of water stress in the UK.

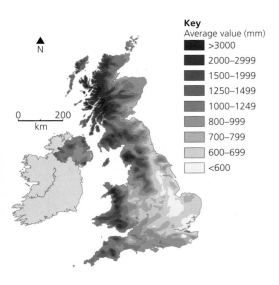

▲ Figure 22.6 Annual rainfall in the UK

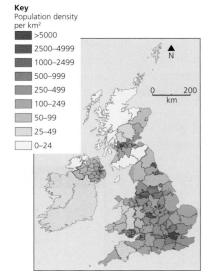

▲ Figure 22.7 Population density in the UK

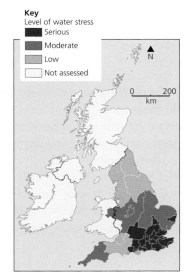

▲ Figure 22.8 Areas of water stress in the UK

Areas of **water deficit** are areas which do not have enough water for the needs of the population and may suffer from shortages. Areas of **water surplus** are places where they have more water than they need. For example, using Figures 22.6 and 22.7, you can see that Wales has a low population density but high rainfall, so it is an area of water surplus.

Areas which suffer from serious water deficits are said to be undergoing serious **water stress**, when the water available is not sufficient to meet the needs of the population, or is of poor quality. UK rainfall is also remarkably unreliable: in some years, the rainfall is well below average, such as in 2006 when water restrictions (hosepipe bans) were put in place, but in other years, the UK has had very wet periods.

What are water transfer schemes and why are they needed?

The British government has considered setting up a national water grid, similar to the national grid for electricity, where the water would flow through pipes from areas of surplus such as Wales to areas of deficit such as London. The idea has been discussed for many years but has not yet been put into practice, partly due to the enormous costs involved.

In addition, there are other concerns, including:

- the impact on the environment of the river basin in the source area
- dams constructed to create reservoirs on rivers (may disrupt the ecology and block migrating species)
- the increased carbon emissions linked to pumping water over long distances
- the displacement of local communities
- potential droughts in river source basins caused by the removal of water to other areas.

However, there are some parts of the country where water transfers do happen on a smaller scale:

- The reservoirs in the mountainous area of North Wales and the Lake District provide water for the densely populated urban areas in the northwest of England, such as Liverpool and Manchester.
- The water from the Kielder Dam in Northumberland (see Section 11.12) is pumped into the North Tyne River. From there the water can be transferred to three other major rivers, the River Derwent, the River Wear and the River Tees. These rivers then supply water to the major cities of Newcastle, Sunderland and Middlesbrough.

→ Activities

1 Explain why the demand for water is likely to rise in South East England in the next 10 years.

2 What challenges does the annual variation in rainfall cause?

3 Explain what is meant by the term water transfer.

→ Going further

Research the Kielder water transfer scheme and explain the advantages (opportunities) and disadvantage (challenges) it has brought to: (a) the local people (b) the economy (c) the local environment.

See Section 11.12 for some information on the Kielder scheme.

Geographical skills

1 a) Using Figures 22.6 and 22.7, calculate the difference in the average annual rainfall in an area of serious water stress and an area of low water stress.

b) Use Figures 22.6 and 22.7 to name two areas of the UK which have a high population density and a low annual rainfall.

c) Discuss the problems challenges that this may cause for these areas.

➤ The causes of water pollution in the UK

➤ How water pollution affects the UK

➤ How water quality is managed in the UK

Keeping it clean

What are the causes of water pollution?

Today rivers, lakes, and coastal waters in the UK are cleaner than they have ever been since before the industrial revolution. The improvement in water quality has seen wildlife, including salmon, otters and birds, returning to live in these habitats. However, water pollution still exists. In the UK only 27 per cent of our water is classified as being of a 'good status' under the EU Water Framework Directive which looks at **water quality** throughout the EU. This means that 73 per cent of our water in the UK could be polluted or of poor quality. Rivers, lakes and coastal waters are polluted by a wide variety of items as seen on Figure 22.9.

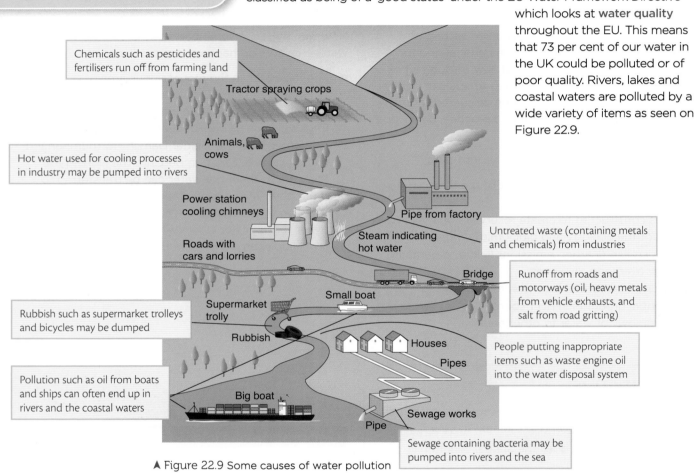

Chemicals such as pesticides and fertilisers run off from farming land

Tractor spraying crops

Animals, cows

Hot water used for cooling processes in industry may be pumped into rivers

Power station cooling chimneys

Pipe from factory

Steam indicating hot water

Untreated waste (containing metals and chemicals) from industries

Roads with cars and lorries

Bridge

Runoff from roads and motorways (oil, heavy metals from vehicle exhausts, and salt from road gritting)

Small boat

Rubbish such as supermarket trolleys and bicycles may be dumped

Supermarket trolly

Rubbish

Houses

Pipes

People putting inappropriate items such as waste engine oil into the water disposal system

Pollution such as oil from boats and ships can often end up in rivers and the coastal waters

Big boat

Sewage works

Pipe

Sewage containing bacteria may be pumped into rivers and the sea

▲ Figure 22.9 Some causes of water pollution

How does water pollution affect the UK?

Water pollution can have a number of serious effects, with long-term and short-term consequences.

- Toxic waste can poison wildlife. Sometimes the toxins can be transferred to humans if they eat the shellfish or fish, leading to birth defects and, in some cases, cancer.
- The supply of drinking water can be poisoned.
- Increased water temperatures can lead to the death of wildlife and disrupt habitats.
- Increased fertilisers can increase nutrients in the water, speeding up the growth of algae and leading to eutrophication. This means there may not be sufficient oxygen in the water, so other wildlife will also die. The increased algae may also block the sunlight to other water plants.
- Pesticides can kill important parts of the ecosystem.
- The microbacteria in sewage can cause the spread of infectious diseases in aquatic life, animals and humans.
- People whose livelihoods depend on a clean water supply, for example fishermen or workers in the tourist industry, may suffer.

How is water quality managed in the UK?

Legislation – The UK and EU have strict laws which ensure that factories and farms are limited in the amount and type of discharge they put into rivers. Water companies which provide our drinking water and sewage systems have very clear regulations and penalties.

Education campaigns – These inform the public about the damage caused by putting inappropriate items into the sewage systems, such as engine oil and baby wipes, and advise how to dispose of them correctly.

Waste water treatments – Local water treatment plants remove suspended solids such as silt and soil, bacteria, algae, chemicals and minerals, to produce clean water for human consumption. They use a number of processes to do this (Figure 22.10).

Building better treatment plants and investing in new infrastructure – Better sewers and water mains can prevent spills and accidents, but can lead to higher water and sewage bills to pay for the investment. For example, Thames Water in London is investing heavily in its sewage works, and new tunnels to prevent the overflow of the current sewers.

Pollution traps – For example, when new roads and motorways are built close to rivers and watercourses, pollution traps such as reed beds are often installed to 'catch' and filter out the pollution.

Green roofs and walls – in cities, new buildings often have green roofs, which filter out the pollutants naturally in rainwater. Green roofs also offer excellent sustainable water management. This reduces the risk of flooding by reducing runoff from the roof. Green roofs can also help to combat climate change by increasing the absorption of CO_2 from the atmosphere.

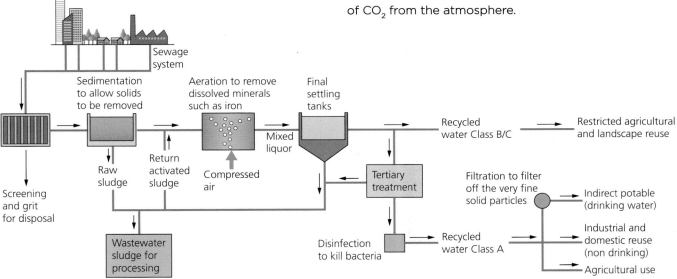

▲ Figure 22.10 The waste water treatment process

→ **Activities**

1 a) Describe three causes of water pollution.
   b) Explain the impacts of these three causes of water pollution.

2 Choose one of the ways given for managing water quality and describe the advantages (opportunities) and disadvantages (challenges) it has.

3 Google the headline '10-tonne fatberg removed from west London sewer' and read the article that comes up about this. Use this and your own research to design a leaflet advising householders how to get rid of waste such as wet wipes and cooking fat.

→ Going further

Choose a method of managing the water quality in the UK and research how this is operating in your local area. A good starting point is your local water company.

Energy resources in the UK

How is the demand for energy changing in the UK?

Energy is important in everything we do. It powers our cars and other transport; it heats our homes, schools and offices; it powers the machines that produce our clothes and food; and it provides the electricity we use to watch TV and use computers.

Today, we actually consume less energy than we did in 1970, despite there being more than an extra 6.5 million people living in the UK. The average household uses 12 per cent less energy while the decline of heavy industry has led to this sector using 60 per cent less. There has, however, been an increase in the amount used by the transport sector. The number of cars on the road has increased dramatically – in 1970 there was 10 million but today there is over 27 million. The increase in air travel has also contributed to this rise.

This reduction in domestic energy consumption can be explained by:

- the introduction of energy-efficient devices, such as light bulbs and washing machines
- the increasing awareness of the public that they must save energy
- the increased cost of energy leading to lower consumption.

What is the UK's energy mix?

The **energy mix** of the UK refers to the different sources of energy used by households, industry and other commercial users, such as shops and offices. Most of the energy we use in our homes is in the form of electricity. Electricity can be generated by burning fossil fuels such as oil and coal, or by using renewable energy sources such as wind or water.

Fossil fuels

Coal, oil and gas make up the three fossil fuels used today in the UK. These were formed over many thousands of years and as they take so long to be replaced they are regarded as non-renewable. In other words they will run out. When burnt, fossil fuels release carbon dioxide, along with other greenhouse gases. Fossil fuels can be burnt directly to produce heat. They can also be used to generate electricity in power stations and finally they can be used to power vehicles and machinery.

Nuclear energy

Nuclear energy uses uranium to produce heat in a nuclear reactor. This heat is then used to drive a turbine to make electricity. It is not a fossil fuel, but is considered non-renewable as the supplies of uranium are finite.

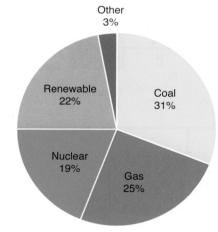

▲ Figure 22.11 The UK's energy mix 2015

Renewable energy

Renewable sources of energy are sources that can be used again and again and will not run out. They include the Sun, wind, waves, the tides, running water in rivers and geothermal heat created deep underground. Methane produced from landfill sites is also burnt to generate electricity from vegetation (biomass). The main problems with these energy sources are the cost of the technology and the relatively small amounts of energy produced. However, they are seen as 'clean' and non-polluting.

How is the UK's energy mix changing?

Until recently, the UK produced enough energy to power homes and industry. The UK had large reserves of oil and gas, but a reduction in these reserves and in the production of coal has led to an increasing reliance on imported fossil fuels.

The production of coal, gas and oil has declined for many reasons (see Figure 22.18). However, there are still significant supplies that could be exploited, often in less accessible areas. One such area is the remote Mariner Oilfield, 150 km east of the Shetland Islands, which started production in 2016.

In addition, policies introduced both nationally and internationally can have an effect on production and the mix of energy used. For example, the use of coal increased in 2011 as older coal-powered stations worked to full capacity, in the knowledge they were soon to be closed due to EU regulations on emissions.

To reduce the reliance on imported fuels, and to reduce the carbon emissions generated by burning fossil fuels, the British Government is encouraging investment in renewable energy sources such as wind and **solar energy**.

Figure 22.12 shows the reduction in the use of fossil fuels in the UK to generate electricity in recent years. However, it also shows the small percentage of energy production from renewable sources, compared to that of fossil fuels. The reliance on fossil fuels to produce our energy will continue into the future mainly due to the cost and unreliable nature of renewable energy sources. However, a reduction in subsidies for solar and wind energy provision may reduce the expansion of this energy production method.

▲ Figure 22.13 Renewable sources of energy

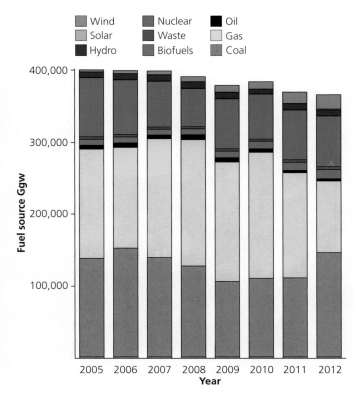

▲ Figure 22.12 Energy production by fuel source, 2005–12

Legend for Figure 22.12:
- Wind
- Solar
- Hydro
- Nuclear
- Waste
- Biofuels
- Oil
- Gas
- Coal

→ Activities

1. Explain why the consumption of energy in recent years has reduced.
2. Make a list of:
   a) non-renewable sources of energy
   b) renewable sources of energy
3. Why is the energy mix of the UK changing?

Geographical skills

1. Study Figure 22.12.
   a) Estimate the increase in the use of wind to generate electricity between 2005 and 2012.
   b) Explain the reasons for this increase.

Economic and environmental issues of energy production

Exploiting new or existing sources of energy comes with opportunities and challenges. Some of these are related to the economic costs and some to the environmental impacts of developing or expanding the energy source.

What are the economic and environmental issues of different types of energy production?

Fossil fuels

Some estimates say that there is sufficient coal in the UK for at least another two to three hundred years. So why do we not dig it up rather than importing coal from other countries? Or, why do we not drill in new locations to find oil or gas?

▲ Figure 22.14 Gas-fired power station in Pembroke, Wales

Nuclear power

The amount of energy generated in the UK by **nuclear power** has fallen from around 25 per cent in the late 1990s to around 18 per cent today. Nuclear power stations in the UK are all expected to be closed (**decommissioned**) by 2023. However, it was recently decided that new-generation plants should be built and should be working by 2025. Many people see nuclear power plants as a clean form of power production that will enable the UK to meet targets for the reduction in emissions of CO_2. However, the risks associated with the industry are seen as great, given past and recent disasters, and the cost of building the power stations, as well as storing or disposing of the nuclear waste, is often very high.

▲ Figure 22.15 Nuclear power station, in Sizewell, Suffolk

Renewable energy

Renewable energy such as wind and solar power could be seen as the solution to maintain the supply of electricity in the UK. However, there are conflicting views about this, too. Many people are concerned about the visual impact of, for example, wind turbines on the environment. Wind turbines need to be placed in exposed positions to maximise their exposure to the wind, and, very often, these are places of particular scenic value. Recently, a proposal to build a wind farm off the Dorset coast was turned down because of the visual impact it would have on the Jurassic Coast, which is a World Heritage site.

Figure 22.17 is an overview of the associated costs and future costs of energy production.

▲ Figure 22.16 Offshore wind farm

Fuel	Cost now		Future cost		Carbon emissions g/KWh	UK supplies
Coal	very low		moderate		1,100	moderate
Gas	very low		moderate	low	400	poor
Oil	moderate		high		900	very poor
Shale gas	moderate		moderate		500	excellent
Nuclear	low		moderate		20	very poor ***
Wind offshore	moderate		low		11 *	excellent
Wind onshore	low		low		11 *	excellent
Solar	moderate	high	low		45 *	moderate
Tidal barrage	moderate	high	moderate		9	good
Biomass	low		low		40	moderate
HEP	very low		very low		7	poor **

◄ Figure 22.17 The costs of energy (*producing and installing equipment, **lack of mountainous areas, ***no UK uranium)

→ Activities

1 Explain the terms:
   a) economic
   b) environmental.

2 Explain why the costs of energy supplies may change over time. Use examples to illustrate your points.

3 Using the information in Chapter 22 and your own research, consider the future of the UK's energy supply. Which source should we be investing in? Should we be considering a mix of energy sources? If so, which and why? What points should you take into account when making your plan? Write a report to summarise your findings.

4 Is there an alternative to increasing production from renewable energy sources in the UK? Justify your answer.

5 A new wind farm is planned for the south coast of the UK. Think about the points that might be made both for and against it. Include viewpoints from local residents, the wind farm company, local fishermen, local leisure sailors, tourist chiefs, the local council and an environmental organisation.

Economic and environmental issues of energy production

▼ Figure 22.18 Challenges and opportunities for different types of energy

	Economic Challenges	Environmental Challenges
Fossil fuels	• Much of the remaining coal is in hard-to-access areas, often deep underground, which is very expensive to mine. • With the UK's last coal mine closed in 2015, coal must be imported from countries like South Africa. • Mining coal causes environmental problems such as waste or spoil heaps, which are expensive to clean up. • Miners often suffer from diseases related to their jobs, which will incur a cost to the health service. Emissions from fossil fuels can also cause respiratory diseases, again incurring a cost. • Costs of exploring more remote and inaccessible areas in the North Sea, or costs of drilling in heavily populated areas (Sussex) or sensitive areas (Dorset). • Costs of climate change for example increased flooding requiring flood defences (see Chapter 11).	• The burning of fossil fuels creates greenhouse gases which may contribute to climate change (Chapter 4) and causes acid rain. • Waste heaps from coal mining can create visual pollution. • Opencast coal mines are unsightly and create dust and noise, disturbing local people and wildlife. They also use huge areas of land. • Access roads and support industries for all of these can destroy wildlife habitats and impact on land visually. • Gas and oil coastal terminals take up space and can mean the digging up of large areas of land for pipelines to transfer to the power stations. • Fracking (see opposite) presents its own challenges.
	Opportunities	**Opportunities**
	• Creation of jobs directly, in support industries and in the manufacture of equipment. It can bring money and jobs in to an area – a multiplier effect.	• Carbon captive storage (CCS) is more efficient but expensive (see Section 4.5).
Nuclear	**Challenges** • The costs of building nuclear power stations are huge. • There are enormous costs to store and transport nuclear waste. It is expensive to **decommission** power stations.	**Challenges** • The waste from nuclear power stations must be stored safely for many years to avoid contamination. • The environment can be considerably more dangerous if an accident occurs. Nuclear accidents can lead to the release of radiation into the atmosphere which can have a long-term detrimental impact on wildlife and people.
	Opportunities • Creates jobs in research and development for new technologies in the nuclear power industry. • After the initial investment, energy generated by nuclear power is seen to be cheaper.	**Opportunities** • Nuclear power is considered cleaner and less polluting than energy generated by fossil fuels.
Renewables	**Challenges** • High set-up costs of renewable energy sources, such as wind turbines, solar farms and tidal power stations, especially in the remote areas suitable for this type of energy generation. • The impact on the visual environment can impact on tourism and reduce income and jobs. • Low profitability is also a concern.	**Challenges** • Evidence shows that wind turbines can effect bird migration patterns and bat life in the area. Turbines located at sea are believed to impact on sea currents and on fish and birdlife. • Many people consider wind turbines ugly. • Wind turbines and the associated access roads can impact on untouched land such as the Highlands of Scotland. • Turbines are also noisy and can disturb people and wildlife living nearby. They can also block TV and phone signals.
	Opportunities • Many jobs are created in the manufacture of solar panels and wind turbines along with jobs in research and development.	**Opportunities** • They produce much lower carbon emissions. • Land used for siting wind turbines can also support other uses such as farming and leisure activities. • Offshore wind turbines can act as an artificial reef, creating habitats for marine wildlife.

Fracking: a new energy production issue in the UK

Fracking or hydraulic fracturing is currently being explored in the UK as a means to extract gas that is locked in rocks thousands of metres below the Earth's surface.

A hole is drilled deep into the rock, and a mixture of sand, water and chemicals are injected into it at high pressure, which splits the rock and releases the gas. This technique has been used in the North Sea oil and gas fields for many years.

Areas where this is being considered include Lancashire, Yorkshire, the East Midlands and some areas in southern England, such as Sussex and Surrey.

The economic and environmental challenges and opportunities of fracking

- Fracking for shale gas can lead to pollution of ground water, which in turn can lead to contamination of drinking water with hydraulic fracturing fluids
- It also requires the use of large quantities of water and can impact water supplies in some areas
- Additionally, the process of fracking has been linked to low-level earthquakes
- However, fracking can bring economic benefits in terms of more government revenues and more jobs for people.

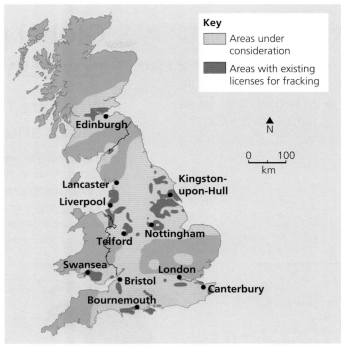

▲ Figure 22.19 Shale gas sites in the UK

→ **Activities**

1 Using Figure 22.19, describe the pattern of areas where fracking may be carried out.

2 Using Figure 22.20, draw a flowchart to explain the sequence of actions used to 'frack' for gas.

3 Classify the bulleted list to the left into economic and environmental, challenges and opportunities.

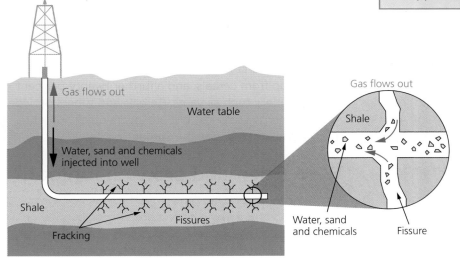

◄ Figure 22.20 The fracking process

23 Food

Global patterns of food consumption

What are the differences in global calorie intake and food supply?

Food security was defined by the World Food Summit in 1996 as 'when all people at all times have access to sufficient, safe, nutritious food to maintain a healthy and active life'. Many people do not have food security (as you may remember from Figure 21.7 on page 308) and are suffering from undernourishment. There is a great difference in calorie intake in different parts of the world (Figure 23.1).

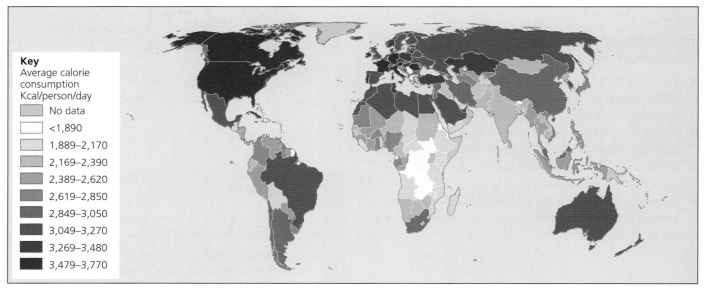

Key
Average calorie consumption Kcal/person/day
- No data
- <1,890
- 1,889–2,170
- 2,169–2,390
- 2,389–2,620
- 2,619–2,850
- 2,849–3,050
- 3,049–3,270
- 3,269–3,480
- 3,479–3,770

▲ Figure 23.1 Average calorie consumption (Kcal/person/day) by country, 2010

The world as a whole produces enough food for everyone, but not everyone has equal access. If we distributed the world's food evenly, there would be enough for every person to receive around 2,720 calories per day, plenty for everyone to live a healthy life. As Figure 23.1 shows, the countries suffering from the lowest calorie intake are almost all in sub-Saharan Africa. Another issue is the composition of the average diet:

■ In HICs, over a quarter of the diet is made up from meat, fish, eggs, milk and cheese and approximately a quarter from cereals.

■ In the countries in Africa and other LICs suffering from the worst **food insecurity**, over half the diet is made up from cereals, and a further twenty per cent from tubers such as yams. Cereals and tubers provide food energy (calories) but are low in other nutrients, which leads to undernourishment.

Why is demand growing for food?

The demand for food has grown over time, as we develop new processes and new products and change our way of life. As LICs and NEEs develop industrially and economically, their demand for different food products increases as well.

Population growth

Global population growth is the greatest pressure on our essential resources such as food. In the past ten minutes, while you have been reading this, the world's population has grown by approximately 1,400 people, so, in the past hour, that's around 8,400 people.

Where are the areas of fastest population growth?

The problem for food supply is that the growth of population is not even. Parts of the world are growing at a faster rate than others (Figure 23.3). There is a huge difference in the rates of population growth. The population in Africa is growing by 2.51 per cent per year, while the population of Europe is growing by 0.1 per cent per year.

Figure 21.6 on page 308 shows the areas of the world where there is a real risk of hunger due to lack of food. If you link this to the graph of population growth shown in Figure 23.3, and Figure 23.1 showing calorie intake, there is a clear correlation between the areas of greatest population growth and the areas which have the highest levels of undernourishment and lowest calorie intake per person.

Economic development

As people in NEEs such as India and China become richer, their diets change. They begin to eat more meat instead of grains such as rice. Grain is fed to animals to produce the meat for people to eat, rather than grain being eaten by people. This means more food resources are needed to feed the people of the country. Additionally, as wealth grows, so does the demand for highly processed and convenience foods, which can also increase the calorie intake. China provides a good example of this, as shown in Figure 23.4.

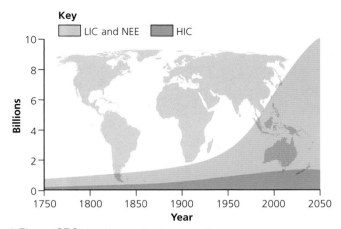

▲ Figure 23.2 World population growth

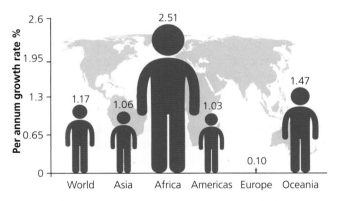

▲ Figure 23.3 Population growth by continent

▼ Figure 23.4 China in 1981 and 2011

Year	Average calorie intake	Percentage of meat in diet	GDP (US$)
1981	2,165	6	193.30
2011	3,073	17	5,574.20

→ Activities

1 Describe the relationship between Figure 23.1 and Figure 23.3.

2 Explain why calorie intake and diet change when GDP changes in a country.

→ Going further

Choose a HIC and a LIC from two different continents, and use the website www.nationalgeographic.com/what-the-world-eats to:

a) draw graphs to show the composition of their diets.

b) describe the differences and similarities in their diets.

Geographical skills

1 Use Figure 23.1 to describe the global pattern of countries that are suffering from food insecurity. Use the information on page 306 to help you establish the average calories per person needed for a healthy diet and to have food security.

2 Use Figure 23.3 to calculate the difference in population growth between Europe and Africa.

Causes of food deficit and surplus

What affects the amount of food we have?

In the parts of the world, mainly LICS , where there is food insecurity, a majority of the population is rural, which means they depend upon agriculture for food and cannot 'pop down to the shops' to buy it. The income they may have will also come from farming. In rural areas, drought and conflict are the main factors that cause problems for food production. However, other factors, such as poverty, and pests, are also important, as explained below.

Climate

This includes climate disasters such as floods (see Chapter 1), tropical storms (see Chapter 3) and long periods of drought.

- Drought is one of the most common causes of food shortages. It causes huge losses not only of crops but also livestock. African countries such as Ethiopia and Somalia suffer especially from droughts that last for many years. Drought can lead to desertification, and salinisation, which can reduce fertile farmland areas.
- Floods, often caused by tropical storms (such as Hurricane Haiyan in the Philippines), can also cause huge losses of crops and livestock.
- Climate change is altering normal weather patterns in many areas; droughts are becoming longer, floods greater and hurricanes more frequent.

Pests and diseases

- LICs often suffer from a wider variety of pests and diseases than HICs, due to the climate and lack of investment. Cattle diseases, such as bovine pleuropneumonia or Rift Valley fever, are prevalent along with insects such as locusts which can decimate a grain crop.
- Other pests such as mice and rats will cause damage after harvest due to insufficient storage facilities for the crops.
- Diseases such as AIDS and malaria worsen food insecurity, as they reduce the available workforce in agriculture.

▲ Figure 23.5 Drought in the Sudan

▲ Figure 23.6 Rotting sacks of grain in India

Technology

Many LICs lack the money to invest in agricultural infrastructure such as: roads to transport their produce, warehouses for safe storage, irrigation systems, and machinery that could increase the crops or land under cultivation. This can reduce yields and therefore food supplies.

Water supply

The level of water stress is more important than the amount of water available. The level of water stress within a country is when annual fresh water supplies drop below 1,700 cubic metres per person per year. Water stress makes food production difficult.

- Water stress can occur where there are floods when the water becomes dirty and polluted; when the water supply (rainfall) is unreliable; and when the population density is high, such as in cities in LICs.
- With the current rate of climate change (see Chapter 4), almost half the world's population will be living in areas of high water stress by 2030, including between 75 million and 250 million people in Africa.
- HICs have the money and technology to manage water stress, using **water transfer** schemes and irrigation to enable them to produce food that LICs do not.

Poverty

People living in poverty are often described as being in the 'vicious cycle' of poverty:

- They cannot afford to buy nutritious food for themselves and their families.
- This makes them weak and unable to work on the land or earn money to support themselves.
- Poor farmers often cannot afford to buy seeds, tools or fertilisers. This can limit the amount of crops they produce, as well as reducing the quality of the crops.

Conflict

There are many ways in which conflict can cause hunger:

- Conflicts and wars can disrupt farming and food production. Fighting forces millions of people to flee their homes, leading to hunger as they find themselves without the means to feed themselves.
- In times of war, food can become a weapon. Food supplies can be seized by soldiers or destroyed.
- Farming areas may be mined to prevent the local people harvesting crops or growing food in the future.
- Wells and water supplies can be deliberately polluted.
- Aid workers are often prevented from reaching the people in need, as travelling through areas of conflict can be dangerous.
- International aid can often not reach the most vulnerable people due to a high level of corruption and political instability in many LICs.

Examples of this can be seen in the conflict in Somalia and the Democratic Republic of Congo in Africa, which has contributed significantly to the level of hunger in the two countries.

→ Activities

1 Use the information on this page to create a mind map showing some of the links between the different causes of food insecurity.
2 Explain how conflict can lead to food insecurity.

Geographical skills

Use a blank world map to annotate the causes of food insecurity in the world. Carefully research the causes, given in this section, of food insecurity and ensure you have a clear place-specific example for each one to add to your map.

⭐ KEY LEARNING

➤ The impacts of food insecurity on people

➤ The impacts of food insecurity on the environment

The impacts of food insecurity

How does food insecurity affect people?

The most obvious problem associated with food insecurity for people is the lack of food.

Famine and undernutrition

Famine is the clearest indicator that people in an area do not have sufficient food. Famine is described as 'a widespread scarcity of food'. Famine:

- not only causes death, but also undernutrition, which weakens immunity and makes people more vulnerable to diseases.
- may also lead to deficiency diseases such as beriberi or anaemia.
- can also hinder the physical and cognitive development of children.

The UN Food and Agriculture Organisation estimates that about 805 million people of the 7.3 billion people in the world, or one in nine, were suffering from chronic undernourishment during the period 2012–14. Almost all of the hungry people, 791 million, live in LICs, representing 13.5 per cent of the population of LICs.

Rising prices

A shortage of food can lead to an increase in prices. Rises in food prices locally can often be caused by global events, such as the poor grain harvests in Russia, Australia, and Pakistan in 2010. This led to a ban on exports of grain from Russia, so causing a reduction in the supply of grain and a rise in global prices. In LICs, the shortage of food can cause prices of basic food stuffs such as rice and maize to rocket and become out of reach for the average family. While not suffering from famine, people are suffering from undernutrition and the diseases associated with poor diet, as they cannot afford to access nutritious food.

Conflict and social unrest

The increased competition for scarce food resources can lead to conflict in both local and international communities.

The need for water for farming can lead to international disputes over ownership of the water sources (see page 342). For example, the River Nile runs through several countries and all of them need the water from the river to help feed their population. There have been long negotiations over how the water is shared:

- The Ugandan government wants to dam the river, which will restrict the flow downstream.
- The Sudanese government would like to use more of the water for irrigation to help grow more food.
- The Ethiopian government currently uses the water from the Nile for irrigation to support its coffee industry, which brings in much-needed income.

- Egyptian farmers use the water to grow vast amounts of food, such as avocados and fruit, for the global and domestic markets.

Who has the right to the water?

Additionally, social unrest in local communities is often related to rising prices of staple food stuffs. Between 2008 and 2011, 60 riots linked to rising food prices or food shortages were recorded around the world, especially in North Africa and the Middle East. There were five days of rioting in Algeria in 2011, when flour and cooking oil prices doubled.

▲ Figure 23.7 A protest against rising prices in Algeria, 2011

How does food insecurity affect the environment?

In LICs, the best land is often used to grow cash crops for export to HICs, such as flowers and green beans in Kenya. This leaves the less suitable land for growing food to feed the local population of the country. This marginal land often does not have sufficient nutrients or water to produce a good harvest and will quickly become infertile. Infertile land cannot support plant growth, which leaves the soil exposed and prone to erosion through wind and water.

Overgrazing by cattle and other farm animals also leaves the soil exposed. Once the soil has no vegetation to hold it in place, it can be blown or washed away, so causing soil erosion.

Where cash crops are grown in both LICs and HICs, the increased use of pesticides and fertilisers to produce a large healthy crop can cause water pollution. It can also increase the demand for water for irrigation and lead to water shortages. Water shortages and pollution can also have an impact on the indigenous wildlife habitat.

▲ Figure 23.8 Overgrazing and soil erosion

→ Activities

1   Using Figure 23.8, suggest reasons why farming can lead to soil erosion in LICs.

2   Use Figure 21.7 on page 308 and Section 23.2 to research the causes of undernutrition in two countries, each in a different continent.

Food, glorious food

How can we increase the amount of food produced globally?

There are many ways in which we can increase the amount of food produced to combat food insecurity. Some are simple and 'low-tech'. They cost very little money but can make a huge difference to many people. Others require huge financial investment and the use of the latest technology but can also help to increase the global food supply.

Irrigation

Irrigation is used to supply extra water to farming areas to increase or maintain production when the water supply is unreliable or low. It can increase crop yields as well as income, and so helps to reduce poverty, a major cause of food insecurity. The source of the water used varies from place to place and includes rivers and lakes.

Irrigation is used a lot in countries with arid or semi-arid climates, including parts of Europe and North America. However, irrigation is not so widely used in sub-Saharan Africa, where the cost of investment along with the very unreliable rainfall can be prohibitive. Here only 3.5 per cent of the land is irrigated; in Asia up to 33 per cent of the farm land is irrigated.

There are many problems with irrigation systems:

- large-scale schemes can push people off the land to be used for reservoirs (see Section 11.12).
- increased waterlogging of soil due to lack of drainage
- salinisation, or the build-up of salts and minerals in soil
- competition for water from irrigation schemes, leading to water scarcity for local subsistence farmers
- the cost involved with setting up the method for delivering water.

Aeroponics and hydroponics

These are new ways of growing plants without soil.

Aeroponics

Aeroponics is the process of using air rather than soil to grow plants. Plants are usually grown suspended in the air, in a closed environment such as a greenhouse. Nutrients and water are sprayed in a fine mist onto the roots and lower stem of the plant every few minutes.

The plants tend to grow faster as the roots are exposed to more oxygen, and it is easier to ensure that the plants have all the nutrients they need. The plants' roots have no direct contact with water or soil.

Figure 23.11 shows some of the advantages and disadvantages of aeroponics.

Types of irrigation

The ways the water is 'delivered' to the crops include:

- large-scale schemes with reservoirs, canals and dams, such as the Aswan Dam in Egypt

- drip irrigation, where water is delivered to the plants' roots through pipes that are full of tiny holes and which are spread across fields. It can be computerised (high-tech) or manual (low-tech). The advantages are that water is not wasted and evaporation is minimised

- flood irrigation, where fields are flooded in a controlled way. Water is pumped by hand, animal or machine along canals controlled by small dams, from a river or well. This is commonly used in rice paddies in Asia

- sprinklers, either permanent or moveable, are often found on farms in HICs, such as maize fields in France. Water is delivered to a central point in the field and then to the plants by sprinklers

- buckets and watering cans.

▲ Figure 23.9 Plants grown in an aeroponics system

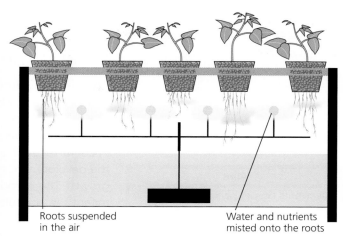

Roots suspended
in the air

Water and nutrients
misted onto the roots

▲ Figure 23.10 An aeroponics system

▼ Figure 23.11 Advantages and disadvantages of aeroponics

Advantages	Disadvantages
Plants are easily maintained.	It relies upon the nutrient misting system. If this breaks down, it can lead to the loss of all the plants.
Less nutrients and water are needed as they are directly absorbed by the roots and not dispersed into the soil.	Specialist knowledge of the exact nutrient mix for the plants, as well as the technology behind the system, is needed.
It requires little space compared to conventional growing methods, as plants can be stacked on top of one another.	The root chambers which hold the plants must be regularly cleaned, as they are very easily contaminated in the warm, moist conditions.
The plants can be easily moved around.	It is very expensive.

▲ Figure 23.12 Lettuce grown using hydroponics

Hydroponics

Plants grown using a **hydroponics** system are grown in water. The roots of the plants are in a nutrient-rich water bath throughout their lifespan. Examples of crops grown this way are lettuce and tomatoes. There are various ways of doing this, using tanks, pumps and wicking systems.

There are various advantages:

- The plants receive more nutrients, which allows them to grow faster.
- They use up less space, as plants can be stacked on top of one another.
- They use less water than plants grown conventionally.
- Crops like lettuce can be shipped alive in the water to consumers, increasing the freshness.
- Diseases found in soil are eliminated.
- The grower has total control over the speed of growth.

However, as with aeroponics, technical expertise is important, and the system is very expensive to set up and run.

→ Activities

1 Start creating a list of strategies to increase global food supply, based on what you have read on these pages. Colour-code them to show which are 'hi-tech' and which are 'low-tech'.

2 Explain how irrigation can damage the environment.

More food, glorious food

What is the new green revolution?

The green revolution occurred in the 1960s, when scientists developed new strains of seeds which produced higher yields of crops, mainly grains such as rice and maize. These were known as high-yielding varieties (HYV). In five years, the yields of rice, maize and wheat rose by 40 per cent in Asia, in countries such as India and Bangladesh. Along with the new seeds came the development of chemical pesticides, herbicides and fertilisers to aid their growth. The introduction of these, and the method of multiple cropping, meant the agricultural industry was able to produce much larger quantities of cheaper food. This increase in productivity made it possible to feed the growing human population.

However, since the green revolution, the global population has continued to grow and crop yields are not growing with it. So there is now a new green revolution, focusing on Africa instead of Asia (see Figure 23.13).

A major part of this new green revolution is the role of biotechnology.

Biotechnology

Biotechnology is when plants, animals and fish are genetically modified (GM). For crops, this means injecting the genes from one plant into another to give the new plant some of the characteristics of the other one. The most common modifications have been to create plants that are resistant to pests, diseases or herbicides. Resistance to herbicides mean the farmer can spray fields with a herbicide to kill the weeds without killing the crop.

Other modifications have:

- increased vitamins or proteins in rice and potatoes
- developed drought – and salt-resistant crops in order that they can grow in poor conditions
- produced insect-repellent crops, so reducing the use of pesticides
- improved the flavour of foods
- increased the shelf life of foods
- produced disease-resistant crops.

Many people are concerned about the effects of GM crops on the environment as well as on human health. In Europe, GM crops are rare and any GM foods must be clearly labelled. In North America and Australia, there has been less controversy and GM crops are more common. GM crops have many potential benefits for farmers in LICs. Drought- and salt-resistant crops, along with nutritionally enhanced foods, could make a huge difference to food security in these areas.

⭐ KEY LEARNING

➤ Strategies to increase global food supply

➤ The new green revolution, biotechnology and appropriate technology

Aims
• Rethinking farming techniques
• Looking at the whole ecosystem in a holistic way
• Raising yields
• Protecting waterways and soil
• Protecting the wellbeing of farming communities
• Conserving biodiversity
• Prioritising long-term growth over short-term gain

Examples
• Breeding plants which can withstand drought
• Encouraging nutrient cycling through crop rotation
• Using genetic pest and disease control rather than chemical methods
• Increasing crop production in some areas to reduce the pressure on marginal areas
• Promote mixed arable and livestock farming in less fertile areas
• Blending traditional farming methods with up-to-date knowledge of crop and livestock farming methods.

▲ Figure 23.13 The new green revolution

Currently, most GM crops are used as animal feed (corn and soybeans), fabric (cotton) and oil (canola). These crops do not feed the world's poor or contribute to food security, but the increased income from, for example, higher yields of cotton can impact on people's ability to buy more nutritious food. For example, in India, each hectare of GM cotton, increased total calorie consumption by 74 kilocalories per adult per day, by increasing the farmers' income and enabling them to spend more money on food. Also, only eight per cent of households are food-insecure when they are growing GM cotton, as opposed to twenty per cent of households that are not growing GM cotton.

The use of appropriate technology is another strategy for increasing global food supply.

Appropriate technology

Appropriate technology, as mentioned in Chapter 7 (see page 96), is a relatively low-tech strategy. Often, aid projects in LICs have used technology that is impractical for the local people. For example, while tractors and combine harvesters will speed up the production of crops, they also require fuel and specialist skills if they go wrong. Spare parts may be very expensive and unavailable. Those using them must be able not only to operate the machines, but also to repair them.

Therefore, to increase food security, the level of technology used must be accessible to the communities it is aimed at. Aeroponics is an excellent and efficient way of producing salad crops for export, but the financial investment will be beyond the means of most farmers in LICs. However, a simple drip irrigation scheme, where the water is pumped into the pipes by a small diesel engine, would enable the farmers to increase their food production while maintaining the system themselves (Figure 23.14).

▲ Figure 23.14 Drip irrigation

→ Activities

1 Add the additional methods you have now read about to your list of strategies to increase global food supply and colour-code them as before, for 'hi-tech' and 'low-tech'.

2 Suggest reasons why some HICs have chosen not to use biotechnology.

3 Should a country such as Kenya invest in a 'hi-tech' method of increasing food production such as aeroponics? Justify your answer.

> ⭐ KEY LEARNING
> ➤ The advantages and disadvantages of a large scale agricultural development

Almería, Spain: a large-scale agricultural development

The southeast of Spain near Almería has always been an arid area and receives only an average of 200 millimetres of rainfall per year. In the past, film production companies used the region to film spaghetti westerns, as the landscape is so similar to the deserts in the USA.

➤ Figure 23.15 Satellite map of the Province of Almería, Spain, showing greenhouse coverage

In the last 35 years, this area has developed the largest concentration of greenhouses in the world, covering over 26,000 hectares. The greenhouses are owned and operated by a mixture of large businesses and individual farmers. Most of the UK's out-of-season crops, such as tomatoes, lettuce, melons, courgettes, cucumbers and peppers, are grown here. The scheme brings in over US$1.5 billion per year in income, as it delivers over half of Europe's fruit and vegetables. This large-scale farming has developed for several reasons:

- changes in people's diets to eat more vegetables
- the development of suitable plastic to build the greenhouses
- new and fast transport methods, which have lowered shipping costs
- the average temperature in the region is 20 °C with about 3,000 hours of sun per year, meaning that the crops can be grown in the winter without artificial heating, unlike other areas in Europe (the Netherlands), so reducing costs
- low labour costs from immigrants
- funding from the EU and the Spanish government.

The greenhouses are so successful that they have covered the plain of Dalías, and have moved up the valleys of the nearby Alpujarra hills, one of Spain's most unspoilt areas. Almost all the plants are grown using hydroponics. Figure 23.16 shows the impacts on the local economy in Almería.

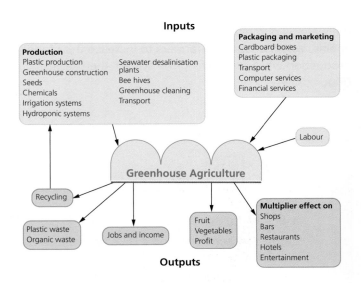

▲ Figure 23.16 The impact of the scheme on Almería

What are the advantages and disadvantages of the scheme?

Advantages

- Large amounts of cheap, temporary labour from North Africa, Eastern Europe, and Central and South America
- The advance of hydroponic growing techniques
- Less water used due to drip irrigation and hydroponics
- A new desalination plant supplying fresh water from sea water to the region
- Low energy costs due to the all-year-round warmer temperatures
- Additional jobs created in packing plants
- Factories producing and recycling the plastic for the greenhouses also provide jobs
- Relatively cheap fresh fruit and vegetables provided all year
- New scientific agribusiness companies have located in the area, providing high-skilled jobs in research and development
- Strict UK regulations on quality have reduced levels of chemicals used and raised production standards

Disadvantages

- The immigrant labour is paid very low wages and often live and work in poor conditions
- There are often clashes between immigrants from different countries
- Many immigrants are working illegally and so have little control over their working conditions
- The local environment has been badly affected – large areas of land have been covered with plastic, destroying the natural ecosystems
- Large amounts of litter are left around, including chemical containers and plastic sheeting
- Plastic is dumped into the sea and is affecting marine ecosystems
- The increased use of pesticides in the area has led to increased health risks for those who work or live near the green houses
- The natural water sources (aquifers) in the area are drying up
- The greenhouses reflect sunlight back into the atmosphere and have contributed to the cooling of the area. Average temperatures have increased in the rest of Spain since 1983, but in Almería they have dropped by 0.3°C per ten years.

◄ Figure 23.17 Greenhouses in Almería

→ Activities

1. Why is Almería in Spain suitable for the intensive production of salad crops?
2. Evaluate the success of the greenhouse industry in Almería, Spain in increasing food production (consider the economic, social and environmental impacts).
3. Discuss the long-term future of the production of salad crops in Almería, Spain.
4. Using Figure 23.16, describe how greenhouse agriculture in Almería contributes to the local economy.

Sustainable food production

What is sustainable food production?

Sustainable food production increases yields but protects land, energy and water resources, in order to maintain food supplies for future generations. It also means increasing food supply and income to poor farmers. The pressure group Sustain defines it as production that:

- contributes to thriving local economies and sustainable livelihoods at home and abroad
- protects the diversity of plants and animals and avoids damaging natural resources and contributing to climate change
- provides social benefits, such as good quality food, safe and healthy products, and educational opportunities.

How can food be produced sustainably?

Permaculture

Permaculture (as in *perma*nent agri*culture*) refers to creating food production systems that co-operate with nature to care for the Earth, for example, using natural predators such as ladybirds to control aphids, instead of chemicals pesticides. It aims to have a minimal negative impact on the natural environment and work in harmony with natural systems.

In practice this means organic farming, growing and buying local food, and eating local, seasonal food (see Chapter 21). While many of these ideas have been popular in HICs for some years, farmers in LICs need support to switch to this way of farming. This may be through education and training, access to better infrastructure such as transport networks and storage facilities, or information and opportunities to invest in their own futures through suitable loans (see page 260).

Organic farming

Organic farming is now practised worldwide, and LICs often produce organic food for export to HICs. Organic farming in LICs not only improves the environment, but can improve the quality of life for the farmers, reducing both their exposure to harmful chemicals and the costs of production. It can, however, increase food miles and energy consumption when the food is shipped to HICs.

Local food

Local food consumption in HICs can benefit people in LICs. It reduces the crops grown by LICs for export, and can release fertile land to be used to produce food for local people, as opposed to marginal land, and so improve food production and quality. The reduction in food miles will also reduce emissions, but if intensive farming methods are used in HICs, their energy consumption will increase (see page 360).

Seasonal food

Eating food that is in season locally can reduce food miles and also reduce the energy used in producing out of season food locally in greenhouses (see Section 23.4).

Urban farming initiatives

This is when gardens are created on unused land in towns and cities. The aim is to increase the connection people have with food production, and contribute to a sustainable system of food production.

In HICs they are often run on a local and voluntary basis, set up by community groups as part of a social movement keen to engage the local community and contribute to a sustainable future. In LICs, however, they form an important part of food security and contribute to income and nutrition by providing fresh fruit, vegetables and meat products.

Many major cities in the USA, such as New York and Detroit, have well-established **urban farming** communities. Other countries, such as the UK, allocate allotments, allowing people with small or no gardens to produce their own food.

▲ Figure 23.18 Urban farming in Chicago

Eating sustainable meat and fish

Meat

There are many views on whether the production of meat for human consumption is sustainable, especially considering the huge cost to the environment. Feeding animals grains and concentrated feeds indoors is less sustainable than the traditional method of grazing animals such as sheep and cattle outdoors on grass (pasture-fed). Grazing livestock outdoors can also be a good way to maintain the landscape, such as moorlands in the UK.

Producing other meat, such as chicken and pork, using free-range methods where the animals are able to go outside for at least part of their lives and exhibit more natural behaviours, is also seen as more sustainable than rearing them intensively indoors. Many countries have developed labelling systems to enable consumers to identify meat produced by more sustainable methods, such as the Freedom Food logo used in the UK.

▲ Figure 23.19 The Freedom Food logo

Fish

Sustainable sourcing of fish and other seafood is when they are farmed or fished in a place where the species can maintain its population indefinitely, and without impacting on other species in the ecosystem by removing their food source or damaging the physical environment. Identifying fish that are sustainable is very difficult due the difficulties in accurately assessing fish

populations. In addition, the methods used to catch the fish can impact on the sustainability of the fish.

Large-scale purse seine net fishing for tuna fish catches many other species, such as turtles, dolphins and sharks, due to the small holes in the nets not allowing them to escape. These other species are often returned to the sea dead and are known as bycatch. A more sustainable fishing method is by using a pole and line.

Seabed dredging or bottom trawling for shellfish can destroy the entire seafloor ecosystem. Diving for shellfish is more sustainable, for example, selecting just the mature scallops and leaving the immature ones to grow and breed, ensuring a future supply.

Sustainable fish and shellfish can also come from fish farms, including mussels and freshwater fish such as trout. However, some large-scale fish farming has caused increased problems for marine ecosystems by introducing diseases to the wild fish population and polluting the surrounding waters with chemicals, antibiotics and vaccines used to control diseases in the intensively farmed fish.

➜ Activities

1   Create a table for the benefits of different types of sustainable food production: at the top include: permaculture, organic farming, local food, seasonal food, urban farming. Make columns for the following, and fill in the benefits:

■   the local environment

■   the global environment

■   people in HICs

■   people in LICs.

2   Why is pasture fed livestock considered a more sustainable form of meat production than grain-fed livestock?

3   Use the websites Greenpeace, The Marine Conservation Society and Hugh's Fish Fight to compile an up-to-date list of sustainable fish.

4   Choose one of the methods of sustainable food production and research it further. Draw a table to show the advantages and disadvantages of the method.

5   Choose one of the methods of sustainable food production and explain why it is sustainable.

➤ How food waste contributes to the problems of food supply
➤ How food waste can be reduced

Reducing food waste

How does food waste contribute to the problems of food supply?

Food loss and food waste are both problems in food supply. Food loss is when food is thrown out due to deteriorating quality before it reaches the point of sale. This may be due to poor storage, transport or packaging. Food waste tends to refer to food thrown away after it has reached the consumer and is mainly a problem in HICs.

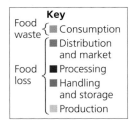

Key

Food waste { ■ Consumption
■ Distribution and market

Food loss { ■ Processing
■ Handling and storage
■ Production

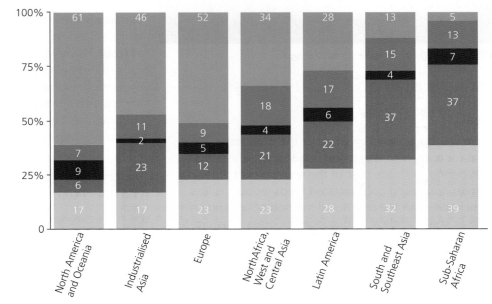

Share of total food available that is lost or wasted

▲ Figure 23.20 The percentage of food lost or wasted globally

Currently approximately one-third of food produced for human consumption is lost or wasted globally. Some sources suggest that the USA throws away as much as 40 per cent of its food supply every year. Imagine putting nearly half your food in the bin every day before eating it. If food waste is reduced, we potentially could have nearly twice as much food as we have today, which would comfortably feed the world. The reduction of food waste is an important part of the problem of food insecurity, along with increasing sustainable production.

Food waste is not only contributing to food insecurity, but also to the landfill sites where this food ends up. The decomposition of the food in landfill sites also contributes to the increase in greenhouse gases. The monetary cost is an issue too. The average family in the UK wastes £700 of food each year. The environmental impact should also be considered – all the water, land and energy that are wasted in the production of that food. Food losses in LICs can severely impact on farmers' incomes and their ability to feed their own families nutritious food.

How can food waste be reduced?

There are many ways to reduce food waste in HICs, mainly due to consumer behaviour and quality regulations imposed on farmers by large supermarket chains.

- Consumers are encouraged to shop more carefully, planning meals and buying exactly what they need, rather than impulse buying. For example, buying the two carrots for a recipe rather than a whole bag.
- Buy odd-shaped vegetables rather than expecting perfect ones so saving tonnes of fruit and vegetables going to landfill due to their appearance. Encourage supermarkets to stock these.
- Look carefully at use by dates and use products before they expire.
- Use left overs rather than throwing them away.
- Ensure food is stored correctly to prevent it going off.
- Learn to preserve products in season, such as by freezing, pickling, canning and making jams.
- Compost peelings and other vegetable waste to ensure they do not go into landfill sites, but return nutrients to the soil.
- In many cases, smaller portions will reduce food waste.
- Supermarkets and governments need to reduce the confusion over 'best before' and 'use by' dates on packaging.

▲ Figure 23.21 Food wastage

How can food loss be reduced?

Food loss occurs most often in LICs where technical problems such as poor handling of produce post-harvest cause deterioration of food. Infestations by pests, such as rats, due to poor storage, lack of 'cold chain' storage and transport and poor marketing can lead to food being thrown away before it is sold. Reduction of food loss in LICs requires:

- investment in infrastructure, such as transport and communications
- education on harvesting, storage and marketing.

This should enable more of the food to reach the consumers whether the food is exported to other countries, sold at local markets or used to feed the farmers and their families.

→ Activities

1   How does food waste contribute to food insecurity?
2   Using Figure 23.20, explain the difference in the percentage of food waste in North America and sub-Saharan Africa.
3   Design an information leaflet for consumers in HICs, encouraging them to do more to reduce food waste.

Geographical skills

Using Figure 23.20, calculate the difference between the percentage of food loss and food waste in Europe and in South and South East Asia. Explain the difference between the two areas.

✪ KEY LEARNING

➤ How food can be produced sustainably

Jamalpur, Bangladesh: a local sustainable food production scheme

How has Jamalpur used a local scheme to increase food sustainability?

Jamalpur district in northern Bangladesh is an area where over 57 per cent of the income is from agriculture. Many farmers are subsistence farmers who grow food to feed their family. The main crops in the area are rice, jute and wheat.

The charity Practical Action has been working with small-scale farmers to increase their income and also to improve the nutrition of the local people. They have been doing this through the use of rice–fish culture, where small local fish are introduced to the paddy fields. The small fish are safely hidden from predators (birds) among the rice plants. The fish provide a natural fertiliser for the rice with their droppings, eat insect pests and help to circulate the oxygen in the water around the rice plants. Using this method rice yields have been increased by ten per cent, and, in addition, the farmers have supplies of fish.

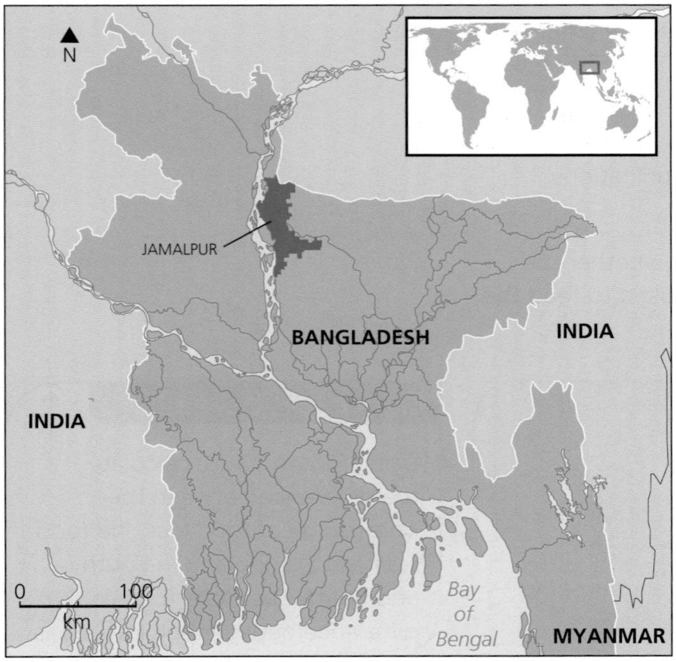

The fish provides a valuable supply of protein for the local people, so improving their health. The increased rice yield not only helps to feed the farmers' families, but also provides a surplus to sell at market, so increasing their incomes. This sustainable method of farming increases food production without the use of increased artificial chemicals or impacting on the local environment.

Construction

First, Practical Action works with a farmer to identify a suitable site: one that is less likely to be washed away should a flood occur. Together, charity workers and the farmer then build a dyke, approximately 60 centimetres high, around the outskirts of the field. This has a dual purpose – to keep the fish in the rice fields and enable vegetable cultivation around the field. The next step is digging a ditch for the fish to live in during the dry season – this is something the whole family gets involved in.

Planting and stocking

The farmer plants the rice in rows that are roughly 35 centimetres apart, and then fills 50 per cent of the ditch with water. The water is purified with a small quantity of lime, and a little organic fertiliser is added. Then, when the rice starts to shoot, the water level across the field is increased to 12–15 centimetres, and small fish or 'fingerlings' are released into the ditch. As soon as they have acclimatised to the rice field water, the farmer releases them into the field and raises the water level as both the fish and rice grow.

▲ Figure 23.22 The Jamalpur district in Bangladesh

Harvesting

At the first harvest, approximately four to five months later, the farmer will harvest the rice first, and then drain the rice field to collect the fish into the ditch, where they can easily be caught.

Every year, Kamrul Barik's family was faced with starvation, but thanks to rice–fish culture he has been able to turn their lives around. Before using the method, he could only produce enough food to last two-thirds of the year. Since adopting the method, he's been able to grow enough food for his family to eat and extra to sell at the local market.

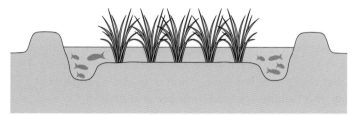

▲ Figure 23.23 Small local fish are introduced to the paddy fields in rice–fish culture

Many other examples can be found of sustainable food production in LICs. The charity Farm Africa is helping people in Kenya, Ethiopia and Tanzania by teaching them how to keep chickens, goats and other livestock. The increased income from the produce can help local people to improve their quality of life through increased income and the increased access to protein, such as eggs and milk, which improve their health.

The most important aspect of all of these schemes is that they are sustainable. They are run by the local people to increase food production. They teach each other and pass on their knowledge once trained. They use crops and livestock suitable for the local conditions and do not use large amounts of chemicals or large scale irrigation, so do not need vast amounts of financial investment. The impact on the local environment is minimal, but the impact on local food supply is great.

'My name is Kamrul Barik and I live in a village in the Jamalpur Sadar district of Bangladesh. It's a small village and I have very little land to live on or to farm – my home is just 15 square metres with 48-square metres of rice fields. I have two sons and one daughter, and my parents are old so they live with us too. In the past it was very hard to grow enough food to feed us all. I could grow enough rice for perhaps eight months, but the other months I struggled to find food. I had to scrape together money by borrowing from moneylenders at a very high interest. I could not afford to send my children to school and we often went hungry for days. Things were very bad.

Then in 2006 Practical Action taught me all about rice–fish culture. I learnt how to choose rice that was more resistant to floods. I also learnt how to protect the fish in my fields. Soon I was able to grow more rice and fish and earn money from selling them – 26,200 taka (£195) from my fish and 18,000 taka (£135) from my rice. Practical Action also showed me how I could farm bananas and vegetables on the dykes, which meant I was able to earn another 10,200 taka (£75).

The extra money has made a very big difference. We can now buy fruit and vegetables at the market so my family can eat better – and of course the fish also give us important nutrients and vitamins. My children are now going to school and I hope they will now be able to find good jobs. I have been able to install a new latrine in our home and I hope to lease another piece of land to start rice–fish culture in another plot. Now the future looks much brighter.'

Source: Practical Action, http://practicalaction.org/rice-fish-culture-3

→ Activities

1 Draw a flow chart to show the sequence a farmer may follow to produce food using rice–fish culture. The first two stages have been done for you:

> Choose a site → Build a dyke

2 Draw an annotated diagram to show how the paddy fields in rice–fish culture work.

3 Explain how rice–fish culture has improved the quality of life for Kamrul Barik and his family.

4 Discuss the ways in which rice–fish culture is sustainable.

★ KEY LEARNING

➤ The global importance of water

➤ The balance between demand and supply

➤ Water security and insecurity

Global patterns of water surplus and deficit

Why is water so important?

Only 2.5 per cent of all the water on Earth is freshwater; the rest is saltwater in seas and oceans (Figure 24.1). It has been calculated that 79 per cent of the world's freshwater is unusable because it is locked up in ice sheets and glaciers. Freshwater is a scarce resource and yet it is essential to all life.

Freshwater is needed for the following purposes (the percentage figures refer to the share of global water consumption):

- domestic use (6 per cent) – for drinking and cooking, bathing and showering, washing clothes and dishes, flushing toilets and watering gardens
- agricultural use (69 per cent) – for irrigating crops and providing drinking water for livestock
- industrial use (20 per cent) – for the production of a whole range of products, from beer and processed food to paper and plastic
- energy use – in the generation of hydroelectric power (HEP) and providing cooling water in thermal power stations. An important point here is that most of the used water is quickly returned to the natural environment
- leisure use – as in sailing and water-skiing on lakes and ponds, sport fishing on rivers, filling swimming pools and watering golf courses.

What is the relationship between water availability and water use?

The global water situation has two important sides to it:

- the demand side – this is the need for, or consumption of, water.
- the supply side – this is the availability of water resulting from the harnessing of various water sources.

The global maps showing the distributions of water demand and water supply do not exactly match each other. If one map were to be laid over the other, we would be able to pick out two types of water balance:

- **water surplus** – where the supply of water exceeds demand
- **water deficit** – where the supply of water is exceeded by demand.

All water

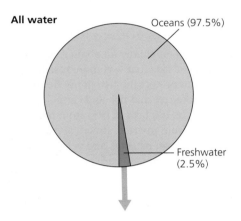

Oceans (97.5%)

Freshwater (2.5%)

Freshwater (2.5% of all water)

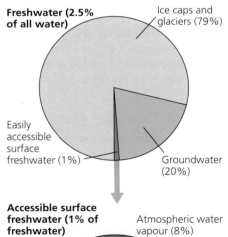

Ice caps and glaciers (79%)

Groundwater (20%)

Easily accessible surface freshwater (1%)

Accessible surface freshwater (1% of freshwater)

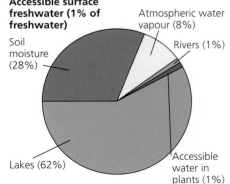

Atmospheric water vapour (8%)

Rivers (1%)

Soil moisture (28%)

Lakes (62%)

Accessible water in plants (1%)

▲ Figure 24.1 The world's stock of water

What are water security and insecurity?

There are two concepts linked to the idea of water scarcity – they are water security and **water insecurity**. Both usually apply at a national level.

Water security occurs in countries where there is a reliable supply of water of acceptable quality. Governments do not have to worry that future increases in water demand cannot be met from known water sources.

Water insecurity occurs where there is an inadequate supply of water of acceptable quality. Governments not only have to find new sources of water to meet present demand, but the pressure to do so will increase with any rise in water demand.

Figure 21.8 (see page 309) looks at the world in terms of water scarcity. There is plenty of freshwater in northern parts of the northern hemisphere. Large areas of Africa and southern Asia suffer water stress. Note that they are located adjacent to areas of low or no water, that is, next to the world's main deserts. In between these two extremes there are areas where water supply and water demand are fairly evenly balanced, as in much of South America and Europe. There it does not need much to tip the situation towards water insecurity. A prolonged drought or sudden rise in the demand for water could easily do this.

▲ Figure 24.2 A symptom of water insecurity – a disappearing reservoir in the USA

Because water is so critical to human survival and well-being, it has the potential to become a source of conflict between countries. The prospect of 'water wars' will become more likely, as the global demand for water will soon be greater than the global supply.

Additionally, both water security and insecurity can exist within the same country. This is particularly true of large countries. Even in England and Wales there is a mismatch between the distributions of water demand and water supply (see Section 22.3). Where there are such mismatches, it is more usual to talk about levels of water stress rather than water insecurity. Clearly, water stress is greatest in South East England and least in Wales and the North.

→ Activities

1. Work out the percentage of the Earth's water that is usable freshwater.

2. Why are no percentage figures given for the use of water by energy production and leisure?

3. Look back at Figure 21.8 on page 309. In which continent do you think a conflict over water might arise? Explain why.

4. Suggest a definition of water stress. Then check your definition with the one at the back of this book.

The consumption and availability of water

It is now necessary to take a closer look at the two components of the water balance: water demand (or consumption) and water supply.

Why is water consumption increasing?

There are two main factors causing water consumption to rise.

Population growth

Although the rate of population growth has been declining for some 50 years, the world's population continues to grow. It is now approaching the 7.5 billion mark. It is hardly surprising that, for example, the consumption of water for domestic use is rising (Figure 24.3).

Economic development

This is happening is nearly all parts of the world. Three particular aspects of it drive the demand for water:

■ the growth of commercial agriculture, with its greatly increased demand for irrigation water. This is by far the largest consumer of water
■ the growth of 'thirsty' manufacturing industries, which need water to make anything from steel and plastics to soft drinks and clothing
■ rising living standards: people have more money to spend on housing, with amenities such as flush toilets, baths and showers. This increases the domestic consumption of water, as does the increasing use of washing machines and dishwashers.

The last two factors behind the rise in water consumption also explain why water consumption levels vary so much from one part of the world to another. Generally, the greater the economic development and the higher the standard of living, the greater the per capita water consumption. Water consumption is therefore lowest in the LICs and highest in the HICs (Figure 24.4). It is interesting to see that NEEs like Brazil, India and China are in the middle of the water consumption rankings.

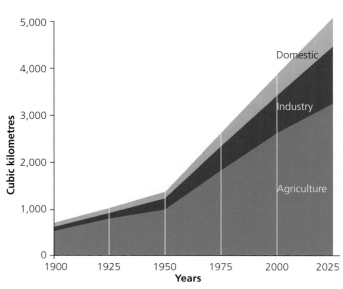

▲ Figure 24.3 The rising global consumption of water (1900–2025)

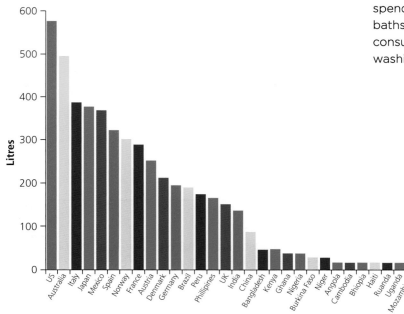

▲ Figure 24.4 Daily per capita water consumption in selected countries

What are the factors affecting water availability?

Let us now look at the other side of the water balance – water supply (also referred to as water availability). In most parts of the world, the water needed to meet the increasing demand comes from three main sources:

■ Rivers and lakes – the water is used either where it is, mainly for domestic purposes such as washing, or is transported by canal or pipeline to the location where it is used.

■ Aquifers – natural underground stores where water collects in porous rocks. The water is extracted by drilling wells or boreholes down into the aquifer. Water is then brought to the surface by buckets, pumps or under its own pressure. This is known as groundwater.

■ Reservoirs – these are artificial lakes usually created by building a dam across a valley and allowing it to flood. River water and direct rainfall are collected, stored behind the dam and, after treatment, distributed to consumers.

There are two main physical factors influencing the availability of water: climate and geology.

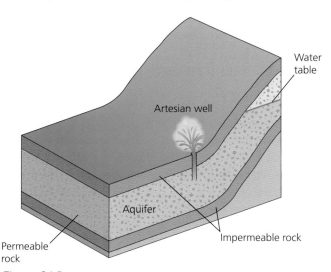

▲ Figure 24.5 Cross-section of part of a syncline

Climate

Globally, this is the single most important factor affecting the availability of water. The areas with most available water are those with tropical or temperate humid climates. Most mountainous areas also receive large amounts of precipitation. A climate that is hot and dry will increase the amount of water that is consumed, for example, in irrigating crops and having frequent showers to keep cool. The high consumption of water in the USA (Figure 24.4) is boosted by its climate.

Geology

This is important in terms of creating aquifers or groundwater stores. Particularly useful are basin-like areas (known as synclines) where a thick layer of porous rock is sandwiched between layers of non-porous rocks. These basins can vary enormously in size.

People also can affect water availability in a number of negative ways.

■ Pollution causes water supplies to become unfit for human use and thus leads to there being less safe water available.

■ **Over-abstraction** occurs when pumping from rivers, lakes and beneath the ground takes place at a rate faster than it is being replenished by rainfall. This leads to sinking water tables, empty wells and higher pumping costs. In coastal areas, over-abstraction can lead to the intrusion of saltwater from the sea which pollutes the groundwater. Sinking water tables can also make rivers less reliable, since many river flows are maintained in the dry season by springs that dry up when water tables fall.

■ A limited infrastructure for the collection and delivery of water will often mean that much water is wasted, for example, from leaking pipes.

■ Poverty often prevents people from having access to safe water. The water available to poor people comes mainly from wells, rivers and standpipes. Pollution is a big risk and the availability of water will vary with the seasons.

→ Activities

1 Why is water so important to commercial agriculture?

2 What is the link between rising standards of living and domestic water consumption?

3 Study Figure 24.4. Calculate the difference between the daily per capita water consumption of Norway and Bangladesh. How do the two countries compare?

4 Why is water insecurity increasing?

5 Look at Figure 24.5. How do synclines help in making groundwater accessible?

KEY LEARNING

➤ Water pollution and waterborne diseases
➤ Food production
➤ Industrial production
➤ Conflicts over water

Impacts of water insecurity

Water insecurity occurs in a country when the demand for water exceeds the supply of acceptable water. See Figure 24.6 for some of the impacts.

Why do water pollution and waterborne diseases occur?

Many rivers multi-task. While they supply water for the three main uses, they also help carry away waste water from homes and industries. As soon as the river flow is unable to carry away all the discharge of waste water, pollution quickly builds up.

The use of hazardous chemicals in manufacturing and agriculture causes severe pollution and makes huge amounts of water unfit for human consumption (see Figure 22.9 on page 316).

The Ganges River in India is one of the most polluted in the world. It contains raw sewage, manufacturing waste, household trash, the ashes of cremated bodies and animal remains.

Health risks will be increased where there is human contact with the polluted waters of lakes and rivers (Figure 24.7). Infectious diseases, such cholera, dysentery, malaria, polio and typhoid, thrive and are readily spread in polluted water. Bilharzia and trachoma are caused by water parasites.

Over-abstraction (see page 345) — Pollution — Safe water shortages

Conflicts — **Some impacts of water insecurity** — Tensions

Waterborne diseases — Rising costs — Lower productivity

▲ Figure 24.6 Some impacts of water insecurity

◄ Figure 24.7 A life put at risk by water insecurity

What is the impact on food production?

Commercial agriculture in many parts of the world is now heavily reliant on irrigation. It is needed to increase crop yields and to top up the supply of water during dry seasons. Livestock quality may also decline if there is not sufficient water for them to drink (Figure 24.8) or for the growth of fodder crops.

What is the impact on industrial output?

Water scarcity has a number of different economic impacts. It will certainly mean that the cost of water goes up. Any price rise is bound to affect the profitability of almost all economic activities. If they have to charge more for their products, this is likely to reduce sales, which in turn will lead to reduced output. Particularly badly affected would be the 'thirsty' manufacturing industries, such as iron and steel, chemicals and textiles.

▲ Figure 24.8 Drinking water for cattle

Water is important in providing much of the energy used by industry. This is obviously the case of hydroelectric power (HEP) generation. Coal, gas and nuclear power stations also require large quantities of cooling water (Figure 24.9).

▲ Figure 24.9 A water cooling tower at a power station

Why might there be conflicts over water?

A world increasingly short of water is fast making water into a valuable commodity. As with oil and other scarce commodities, international competition could so easily lead to international tensions, and possibly even 'water wars'.

Today, tensions over water are inevitable where large river basins and inland waters are shared between two or more countries. For example, India and Bangladesh share the Ganges, but the advantage is with India because it lies upstream from Bangladesh. The Uruguay River acts as part of the border between Argentina and Uruguay. The Aral Sea is now divided between Kazakhstan and Uzbekistan. Part of the boundary between Canada and the USA runs through the Great Lakes. Tensions exist in all these cases, but they only simmer.

We need to look elsewhere in the world for the likely flash points of the first 'water wars'. For example:

- the River Jordan provides the boundary between Israel and Jordan (Figure 24.10). Upstream, the river runs through Syria and Lebanon. All four countries suffer from water insecurity and are competing for the Jordan's water.
- the Tigris-Euphrates river system rises in Turkey and then flows through Syria and Iraq. As such, Turkey holds the upper hand.

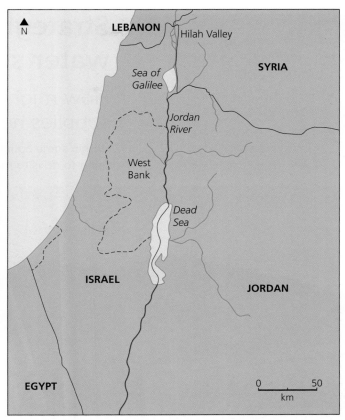

▲ Figure 24.10 Countries competing for the River Jordan's water

→ Activities

1 When does water insecurity occur?

2 Explain why increasing water scarcity often leads to rising water pollution.

3 Study Figure 24.7. Explain how this child's life is being put at risk.

4 Which activity do you think is likely to be most affected by increasing water insecurity?

5 Why do upstream countries have an advantage in shared river basins?

6 Draw your own table with four columns, with the headings 'Environmental', 'Economic', 'Social' and 'Political'. Take the impacts of water insecurity in Figure 24.6 and arrange them underneath the four categories.

→ Going further

Research one of the examples of potential 'water wars' above. Explain how conflicts may arise.

Strategies to increase water supply

How might the diversion or transfer of water supplies help?

Within some countries, water availability varies between regions. The obvious thing to do is to move any surplus water to areas where there is a water deficit. This can be done by pipeline, as in England and Wales. The mountains of mid-Wales, with their heavy rainfall, supply water to the large urban populations of the West Midlands. The River Severn rises in those same mountains (see Section 22.3). Some of its water is diverted near Gloucester and moved by canal, to supply the city of Bristol with half its water.

Rather larger schemes involve taking water from one river basin and transferring it to another (Figure 24.11). One of the best known schemes is the Tagus–Segura transfer project in Spain. There are some even larger and well known schemes in Australia (Murray–Darling), the USA (Colorado) and China (the South–North scheme). The last of these will be looked at more closely in Section 24.5.

Water transfer schemes, both small and large, may help to relieve water shortages in one area, but they do pose challenges:

- They are expensive.
- They often encourage an unsustainable and wasteful use of water in the receiving region.
- There are environmental impacts in the donor region: such as damaged fish stocks and increased pollution.

▲ Figure 24.11 Bolarque Dam pumping station: part of the Tagus Segura transfer project in Spain

What about building more dams and larger reservoirs?

Another strategy is to catch more water by building more and larger reservoirs. They can be created by constructing dams across valleys. On lowlands, the reservoir might simply be an artificial lake created by an encircling embankment. As with water transfer schemes, the reservoir solution to the water supply problem does have its disadvantages:

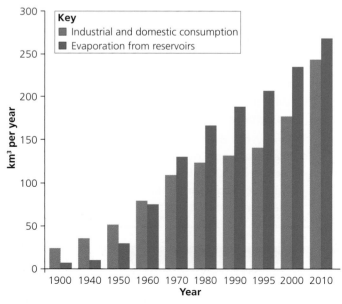

▲ Figure 24.12 More water evaporates from reservoirs than is consumed by people

- Economic costs – Dams are very expensive to build. Water sales, even over many decades, rarely cover the construction costs. Perhaps this might change if the price of water increases in the future.
- Social costs – Large-scale dam and reservoir construction cause much disruption. Where valleys are drowned, people lose their livelihoods and are displaced. They have to start new lives elsewhere.
- Environmental costs – There are many, for example, ecosystems are destroyed by the flooding. (See Section 11.12.)
- Physical costs – A major disadvantage of reservoirs is that there are huge losses of water, not so much by water seeping into the ground, but from direct sunlight and wind causing evaporation from the surface of the reservoir. Figure 24.12 shows how evaporation from reservoirs has increased since 1900.

Is desalination a possibility?

This is the removal of salt and other minerals from seawater. It produces freshwater suitable for human consumption and irrigation. Because of the technology needed, the process has been so costly that only the most wealthy countries have been able to afford it (Figure 24.13). However, the situation is changing. As a result of rising water prices and cheaper technology, more countries are looking to desalination as part of their strategy to increase water supplies.

Today, countries making most use of treated seawater are still Middle Eastern ones, such as Saudi Arabia, the UAE and Kuwait. But they have now been joined by the USA, Spain and Japan. It will not be long before some NEEs are able to turn to desalination.

Having looked at these possible strategies for increasing water supplies, there are two conclusions:

- all strategies have their costs or disadvantages.
- the world must be coming close to over-using or even exhausting its freshwater resources.

Are there other ways of reaching a sustainable balance between water demand and water supply? This question is explored in Section 24.7.

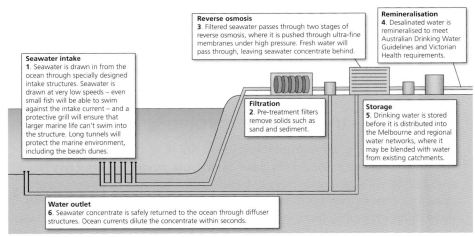

Reverse osmosis
3. Filtered seawater passes through two stages of reverse osmosis, where it is pushed through ultra-fine membranes under high pressure. Fresh water will pass through, leaving seawater concentrate behind.

Remineralisation
4. Desalinated water is remineralised to meet Australian Drinking Water Guidelines and Victorian Health requirements.

Seawater intake
1. Seawater is drawn in from the ocean through specially designed intake structures. Seawater is drawn at very low speeds – even small fish will be able to swim against the intake current – and a protective grill will ensure that larger marine life can't swim into the structure. Long tunnels will protect the marine environment, including the beach dunes.

Filtration
2. Pre-treatment filters remove solids such as sand and sediment.

Storage
5. Drinking water is stored before it is distributed into the Melbourne and regional water networks, where it may be blended with water from existing catchments.

Water outlet
6. Seawater concentrate is safely returned to the ocean through diffuser structures. Ocean currents dilute the concentrate within seconds.

▲ Figure 24.13 A seawater desalination plant, Adelaide, Australia

→ Activities

1 Give some examples of the environmental costs of dam and reservoir schemes.

2 Should we worry much about the social costs of dam and reservoir schemes? Explain your ideas.

3 Explain the increases of evaporation and consumption shown in Figure 24.12.

4 Why is desalination not helping more to increase water supplies?

5 Complete a table listing the advantages and disadvantages of each of the three strategies. What conclusions do you reach?

Example

⭐ KEY LEARNING
➤ The need for the scheme
➤ What the scheme involves
➤ The situation today

The SNWTP, China: a large-scale transfer scheme

China's South–North Water Transfer Project (SNWTP) aims to move huge quantities of water from the humid south of the country to the arid north (Figure 24.14).

Why is this water transfer necessary?

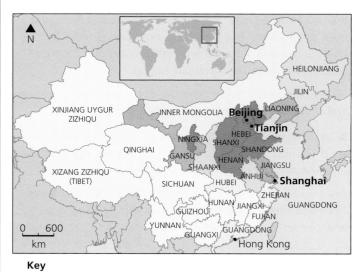

Key

☐ Extreme scarcity ☐ Stress ☐ Adequate
☐ Scarcity ☐ Borderline adequate

▲ Figure 24.14 Water availability in China

The need for this ambitious scheme is the concentration of much of China's amazing economic growth on the North China Plain, particularly around the megacities of Beijing and Tianjin. Here, there is a huge water demand created by over 200 million people, the irrigation of vast areas of farmland, and a vast concentration of thirst-heavy industries.

Given the shortage of rainfall, the North China Plain obtains much of its water from groundwater. The water table below Beijing has been dropping by five metres each year. Some wells around Beijing now have to be drilled to a depth of around one kilometre. Clearly, there is an acute water crisis.

The SNWTP came into being because of two mismatching distributions in China: water availability and water demand.

Key

— Waterway routes

▲ Figure 24.15 The three routes of the South-North Water Transfer Project

What does the project involve?

The idea of moving water from the 'surplus' South to the 'deficit' North was first discussed in 1952. Fifty years later, work started on the SNWTP, a US$62 billion scheme to move 12 trillion gallons of water per year over a distance of more than 1,000 kilometres. The water would be moved along three different routes, from the Yangtze River basin in the south to the Yellow River basin in the north (Figure 24.15), making it an inter-basin water transfer scheme.

Eastern route: this was completed in 2013 and provides water for domestic and industrial use by the cities of Tianjin and Weihai. It mainly uses existing canals, rivers and lakes. All three are already polluted by agriculture and industry. So unless existing water users clean up their act, there is a high risk of transferring polluted water.

Central route: this opened a year later, in 2014, and provides water for twenty large cities, including Beijing. The problem with this route has been the displacement of 300,000 people to create a reservoir at Danjiangkou.

Farmers will not benefit from water being delivered by these two routes. Both routes have also changed drainage patterns and river flows. Wildlife and ecosystems have been badly disturbed.

Western route: this will involve constructing several dams in the Upper Yangtze basin and hundreds of kilometres of tunnels through the Bayankala Mountains. This has been controversial because of its huge construction and environmental costs. The route also crosses an area where there are frequent earthquakes. For these reasons, the project has been put on hold and may never happen.

▲ Figure 24.16 A section of the central route canal

What is the current situation of the project?

After 15 years, the SNWTP has still not been completed. Clearly much has happened in China since the idea was first mooted. Is the project still receiving enthusiastic support? Clearly, the project has its costs and benefits (Figure 24.17). The following suggest that the challenges may be too large:

- Since the 1960s, there have been some severe droughts in the south, so it looks as if it will not be able to supply as much water as was planned.
- Along much of the eastern and middle routes, the water is being transferred in open channels. Evaporation losses must be substantial.
- There are doubts if the benefits of the SNWTP really justify the huge sum of money spent on it. That figure now stands at US$79 billion. The project has been costly, both for Chinese taxpayers and the environment. It is by far the world's most expensive water transfer scheme.

It is increasingly being recognised that there are ways of saving water and not damaging the environment by:

- improving the efficiency of the water distribution networks and reducing water losses
- improving water use efficiency, particularly in agriculture
- reducing the extravagant demand for water by increasing the price
- increasing water re-use by investing in water treatment and tightening controls on pollution
- recharging and conserving groundwater stocks.

The growth of the Chinese economy is slowing down. Maybe this will persuade the Chinese government that it would be better to change its water policy in favour of cheaper and more environmentally friendly ways of tackling the water crisis.

▼ Figure 24.17 Some costs and benefits from the SNWTP

Costs	Benefits
Displacement of huge numbers of people	Provides reliable water supply in the water-deficient north
Wildlife and ecosystems badly disturbed	Improves availability of safe water therefore reduces health risks
Loss of antiquities	
Huge capital investment – taxpayers to pay	Water for industrial growth
Water exports might run the south dry	Water for irrigation
Evaporation losses from canals	

→ Activities

1 Study Figure 24.15. Describe three routes of the SNWTP. Mention the routes of the main river basin invested in the water transfer.

2 Compare Figure 24.15 with Figure 24.14. What potential problems might occur in the south? What about the north?

3 Discuss other alternatives to large-scale water transfer projects. What do you think the Chinese government should do? Give reasons.

➤ The aims of water conservation
➤ Groundwater management
➤ Recycling water
➤ Making use of 'grey' water

Towards a more sustainable use of water

What is water conservation?

Water conservation is about the strategies and actions needed:

- to manage freshwater as a sustainable resource
- to protect the water environment from overuse and pollution
- to meet present and future water demands.

To reach a **sustainable water supply**, it is vital to:

- achieve a balance between water consumption and water supply
- make sure that this balance is maintained in the long term, as water demand inevitably rises.

More specifically, a sustainable use of water lies in using water more efficiently and reducing the wastage of water. Improving the efficiency of water use inevitably means there is less wastage. There is a range of actions that should bring about a more sustainable use of water.

What is groundwater management?

This is usually conducted at local aquifer or drainage basin level. Its aim is to ensure that groundwater supplies do not become heavily stressed by over-use and pollution. Should the former be the cause, then the simplest actions would be:

- to reduce the amount of permitted pumping
- to top up (recharge) the groundwater with either reclaimed or treated grey water (see below).

If pollution is the problem, then clearly the polluters need to be identified and action taken to stop their discharges.

As part of **groundwater management**, the following actions can play an important role.

The challenge with groundwater management is that it needs some sort of authority in overall control of the drainage basin. That authority might be a local community group or a form of public organisation. Whatever form it might take, the authority does need to have legal powers to prosecute those who ignore the groundwater management strategy.

Cutting back on domestic use

People need to become more 'water aware' by cutting down on domestic use, such as:

- reducing the number of showers taken a day and how long they last
- only using washing machines and dishwashers when they are fully loaded.

The aim should be to minimise our individual water 'footprints'.

Stopping leakages

As an example, every day in England and Wales more than 3.3 billion litres of treated water – 20 per cent of the water supply – are lost through leaking pipes. This lost water would meet the daily needs of 21.5 million people.

Improving irrigation practices

Irrigation is one of the largest consumers of water (Figure 24.18). It is also one of the least efficient users of water. Irrigation water needs to be distributed through low-pressure pipes instead of open canals. Drip irrigation of crops is much more water-efficient than spray irrigation. With the latter, much is lost through evaporation and only a small proportion of the water actually reaches the roots of crops. Micro-sprinklers can help here.

The amount of irrigation being applied to the crops needs to be carefully controlled. Too much irrigation can easily lead to the ground becoming waterlogged.

Controlling pollution

In the more developed world, industry is often the largest consumer of water. It is also the largest polluter (see Chapter 21). Too many factories discharge liquid wastes into streams and rivers. Lethal agricultural chemicals also enter watercourses by seepage or runoff. The pollution problem is made worse by the fact that once water becomes polluted, it cannot be safely used for other purposes.

▲ Figure 24.18 Different irrigation methods, flood irrigation for crops such as rice, and sprinkler irrigation

How might recycling water help?

Recycled water is more often referred to as **reclaimed water**. This is sewage water that has been treated to remove solids and impurities. It can be used:

- to irrigate crops
- to meet industrial needs
- for drinking
- to top up rivers where stream flow is low
- to recharge groundwater aquifers.

It is a proven form of water conservation. It is so much more sustainable and environmentally friendly than simply discharging the treated sewage water into a river or the sea.

What about using grey water?

Grey water is used water from sinks, showers, baths and washing machines. Grey water may contain traces of dirt, food, grease, hair and certain household cleaning products, but it can be collected and reused. Figure 24.19 is a simplified diagram of a possible household system for processing grey water. Once processed, the water can be used, for example, to water the garden or flush the toilet. It can certainly be safely returned to groundwater.

At the moment, grey water reuse systems are not common because of:

- the expense – they can cost a lot to install and maintain, and the payback period may be long
- the quality of the water – after some basic treatment, grey water is usually clean enough for flushing toilets. But if is left too long in a grey water storage tank, water quality can decline as bacteria levels rise.

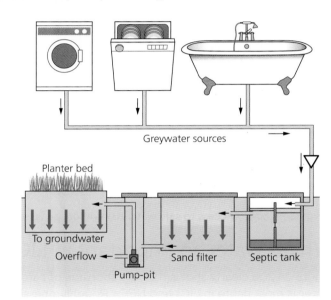

▲ Figure 24.19 Advanced grey water treatment

→ Activities

1 Explain what is meant by 'a more sustainable use of water'.

2 What aspects of groundwater need to be managed?

3 What is the difference between grey water and reclaimed water?

4 Compare the two examples of irrigation in Figure 24.18. (a) Suggest which of these two examples is the most sustainable. Give reasons. (b) Is there a more sustainable method of irrigation than either of these? What is it and why is it more sustainable?

→ Going further

Find out how agricultural chemicals (fertilisers and pesticides) find their way into streams and rivers.

⊛ KEY LEARNING

➤ The need for water in LICs
➤ Strategies used by LICs
➤ The Hitosa scheme in Ethiopia

Hitosa, Ethiopia: a local scheme in an LIC

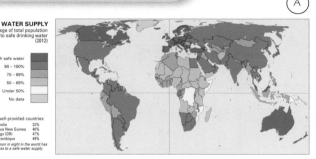

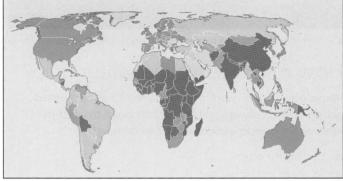

Key
Population with no access to sanitation in % of the total population (2004)

>50%	5–30%	Data not available
31–50%	<5%	

▲ Figure 24.20 (a) Access to safe water (b) Without access to sanitation

Why is water so important in LICs?

Figure 24.20a clearly shows that the shortage of safe water is most acute in LICs and NEEs. Roughly one-sixth of the world's population does not have access to it, and roughly two-fifths do not have access to adequate sanitation (Figure 24.20b). Nearly two million children die each year as a result of diseases caused by unclean water and poor sanitation.

The need for sustainable supplies of water is increased in many LICs because they are often located in water-deficient parts of the world. Modern technology is not available to help them to overcome the water deficit.

What strategies are LICs using?

A range of schemes is now in place in many LICs to help improve the availability of safe water. Most of them are small-scale and local, and make use of appropriate technology. Much of the money and technical know-how comes from international aid organisations, such as WaterAid and Oxfam.

They have done much to promote local WASH (an acronym for Water, Sanitation and Hygiene) projects. They can be found in many countries in Asia and sub-Saharan Africa.

What lessons have been learned from the Hitosa project in Ethiopia?

Hitosa is a largely rural area located 160 kilometres south of Addis Ababa, the capital city of Ethiopia (Figure 24.21). Ethiopia, in the north-east of Africa, is one of the poorest countries in the world. The plains here are hot and very dry, but have been extensively farmed for wheat, barley and oil-producing crops. Prior to the water scheme, the people collected their water from a few shallow, largely seasonal rivers and one spring.

The gravity-fed water scheme began in the 1990s. It involves taking water from permanent springs high on the slopes of Mount Bada, a mountain reaching to over 4,000 metres above sea level. Under gravity, the springwater flows through 140 kilometres of pipeline to over 100 public water points (known as tap stands) and nearly 150 private connections largely related to agriculture.

Successes

- Construction was completed on time and within cost. WaterAid provided over half of the funding and largely designed and supervised the project.
- Twenty years on, it continues to provide a reliable supply of water to Hitosa. Over 65,000 people are supplied with 25 litres of water a day.
- The project is completely managed by local communities.
- People are charged a small amount for the water. The money is used mainly to maintain the physical infrastructure. There has been no misuse of funds.

Problems

- The pipeline, supplied from the UK, may be too costly to replace after its expected lifetime of 30 years.
- The scheme did not include any accompanying education about hygiene and sanitation.
- Hygiene around the tap stands has been neglected, so the risk of disease has increased.
- It has been argued that agriculture is using too much of the water.
- The availability of water has encouraged migration, which means that the scheme is now expected to meet the water needs of well over 65,000 people, threatening the sustainability of the project.

Conclusion

The scheme showed that community management of a large-scale project such as this is possible, but people living close to the springs felt that other people were taking their water. However, the availability of water has had direct economic benefits. Cattle fattening has become one of a number of new businesses in the area. Finally, the time spent collecting water from rivers has been vastly reduced.

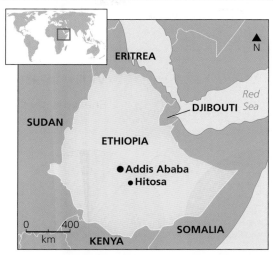

▲ Figure 24.21 The location of Hitosa, Ethiopia

▲ Figure 24.22 Women collecting water at a tap stand

→ Going further

There are five main ways of obtaining water in LICs:

- hand-dug wells
- tube wells and boreholes
- rainwater harvesting
- gravity-fed schemes
- recycling schemes.

Research two of these ways and compare their advantages and disadvantages.

→ Activities

1. From what is 'safe water' meant to be safe?
2. Explain why proper sanitation is so important to human health.
3. Has the Hitosa scheme has been successful or unsuccessful? Justify your answer.
4. Compare the local water scheme at Hitosa with a large-scale water transfer scheme, like the SNWTP in China (Section 24.5). What are the advantages and disadvantages of the two schemes? Do you think a large-scale scheme would be appropriate in Ethiopia, or a local scheme in China? In each case, give reasons.

Global patterns of energy consumption and supply

Why is energy so important?

Energy is needed for a whole range of everyday activities:

- transport
- producing food
- manufacturing
- public utilities
- lighting, heating and cooling in the home
- communication (Figure 25.1).

Chapter 21 gave an overview of the sources of energy (see pages 306–9). As countries become more developed, they consume more energy. Access to cheap and reliable energy sources is important to the health of a country's economy. As a country becomes more developed, it consumes more energy. This includes all the energy we use in our homes for lighting, heating, cooling and a range of household appliances.

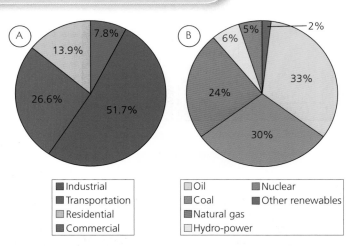

▲ Figure 25.1 a) Global energy consumption; b) Global energy supply

What is energy security and energy insecurity?

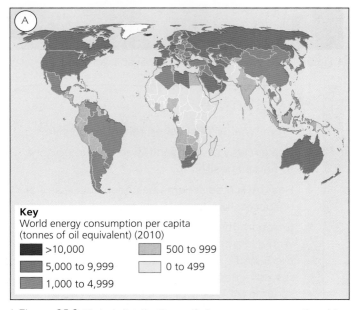

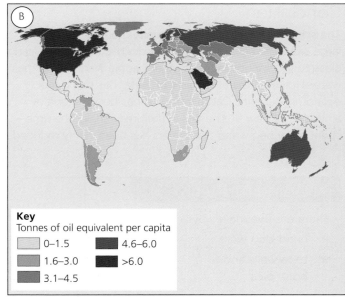

▲ Figure 25.2 Global distributions of a) energy consumption; b) energy supply

Figure 25.1 b) shows that nearly 90 per cent of the world's energy supply comes from three non-renewable fossil fuels: oil, natural gas and coal. The other renewables category is made up of geothermal wind and solar energy, biofuels, biomass and waste. All are mainly used to generate electricity. If we compare the global maps of energy consumption and energy supply in Figure 25.2, we note two significant things:

- the world's major consumers of energy are also the major producers of energy; they are the mainly the high-income countries (HICs). They mostly have energy security, where energy is available for all at an affordable price.
- energy production is low in those countries where the demand for energy is also low; they are mainly low-income countries (LICs). They mostly have **energy insecurity**.

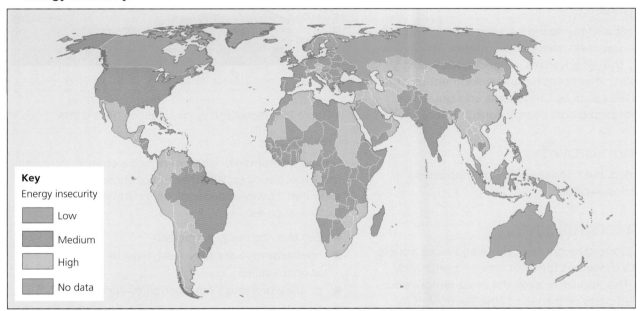

Key
Energy insecurity

Low

Medium

High

No data

▲ Figure 25.3 Risk of energy insecurity

The parts of the world with the lowest risk of energy insecurity include Canada, Russia, most Middle Eastern countries, Indonesia, and Australia (Figure 25.3). High-risk countries occur in Africa as well as in parts of Asia and Latin America. But in some of these countries there is no need to import large quantities of energy as their demands are small.

What is the energy gap?

An increasing number of countries face an **energy gap**. In the UK and elsewhere, the energy gap is becoming wider. The use of fossil fuels, particularly coal, is being deliberately phased out. But the resulting loss of energy is greater than the amount of energy being produced from alternative renewable sources. These countries therefore have energy balances increasingly in deficit. In short, energy insecurity is increasing, particularly when there is more reliance on imported supplies of energy.

→ Activities

1. Study Figure 25.1a. In what ways does the commercial sector consume energy?

2. Study Figure 25.3. Identify a European country that has a low risk of energy insecurity. Explain why it is low-risk.

3. Why does energy consumption rise as a country develops?

Factors affecting energy consumption and supply

Why is the global consumption of energy increasing?

Global energy consumption has nearly tripled over the last 50 years (Figure 25.4). Three factors explain much of this rise.

Development

Development and the accompanying rise in living standards, increases energy consumption. This is particularly true of advances in agriculture, industry and transport. The recent rapid development of large, populous NEEs such as China, India and Brazil, has certainly boosted global energy consumption.

Population growth

As mentioned, the huge rise in global population increases consumption.

Modern technology

Thanks to modern technology, more and more people have access to various forms of energy, particularly electricity. This applies to even the most remote and undeveloped parts of the world (see Section 25.7). Technology is also helping to lower energy costs. This increases the availability, affordability and consumption of energy.

What are the physical factors affecting energy supply?

Just as energy consumption has increased, so too has energy supply. The following are some physical factors that affect energy supply.

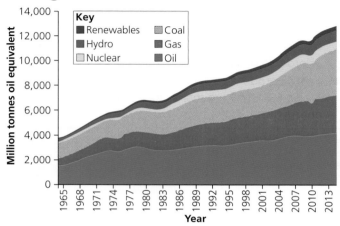

▲ Figure 25.4 Global energy consumption (1965–2013)

Geology

Geology determines whether or not a country contains fossil fuel deposits. Whether or not those deposits are exploited, however, depends on a number of other factors, such as:

■ can they be easily extracted?
■ how extensive are they – can they be worked economically?
■ do they promise a long period of energy supply?
■ does the country have the capital and technology to exploit them?

Climate

With efforts being made to make greater use of renewable energy sources, climate can become an important resource. The three relevant features are rainfall, sunshine hours and windy days. The more there are of these, the better chance there is of generating electricity by hydroelectric power, photovoltaic cells and wind turbines.

Environmental conditions

In some locations, the accessibility of energy sources will be affected by other factors, such as environments with harsh climates and difficult terrain, as in the oilfields of Alaska in the North Sea (Figure 25.5). The more difficult the environment, the greater the costs involved in overcoming the physical difficulties.

What are the human factors affecting energy supply?

Costs of exploitation and production

There is an important link between energy supply and energy consumption (Figure 25.6). Low production costs will mean a relatively cheap energy supply. Cheap energy, in its turn, will encourage high energy demand.

Cheaper energy sources are likely to be the most **exploited**. The cost of conversion into electricity also varies from one energy source to another. This will affect which source is the preferred one to generate electricity.

Movement in the price paid by the consumer is another link between energy supply and consumption:

- A rise in the price of energy may give energy producers a larger profit margin to exploit more difficult or expensive energy sources.
- A fall in the selling price is likely to persuade energy producers to close down their more expensive operations – at least, until the price picks up again.

Technology

Advances in technology often result in:

- the discovery of new sources of energy and the invention of new ways of generating energy
- the development of new ways to exploit more inaccessible sources of energy – for example, fracking (see Chapter 22).

Political factors

Energy security lies in a country producing as much as possible of its energy supply from within its borders. An energy-importing country needs to choose its foreign suppliers carefully. Above all else, they need to be:

- politically stable countries where an uninterrupted supply is guaranteed. For example, exports of oil from Libya have been almost cut off by the civil war there.
- politically 'friendly' countries, namely in the same geopolitical camp, such as the EU, NATO or another international alliance.

→ Going further

Find out why is there so much opposition to fracking in the UK.

▲ Figure 25.5 Oil rig in the challenging North Sea

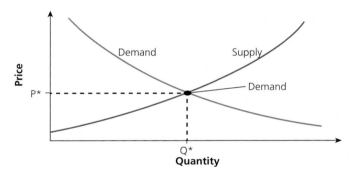

▲ Figure 25.6 Energy supply and demand curves

→ Activities

1. Look around your home and make a list of all the things that consume energy. Do you know where the energy for each thing comes from? If not, try to find out.

2. Give three reasons for the rise in energy consumption.

3. Explain why and how advances in agriculture, industry and transport increase energy consumption.

4. What is meant by the accessibility of energy?

5. Explain the link between the costs of energy production and energy consumption.

6. Why is it important for energy-supplying and energy-importing countries to be in the same political alliance?

⭑ KEY LEARNING

➤ Environmental impacts of energy insecurity

➤ Economic impacts of energy insecurity

➤ Impacts of energy insecurity on people

➤ Potential conflicts over energy

The impacts of energy insecurity

A shortfall in the supply of energy can have a number of different impacts on an energy-importing country.

What are the environmental impacts of energy insecurity?

A shortfall in the supply of energy will persuade a country to take risks with the environment, including:

- clearing forests for wood and then using the land to grow biofuels as in the tropical rainforests of Brazil (see Section 6.4)
- drilling for oil and gas in environmentally sensitive areas, as on Alaska's North Slope, where great damage has been done to the tundra (Figure 25.7)
- flooding valleys to generate more hydroelectric power, as in the Three Gorges project in China, displacing people and depriving them of their livelihoods
- building wind and solar farms in areas of scenic beauty, as in the Scottish Highlands and Cornwall.

▲ Figure 25.7 Pipelines cutting across the Alaskan tundra

What are the economic impacts?

The pressure to use energy sources in economically marginal situations will add to the costs of energy production and therefore will increase the price to the consumer.

Food production

Modern, large-scale agriculture uses vast amounts of energy, from irrigation rigs, milking machines and feed mills to tractors and combine harvesters. Farming and food production are therefore vulnerable, particularly to any rises in energy prices demanded by foreign energy suppliers. Such rises not only make home food production more expensive, but they are likely to sharpen competition from foreign food suppliers in countries that are more energy-secure.

Industrial output

Industrial output is even more vulnerable to the possible impacts of energy insecurity. There is also the same risk of losing out to foreign competition and the consequences of that – a reduced output and labour made redundant.

Energy insecurity not only affects production costs and consumer prices. It might also result in the supply of energy becoming unreliable, leading to power cuts and rationing.

What are the impacts on people?

Energy insecurity impacts on people in a number of different ways:

- a rise in the price of energy will increase the cost of living – running the home, food, manufactured goods and travel will become more expensive
- jobs may be put at risk as sales of goods and services fall – cutting jobs would be one way of offsetting falling sales
- people will be inconvenienced by power cuts, which deprive them of light, heating and the use of electrical appliances (Figure 25.8).

What is the potential for conflict?

At least three different types of conflict might result from growing energy insecurity:

▲ Figure 25.8 A power cut hits a home

- Conflict between the main energy consumers – people, agriculture, industry and transport – as they compete for what energy is available and push up the price. In the UK, it is likely that agriculture and poor households will be the losers.
- Conflict between home-produced goods and imported goods, particularly if those foreign goods come from countries where energy costs are less. Chinese manufacturing has access to energy that is much cheaper than in most of Europe. This, plus cheaper labour, means that China's manufactured goods compete on price.
- Conflict between those countries with sufficient energy supplies and those without. For example, it is thought that Argentina's claim on the Falkland Islands, owned by the UK, may be to do with the untouched reserves of oil and gas supplies located within the territorial waters of the islands. Where pipelines or power cables cross, countries located between the supplier and consumer might be a conflict flashpoint, for example, the dispute over the pipelines that move Russian natural gas across Ukraine to its markets in Europe. Current conflicts in the Middle East are partly rooted in ensuring access to the region's rich oilfields and achieving energy security.

→ Activities

1. In your own words, write a definition of energy insecurity.
2. Study Figure 25.7. What are the problems caused by these pipelines in Alaska?
3. What are the environmental impacts of either wind or solar farms?
4. What would be the impact of rising energy prices on either food production or industrial output?
5. (a) What would be the impact of a power cut on you and your family?
   (b) Now, think about the wider impact on your community. What would be the impacts on the infrastructure and services you use, e.g. your school?

→ Going further

Research which HICs are most dependent on Middle Eastern oil and gas. What other energy options are available for them to exploit within their own borders?

⊛ KEY LEARNING

➤ The energy supply challenge
➤ Future reliance on fossil fuels
➤ Producing more renewable energy
➤ The nuclear option

Strategies to increase energy supply

What is the energy supply challenge?

The challenge facing the world today is how to increase energy supply in ways that:

- do less damage to the environment
- are affordable
- suit the available technology (see Section 25.7)
- do not increase energy insecurity.

Do fossil fuels have a future?

Growing concern about global warming and climate change has persuaded many governments to reduce their use of coal, oil and gas. The burning of these fossil fuels is pumping huge quantities of carbon dioxide and other greenhouse gases into the atmosphere. These fuels account for nearly all the global energy supply. Figure 25.9 shows that two-thirds of all electricity comes from the burning of fossil fuels.

To combat this, people could:

- rely more on alternative sources of energy (renewables)
- make more use of natural gas, the 'cleanest' of the fossil fuels (see Section 25.5), as the overall reliance on fossil fuels is gradually reduced
- find ways of reducing the emissions of carbon dioxide and make fossil fuels a little more environmentally friendly (see Section 25.6).

Can energy supply ever be 'green' enough?

Despite huge efforts over the last two decades, less than ten per cent of the world energy supply comes from renewable sources (see Figure 25.1b on page 356). To date, the main renewables have been hydro and wind power, and biofuels (Figure 25.10).

Biofuels (also referred to as biomass) are often thought to be newcomers to the world of energy. The term is a broad one and includes the earliest fuel known to humans – wood, which is still widely used throughout many LICs. Today's biofuels include:

- biodiesel and ethanol made from crops such as sugar cane and soy beans
- methane captured from the decomposition of rubbish in landfill sites.

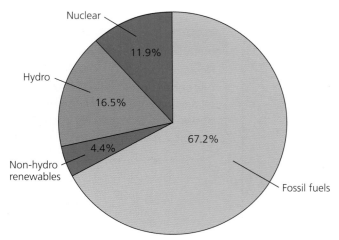

▲ Figure 25.9 Fuels used to generate electricity

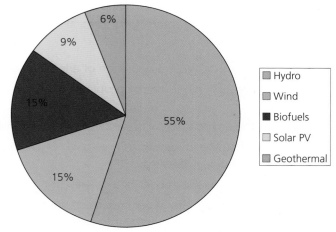

▲ Figure 25.10 Global sources of renewable energy

Other renewables include geothermal energy, tidal and wave power. It is hoped that new technologies will allow these sources, particularly the last two, to make a bigger contribution to energy supply.

Most people say they want our energy supply to become greener. However, proposals to build a new reservoir, wind or solar farm are usually greeted by public outcry and protest. Even when a wind farm project is to be located well offshore, the tourist industry claims that it will ruin the tourists' view and their business. Renewable energy is unlikely to be able to supply all the energy needed globally, and of course, much of the world would need to be covered by reservoirs, wind, solar and biofuel farms.

Is there a nuclear option?

Today, five per cent of global energy comes from nuclear power (see Figure 25.1b on page 356). All this power is used to generate electricity (15 per cent of the world's supply). Uranium is the fuel used in a process known as nuclear fission. The process creates great heat, which is then used to raise the steam that drives the electric turbines.

Because nuclear power needs the mineral uranium, it is classified as a non-renewable source of energy. However, the quantity of uranium needed is very small.

Will nuclear power be part of strategies to increase the supply of energy? Not all HICs are in agreement on this issue. For example, Germany has indicated that it favours a non-nuclear energy strategy. Japan is thinking about abandoning its nuclear power stations. See Figure 22.18 (on page 322) for challenges and opportunities.

A key feature of any supply-increasing strategy must be to find the right energy mix. The mix should:

■ meet present and future energy demands
■ make the best use of energy sources within the country
■ understand the strengths and weaknesses of available technology - the nuclear power option would not be suitable for an LIC, but mini HEP schemes would (see Section 25.7).

▲ Figure 25.11 A tidal power hub

Other renewable sources that might make a greater contribution to energy supply in the future include:

Geothermal energy is thermal energy that is generated and stored in the Earth. It is tapped by drilling holes deep into the Earth's crust. It generates electricity and provides hot water to residential areas. It is cost-effective, reliable, sustainable and environmentally friendly. So far its exploitation has been limited to areas near tectonic plate margins.

Tidal power converts the energy from the tides into electricity (Figure 25.11). The world's first large-scale tidal power plant became operational in France in 1966. Over the years, proposals have been made to harness the tidal power in the River Severn's estuary. Technologically this is possible, but there have always been objections, mainly related to a negative impact on the environment and wildlife. A rather different experimental site is the Race between Alderney and the Channel Islands and the Cotentin Peninsula of France.

Wave power is potentially a great source of energy for conversion into electricity. The energy comes from the up-and-down motion of passing waves. Research is underway to develop an efficient wave energy converter.

These last two energy sources are sustainable, but they are only possible options for countries with coastlines. But not all of those countries will have either the tidal range or the ocean swell to make them viable options.

→ Activities

1 Do you think fossil fuels have a future? Justify your answer.

2 Which of the possible sources of renewable energy shown in Figure 25.10 has the most to offer the UK? Explain your answer.

3 How do you persuade seaside hoteliers that offshore wind farms will not ruin their tourist trade?

4 Do you think that the world can manage without electricity generated by nuclear power? Justify your answer.

⭐ KEY LEARNING
➤ The occurrence, extraction and use of natural gas
➤ The advantages and disadvantages of natural gas
➤ The overall verdict

The dash for natural gas

What should we know about natural gas?

Natural gas provides 24 per cent of the world's energy supply. It was around the middle of the twentieth century that oil took over from coal as the main source of energy (Figure 25.4). It is only since the 1970s that natural gas has begun to challenge the other two fossil fuels.

Natural gas is formed from decaying animal and plant matter that lived millions of years ago. Today, it is found underground, mainly trapped in deep shale rock formations. Wells are drilled and the gas comes to the surface, either under its own pressure or forced up by pumped water. Once at the surface, the gas is pumped through pipelines to where it is used. This transfer often involves pumping the natural gas to ports, where it is shipped in huge tankers. At some point in its transmission, the gas will be refined to remove unwanted chemical impurities.

Figure 25.12 shows the fifteen countries that between them hold over 80 per cent of the world's proven reserves of natural gas. The rankings are dominated by Russia and some Middle Eastern and Asian countries. The USA is well placed, but Russia is the only European country shown. The largest gas field in Western Europe is at Groningen in the Netherlands.

Trillion cubic metres (tcm) — vertical axis: 0, 5, 10, 15, 20, 25, 30, 35, 40, 45

Countries (horizontal axis): Russia, Iran, Qatar, Turkmenistan, United States, Saudi Arabia, UAE, Venezuela, Nigeria, Algeria, Australia, Iraq, China, Indonesia, Malaysia

▲ Figure 25.12 The world's largest natural gas reserves

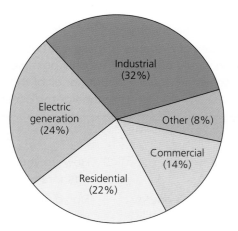

Industrial (32%)
Electric generation (24%)
Other (8%)
Commercial (14%)
Residential (22%)

▲ Figure 25.13 Natural gas use by sector

Figure 25.13 shows that natural gas has a wide range of uses. Most natural gas is burnt to create heat that is either used directly in factories, the home, commercial premises and transport or is converted into electricity.

What are the advantages and disadvantages of natural gas?

Figure 25.14 summarises the main advantages and disadvantages of natural gas.

	Advantages	Disadvantages
Environmental	• A cleaner fuel producing less carbon emissions - 45 per cent less than coal and 30 per cent less than oil • Does not produce waste, such as coal ash • The extraction infrastructure causes less damage to the ground surface. • It is lighter than air and therefore disperses quickly in the case of leakages	• Leakages can be very dangerous, causing explosions and fire. If inhaled, the gas is very toxic • Burning releases greenhouse gases into the atmosphere, but less so than coal or oil • Ground subsidence and earthquakes caused by pumping of gas and fracking
Practical	• Can be used for many different purposes • An economic and instant fuel for heating water and large areas, as well as for cooking • Ideal because it allows precise control and quick results • More abundant than other fossil fuels with large proven reserves • Easy to distribute via pipelines	• Is odourless and leaks cannot be detected unless some odorant is added to the gas
Economic	• Cheaper than electricity • Also produces competitively priced electricity	• The infrastructure for extraction and distribution is fairly expensive • As motor vehicle fuel gives less mileage than petroleum (refined oil)

▲ Figure 25.14 The advantages and disadvantages of natural gas

What is the overall verdict on natural gas?

Natural gas is not a perfect source of energy. There is not one known to us at the moment. But the above evaluation shows that there is much to commend it: the large proven reserves; its relative cleanness and its versatility. Overall, it certainly compares more favourably with the other two major fossil fuels, coal and oil, but of course there are issues with fracking as a method of extraction (see Chapter 22, page 323).

→ Activities

1 Explain why natural gas is challenging coal as a leading energy source.

2 Study Figure 25.13. Suggest possible uses of natural gas that make up the 'other' sector.

3 What do you think is the greatest advantage of natural gas? Give your reasons.

→ Going further

Find out whether the UK has any natural gas sources of its own and where they are located.

⊗ KEY LEARNING

➤ Towards a sustainable energy future
➤ Ways of achieving energy conservation
➤ Making the use of fossil fuels more efficient

- Switch off lights, power sockets, phone chargers and televisions when not in use
- Use energy-efficient light bulbs and rechargeable batteries
- Only use the washing machine or dishwasher when they have a full load
- Use curtains and blinds to provide insulation - from heat in summer and from heat loss in winter
- Wear warm clothing indoors in winter and turn down the central heating
- Walk and cycle more and become less reliant on transport over short distances
- Spend less time on the internet, playing electronic games and texting friends

▲ Figure 25.15 Some ways of reducing personal energy use

Towards a sustainable energy future

What does a sustainable energy future involve?

We need to make sure that today's sources of energy supply will be available for future generations to use. As global energy consumption increases and even greater pressure is put on energy sources, three priorities have become clear:

- Reliance on fossil fuels must be reduced and every effort made to reduce emissions of carbon dioxide to a minimum.
- More use must be made of renewable sources of energy. Research into new sources of renewable energy should also be encouraged.
- Energy efficiency must be improved. This means getting more from the energy we use and eliminating wastage.

What can we do by way of energy conservation?

Energy conservation is all about minimising the wastage of energy and using it as efficiently as possible. The following includes some actions we can take to do this.

Reducing our individual use of energy and our carbon footprints

Energy conservation begins with us as individuals, both in our homes and in our lifestyles. If we all contribute, then the impact on energy conservation can be massive. Figure 25.15 details some simple actions that we all should take.

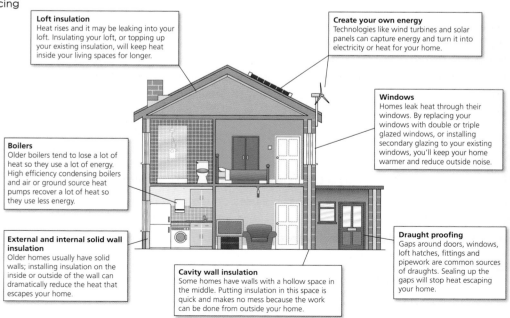

➤ Figure 25.16 Making the home more energy-efficient

In the UK, as in many other countries, most of our energy still comes from the burning of fossil fuels. This means that any saving in our consumption of energy will mean a reduction in our carbon footprint.

Designing homes and workplaces

There are a number of things that can be done to make both existing and new buildings more energy-efficient. Figure 25.16 shows some design features that would make homes more energy-efficient. The same features should also be adopted in the design of new workplaces. Figure 25.17 has some suggestions for energy-saving actions in the workplace (including schools and colleges).

Transport

In this age of globalisation, it does not look as if there is much hope of reducing either the use of transport or its consumption of energy. But as responsible citizens, there are some actions that we can or take that collectively would contribute to a more sustainable use of energy (Figure 25.18).

Demand reduction

The energy conservation actions just described all have the same outcome. They will, to varying degrees, reduce the demand for and consumption of energy. Any such reduction is an important step towards energy conservation. It means that there will be more energy left for use tomorrow.

Can modern technology help us to use fossil fuels more efficiently?

Carbon emissions from the burning of fossil fuels continue to threaten global warming and climate change. Can modern technology help at all?

Chapter 3 mentions carbon capture as one technology (see page 52). Another is combined heat and power (CHP). This generates heat and electricity from a single fuel source, most often oil or natural gas. It uses the heat created during the generation of electricity to provide a supply of hot water for the heating and air conditioning of housing schemes, shopping centres and hospitals. Its overall fuel efficiency is 85 per cent, compared to 52 per cent for a traditional thermal power station. This better efficiency means that consumption of fossil fuels and carbon dioxide emissions are much less. These are two just steps in the right direction. But who can predict what better solutions technology might devise in the future?

- Ensure temperatures are set at no more than 19 °C in offices and classrooms.
- Keep doors and windows closed when the heating is on.
- Avoid heating unused spaces such as corridors and storerooms.
- Don't leave electronic equipment (computers and printers) in standby mode.
- Ensure that inbuilt energy-saving software is activated.
- Use daylight where possible.
- Use low-wattage, compact fluorescent bulbs rather than tungsten ones.

▲ Figure 25.17 Some ways of energy saving in the workplace

- Use public transport rather than private cars.
- Use smaller, more energy-efficient hybrid cars.
- Use alternative, cleaner fuels, such as electricity.
- Car-share when commuting to school or work.
- Reduce the number of aircraft journeys taken, especially short-haul flights.
- Cut down on the number of holidays taken abroad.

▲ Figure 25.18 Some ways of energy saving on transport

→ Activities

1 Explain what is meant by energy conservation.
2 Why should transport be a top target for energy conservation?
3 Can you suggest other energy-conserving actions?

→ Going further

Investigate ways in which the UK government is encouraging a more efficient use of energy.

✪ KEY LEARNING

➤ The energy situation in Nepal

➤ Micro-hydro plants in Nepal

➤ Other possible sustainable sources of energy in Nepal

A local scheme in an LIC: sustainable energy in Nepal

What is the present energy situation in Nepal?

In the Himalayan kingdom of Nepal (Figure 25.19), the present demand for energy is relatively small, but the demand is growing as the country tries to develop and its people struggle for a better quality of life. Wood has been the traditional source of energy for heating and cooking. This had led to widespread deforestation. Crop and animal waste continue to be used as a fuel for cooking.

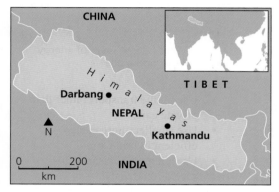

▲ Figure 25.19 Map of Nepal

The wish to supply electricity to its population of 28 million is held back because Nepal has no significant deposits of coal, oil or natural gas of its own. As Nepal is a land-locked and mountainous country, importing fossil fuels is difficult. There is an electric grid system covering part of the country. But power cuts lasting an average of ten hours a day are common. They are so common that the Nepal Electricity Authority even publishes a timetable of power cuts!

What are micro-hydro plants?

The government of Nepal, with support from the World Bank, is helping to create micro-hydro plants across rural Nepal. The plants are built and run by local communities. They are sustainable and bring much-needed electricity for use by local industry, agriculture and commerce, as well in the home.

The micro-hydro schemes in Nepal, as elsewhere, are of the 'run-of-the-river' type (Figure 25.20). They do not need a dam or reservoir to be built, but instead divert water from a stream or river. This water is then channelled to a forebay tank. This is a settling basin that helps to remove damaging sediment from the water before it falls to a turbine via a pipeline called a penstock. The small turbine drives a generator that provides the electricity to the local community.

By not requiring an expensive dam and reservoir for water storage, run-of-the-river systems are a low-cost way to produce power. They also avoid the damaging environmental and social effects that larger HEP schemes cause, including a risk of flooding.

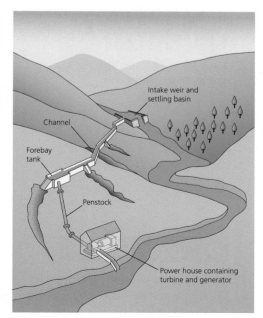

▲ Figure 25.20 Diagram of a micro-hydro scheme

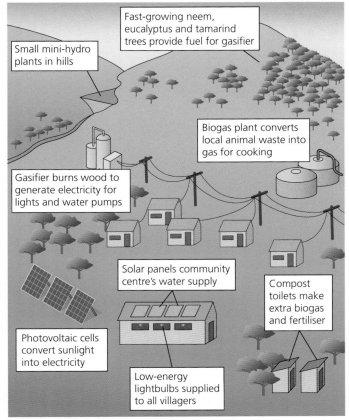

▲ Figure 25.21 Possible energy sources in Nepal

Labels on figure:
- Small mini-hydro plants in hills
- Fast-growing neem, eucalyptus and tamarind trees provide fuel for gasifier
- Biogas plant converts local animal waste into gas for cooking
- Gasifier burns wood to generate electricity for lights and water pumps
- Solar panels community centre's water supply
- Compost toilets make extra biogas and fertiliser
- Photovoltaic cells convert sunlight into electricity
- Low-energy lightbulbs supplied to all villagers

Are there other sustainable sources of energy?

The harnessing of solar energy either by solar panels (for heating) or by voltaic cells (for electricity) offers other sustainable sources, but there are some possible limitations:

- the number of sunshine hours in this mountainous country
- the costs of the panels and cells
- the technology required to maintain the panels and cells.

Three other ways shown in Figure 25.21 are linked to low-tech ways of generating energy that are well-suited to an LIC such as Nepal, and are sustainable.

→ Activities

1 Why is it difficult for Nepal to import oil and natural gas?

2 Why is Nepal well-suited to development of micro-hydro plants?

3 How does a micro-hydro plant benefit agriculture?

4 Why might the use of solar energy be a less attractive option for Nepal than micro-hydro power?

5 Which one of the three biogas energy possibilities shown on Figure 25.21 do you think is best? Give your reasons.

→ Going further

Make a case for and against building mini-hydro plants in the UK.

Local communities are gradually becoming aware of the benefits that micro-hydro plants bring. Take the example of the rather inaccessible settlement of Darbang, located between Kathmandu and the Tibetan border (Figure 25.19). This now boasts a number of new industries. These include a metal and several furniture workshops, a cement block maker, a noodle factory, poultry farms and dairy farms. All these activities have sprung up since the Ruma Khola micro-hydro power plant came into operation in 2009.
The 51-kilowatt micro-hydro supplies electricity to 700 households in five villages, including Darbang.

Over 1,000 micro-hydro plants have been built so far in 52 districts. The Nepalese micro-hydro project is meeting the energy needs of rural communities. In doing so, it is encouraging economic development. It is also making the point that even poor rural communities can enjoy clean renewable energy. The project promises a cleaner and more prosperous future.

Question Practice

Unit 2 Section C

1 Using Figure 21.4 (page 306), state the percentage of water that people in the UK use for washing and cleaning. Select **one** answer only.

    A 54%   C 33%

    B 84%   D 41%   **[1 mark]**

2 Using Figure 22.8 (page 314), describe the distribution of areas of water deficit in the UK.   **[2 marks]**

3 a) Create a line graph showing the changes in organic sales 2005–14 using the data below.

Year	Sales (£ billion)
2014	1.86
2013	1.79
2012	1.74
2011	1.67
2010	1.73
2009	1.84
2008	2.10
2007	2.10
2006	2.00
2005	1.60

> Make sure you are confident in basic graphicacy skills such as drawing graphs.

▲ Figure 1 Changes in organic sales 2005–14   **[1 mark]**

   b) Calculate the rate of increase in organic sales from 2011 to 2014.   **[1 mark]**

4 Using Figures 22.6 and 22.7 (page 314), suggest why there might be a need for a water transfer scheme in the UK.   **[3 marks]**

> You should aim to read the question carefully, especially when more than one explanation or factor is being asked for.

> Look at the marks available and use that to decide how many points to make.

5 Using Figures 22.14 and 22.20 explain why the exploitation of energy resources in the UK can damage the environment.

▲ Figure 22.14 Gas-fired power station in Pembroke, Wales

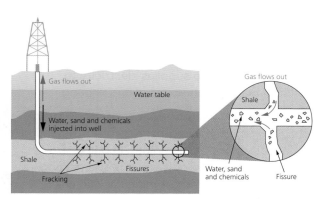

▲ Figure 22.20 Fracking

                                                        [6 marks]

Food

6   Study Figure 23.1 (page 324). How many calories per person per day did the population of the USA consume in 2010? Select only **one** letter.

    A 3,480–3,770

    B 3,270–3,480

    C 3,050–3,270

    D 2,850–3,050   [1 mark]

7   Describe the global distribution of calorie intake.   [2 marks]

8   Give **two** reasons why farmers in LICs may encounter problems with using irrigation.   [2 marks]

9   Explain how irrigation in an LIC such as Kenya can help both the people and the economy.   [6 marks]

> It could be helpful to learn common command words such as discuss, explain and describe.

Water

10 Study Figure 21.8 (page 309). Which **one** of the following statements is the most accurate concerning global water insecurity? Select only **one** letter.

    A Water insecurity is greatest in the most densely populated areas.

    B Australia has the least water insecurity.

    C The whole of Africa suffers from water insecurity.

    D The southern half of Eurasia contains the largest area of water insecurity.   [1 mark]

11 Using Figure 24.14 (page 350), describe the distribution of water availability in China.   [2 marks]

12 Give two reasons for the global rising consumption of water.   [2 marks]

13 Explain how the use of water in the UK can be made more sustainable.   [6 marks]

Energy

14 Study Figure 21.9 (page 309). Which **one** of the following statements is the most accurate concerning global energy insecurity? Select only **one** letter.

    A It is highest in developed countries.

    B It is highest in oil-producing countries.

    C It is highest in the temperate parts of the world.

    D It is highest in the tropical parts of the world.   [1 mark]

15 Using Figure 25.2a (page 356), describe the distribution of energy consumption.   [2 marks]

16 Give two reasons for the global rising consumption of energy.   [2 marks]

17 Explain how our use of energy can be made more sustainable.   [6 marks]

26 Issue evaluation

❂ KEY LEARNING

➤ What a geographical issue is

➤ What evaluation is

➤ Issues for the exam

➤ The pre-release booklet

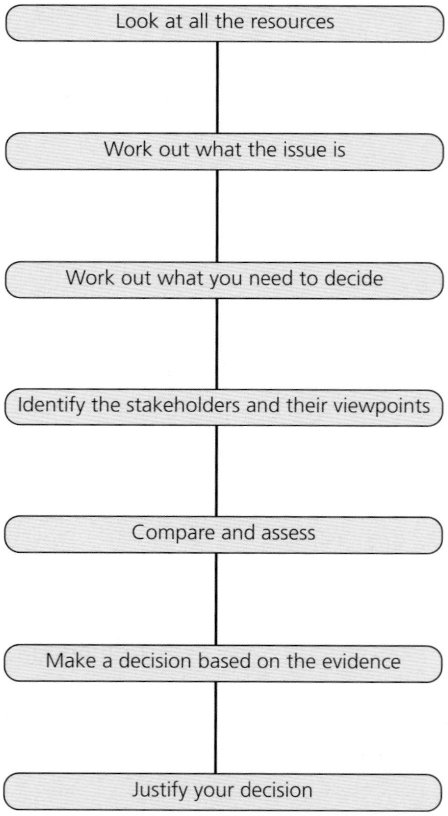

| Look at all the resources |
| Work out what the issue is |
| Work out what you need to decide |
| Identify the stakeholders and their viewpoints |
| Compare and assess |
| Make a decision based on the evidence |
| Justify your decision |

▲ Figure 26.1 Steps in issue evaluation

Introducing issue evaluation

What is issue evaluation?

Issue evaluation involves critical thinking and problem solving. You need to:

■ apply your knowledge and understanding to a particular issue

■ show your ability to interpret a variety of secondary sources (maps, diagrams, tables and text) that illustrate different aspects of that issue

■ evaluate different viewpoints, options or outcomes.

What are geographical issues?

During your AQA GCSE Geography course, you will come across quite a few geographical issues. Most often, they will relate to topics on which different stakeholders hold conflicting views. For example, a local authority may not want to invest money in a coastal protection scheme, but people who live close to the retreating cliff – and whose homes are at risk – will most likely see the scheme as essential.

You should also consider what the financial costs are. Can spending public money really be justified when only some groups benefit?

What is evaluation?

Evaluation means making an assessment of a situation.

It requires collecting and analysing information about an issue and then reaching a decision about it. When you bought your last smartphone you no doubt went through a process of evaluation:

■ weighing up different things such as price and screen size.

■ mentally ranking those features in the order of their importance to you

■ then making your final decision on the basis of your ranking.

The same sort of process goes on in issue evaluation (Figure 26.1).

The geographical issue appearing in the exam could be a controversial one. You will be asked to weigh up all the evidence given you in a resource booklet. From that you will reach and justify a conclusion.

What issues might appear in the exam?

It would be helpful to become familiar with the core compulsory topics (Figure 26.2). The issue will be synoptic: it will draw on your knowledge and understanding of different parts of the core topics. For example, the airport issue in Section 26.4 is rooted in the 'Changing economic world' topic, but it also needs you to apply what you have learned in parts of two other compulsory topics: 'Urban issues and challenges' and 'Resource management'.

▼ Figure 26.2 Possible issues for evaluation

Compulsory topic	Possible issue
The challenge of natural hazards	Tropical storms – reduce risks or face the consequences?
	Rising sea levels – protect or retreat?
	Responses to tectonic hazards – take the risk?
Ecosystems and tropical rainforests	Deforestation – how to reduce the rate?
	Indigenous people – protect or abandon?
	Sustainable management – dream or reality?
UK physical landscapes	(Though this is compulsory, it is a very short section that is unlikely to yield suitable issues)
Urban issues and challenges	Waste disposal – bury, burn or recycle?
	Urban growth – greenfield or brownfield?
	Urban regeneration – how?
Changing economic world	Closing the development gap – how?
	TNCs – good or bad?
	Airports – expand existing or build new?
Resource management	Food miles – are they really necessary?
	Water pollution – who are the culprits?
	Fossil fuels – to burn or not to burn?

As you study the core topics, your teacher may point out and explore some issues. Figure 26.2 lists some possibilities, but remember that the issue will be a broad one and draw on other core topics (see the Specimen Paper 3 on the AQA GCSE Geography website).

Do not worry too much about the actual issue you will have to face. The pre-release resource booklet you receive before the exam will give you a fairly good idea. Your teacher will also be on hand to provide help and guidance during those weeks. So you will know the resources, but you will not know the actual questions based on them.

There is still an element of surprise! What you do know is that you will need to think synoptically across a range of information, evaluate and justify your conclusions. You will also have to use a wide range of skills.

The pre-release resources booklet

You will be given a resource booklet twelve weeks before the exam, containing a mix of resources (see Sections 26.2 and 26.3): maps, diagrams, tables and data.

During these twelve weeks, you need to:

- broadly identify the likely issue and what you expect you will have to evaluate. Will it be choosing between different action options? Or will it be supporting the viewpoint of one stakeholder against those of others?
- find out more about the issue. Look at the relevant parts of this book. Look online at reputable websites for more information. Ask your teacher for some suggestions.
- make notes about each of the resources (see Sections 26.2 and 26.3). What is the important message contained in each?

→ **Activities**

1 Identify an issue of your own that relates to one of the core topics.

2 Check that you understand what a stakeholder is.

3 Which of the issues listed in Figure 26.1 would you most like to be in your exam? Give your reasons.

4 If the issue were the plight of the indigenous people in the tropical rainforests, who would be the other stakeholders? (It might help if you look at Chapter 6.)

5 Identify the different types of resource in the specimen resource booklet on the AQA website.

6 Have a go at completing the questions in the specimen exam paper on the AQA website.

7 What distinguishes a levels-marked question from a point-marked one?

Textual resources

Using the resources booklet

Making a sensible decision needs a sound knowledge and understanding of the situation. Much of this should come from your analysis of what is in the pre-release resource booklet. It is recommended that you study each resource in turn. What is the basic message of each resource? What is it telling you? What light is it shedding on the issue to which all the resources relate?

It would be useful to make some notes on each, which will help you when the time comes for you to revise. Unfortunately, you will not be allowed to take those notes into the exam room. However, once there, you will be given a new copy of the resource booklet. The challenge now is to remember the points you made in your notes!

The resources you will have to deal with fall into two broad types:

1   textual – much of the pre-release booklet contains text in various forms, particularly in fact files or information panels (Figure 26.3)

2   visual – information presented in the form of maps, diagrams and photos. Here and in Section 26.3 we will look at the range of resources you might be faced with. You will also be shown the sort of revision notes you might write. Some of the resources here relate to the issue you will evaluate in Section 26.4: should a new major runway be built at London's Heathrow airport or at Gatwick?

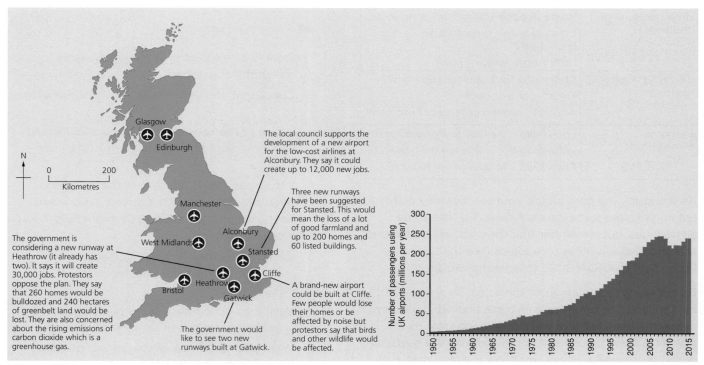

▲ Figure 26.3 Fact file: Possible sites for airport expansion in the UK

Fact file or information panel

This is a popular way of introducing an issue (Figure 26.3). It may be mainly text, but will include some visual information too. What do you think caused passenger traffic to slump in 2009–11?

Issue

- *Where should the new runway be built?*

Notes

- *Shows a steady growth in UK airport traffic*
- *Shows nine airports as possible sites for expansion*
- *Airport expansion has both supporters (business, tourism) and critics (environmentalists worried about carbon dioxide emissions from aircraft)*

Newspaper extracts or headlines

The banner headlines of newspaper can often sum up situations or point to strengths and weaknesses (Figure 26.4).

New Gatwick runway will create much-needed jobs

Heathrow runway will cause noise pollution levels to skyrocket

▲ Figure 26.4 Contrasting newspaper headlines

Notes (continued)

- *The headlines point to two important issues: noise and jobs*
- *Which is the more important?*
- *More people will be affected by noise than will find work at the airport*

Stakeholder speech bubbles

These are often used when different views on an issue need to be illustrated.

We are outraged that this has dragged on for ten years – for a whole decade our community has felt like it's on borrowed time. We will continue our struggle; we will take our fight against these plans to the courts.

Local resident, close to Heathrow

You can't increase air traffic in the UK while even remotely hoping to meet the government's reduced carbon emissions target. We think railways should be invested in, to encourage people to use them, and all internal flights be scrapped. We also believe there is no need for a third runway at Heathrow and these plans should be binned.

Climate change protester

We need better airports so that more tourists visit the UK, and continued growth of air travel is essential. We must expand our airports. We can offset any carbon dioxide emissions from aircraft in the future, as better technology becomes available for cleaner and more efficient aeroplanes.

Government minister

▲ Figure 26.5 Opposing views on the growth of air travel and the expansion of UK airports

Notes (continued)

- *Only three stakeholders: one for and two against*
- *What about other important stakeholders, such as businesses, people of working age, and so on?*

→ Activities

1. With reference to Figure 26.3, give reasons why so many airports in the South East are shown as possible sites for expansion.

2. Figure 26.4 shows one major cost and one major benefit. Which do you think is more important? Give your reasons.

3. Look at Figure 26.5. Suggest some more stakeholders involved in this airports issue.

4. Trial question for 1 mark: Name one UK international airport located outside England.

Visual resources

As far as Paper 3 is concerned, the visual resources fall mainly into three categories:

1 maps
2 diagrams
3 photos.

The aim here is to show a range of visual resources for various issues. For each resource, as in Section 26.2, there will be notes showing:

● the issue to which it relates
● the sorts of notes you might make during the 12 weeks running up to the exam.

Maps

Maps come in many different forms. They range in scale from local to global. They also vary in type. There are topographic maps, like those produced by the UK's Ordnance Survey (see Chapter 11 pages 155 and 161 of this book for examples of OS maps).

There are also thematic maps, which show the distributions of population, volcanoes, and major cities. Figure 26.6 shows the global distribution of Samsung's operations. Samsung is the world's largest producer of smartphones. There are four main aspects to its business: production, sales, research and others (mainly professional support).

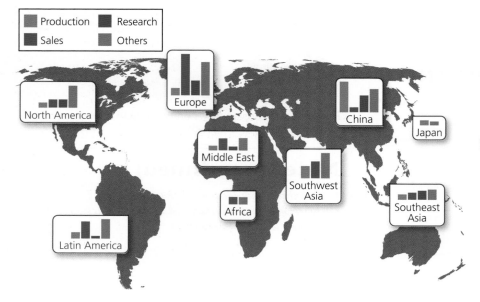

▲ Figure 26.6 The global distribution of Samsung's operations

Issue

• *Is the global shift in manufacturing really helping developing countries?*

Notes

• *Research is located mainly in China and Europe*
• *China is the leading producer*
• *Sales are highest in Europe, Latin America and Southwest Asia*
• *Professional support is well provided in all areas except Africa and Japan*
• *Developing countries, particularly in Africa have benefited little from Samsung's operations. The same is true for Australia*
• *Emerging countries, such as Brazil (Latin America), China and India (Southwest Asia) have done well.*

Diagrams

Again, there are many types of diagram, but three in particular are likely to figure frequently in the resource booklet: graphs, bar charts and pie charts.

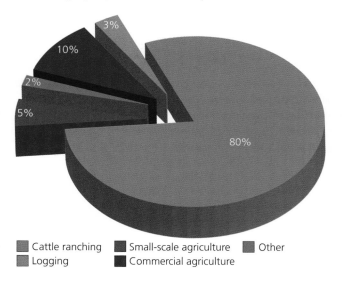

Cattle ranching ◼ Small-scale agriculture ◼ Other
◼ Logging ◼ Commercial agriculture

▲ Figure 26.7 The main causes of deforestation in the Brazilian Amazon

Issue

- *How can the loss of tropical rainforest be reduced?*

Notes

- *Identifies the main culprits*
- *Cattle ranching is by far the greatest cause of deforestation (80 per cent)*
- *This is followed by commercial agriculture (10 per cent)*
- *Logging is shown as small contributor (3 per cent). Is this really so small?*
- *It may be that once the best trees have been felled, the forest is left to recover, so it does not count as deforestation.*
- *Clearly, it is the two forms of farming that need to be controlled*

Photographs and satellite images

Photographs will certainly appear in the resource booklet. They are likely to be either taken from the air or at ground level.

▲ Figure 26.8 The proposed Garden Bridge in London

Issue

- *Should the Garden Bridge be built?*

Notes

- *Whose money will be spent in creating the bridge?*
- *Who will actually use this bridge?*
- *Public access for leisure and recreation? Or more for tourists?*
- *What environmental benefits are there?*

→ Activities

1 Add another note to each of the resource boxes.

2 How do you evaluate these resources? Which do you think is most effective? Give your reasons.

3 What do you think might be the disadvantages of visual resources when evaluating issues?

4 Trial question for 6 marks: Using evidence from Figure 26.6, suggest how globalisation exploits the developing world.

5 Trial question for 6 marks: At a recent meeting, an environmental manager said, 'The disposal of waste paper is no longer an environmental problem.' Do you agree with this view? Give reasons for your decision.

KEY LEARNING

➤ Understanding the issue
➤ Weighing the evidence
➤ Reaching a decision

Evaluation

The issue

Remind yourself of Section 26.2 and the issue of airport expansion. The need for more airport capacity in the UK is widely accepted. But where should a new runway be built – near London, Birmingham, Manchester or elsewhere?

In 2012, the Airports Commission was asked to advise the UK government on where this additional runway should be built. The government also stated that it was keen to maintain the UK's status as an international leader in aviation.

The Commission published its report in 2015. In this it focused on a shortlist of three options. The option it finally recommended was to build a third runway at Heathrow rather than a second runway at Gatwick. It is likely, if this were your real exam paper, that there would be OS maps of both areas for you to look at (Figure 26.9).

▼ Figure 26.9 New runway proposals at (a) Heathrow and (b) Gatwick

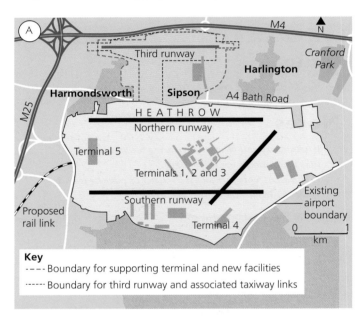

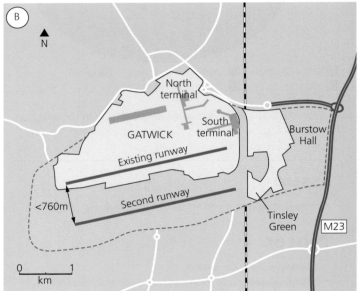

The government has yet to give the go-ahead to the Heathrow proposal. Do you agree with the recommendation, or would you support the expansion of Gatwick instead?

In making your evaluation, you should pay particular attention to:

- financial costs
- environmental impacts
- employment
- accessibility.

Weighing up the evidence

Some evidence follows, but you might also look back at Section 26.3 for more. Remember that you are comparing the two airports on only four different items. Clearly, there are many more that the Airports Commission considered.

Before looking at the evidence, it is important to understand that Heathrow is a much bigger airport than Gatwick (Figure 26.10). Do you think this is an advantage or disadvantage?

	Heathrow	Gatwick
Passengers in (millions)	73.4	37.8
Airlines using	80	45
Destinations	185 in 84 countries	200 in 90 countries
Average number of daily flights	1,290	908
Total flights in 2014	470,695	255,711

▲ Figure 26.10 A statistical comparison of Heathrow and Gatwick

Financial costs

Heathrow's third runway will cost much more than Gatwick's second: £12.8 billion compared with £7.8 billion.

Environmental impact

Aircraft generate noise pollution and atmospheric pollution. These environmental impacts are thought to be proportional to:

- the number of flights
- the number of people living close to the runways; Gatwick claims that 320,000 people will be affected around Heathrow.

Also under the heading of environmental impact is the question of how many homes will have to be cleared to make way for the new runways and terminal buildings.

Employment

Being a larger airport, it is not surprising that the expansion of Heathrow will lead to a greater number of jobs – more than 1,000 more than at Gatwick. Of course, there will be more jobs in companies establishing sites near the airport as well.

Accessibility

Figure 26.11 shows the transport connections of both airports, but you need also to take into account their location relative to London and the rest of the country. Accessibility is about distances to be covered and the time it takes to travel those distances.

> **→ Activity**
>
> 1 Make notes on each of the resources.
> 2 Rank the four criteria (financial costs, environmental impact, employment and accessibility) in order of importance.
> 3 Give your reasons (use your resource notes).
> 4 Compare the airports against each of the four criteria. Tick the airport that you think scores highest on each criteria to reveal your choice.
> 5 Now answer the following 9-mark question: Should the new runway be built at Heathrow or Gatwick? Explain why you have reached this decision.

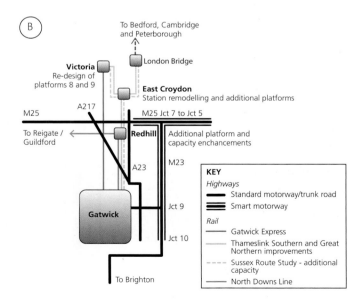

▲ Figure 26.11 The transport links of (a) Heathrow and (b) Gatwick

27 Fieldwork and geographical enquiry

KEY LEARNING

KEY LEARNING

➤ What a geographical enquiry involves

➤ What makes a good question for a geographical enquiry

➤ How fieldwork risks are assessed

Preparing for a geographical enquiry

What is involved in a geographical enquiry?

Fieldwork is the bit of geography that everyone enjoys the most. Years after people leave school, they always remember their geography field trip!

You have to do at least two geographical enquiries in different environments as part of your fieldwork. One must be based in a physical environment and the other in a human environment. Each enquiry, whether physical or human, is likely to follow a similar outline (Figure 27.1). At least one of these enquiries should include the interaction of physical and human geography, for example, investigating the impacts of migration or economic development on a habitat or ecosystem.

▼ Figure 27.1 Outline for a geographical enquiry. You could also try making your own outline

1   Think of a suitable question or hypothesis to investigate.

2   Choose one or two fieldwork methods to collect data that will help you to answer the question. Then carry them out in a fieldwork location.

3   Select two or three ways to process and present the data in the form of graphs, maps, images or tables.

4   Describe, analyse and explain the data you have presented.

5   Draw conclusions to answer the original question or hypothesis you set out to investigate.

6   Evaluate your enquiry to say how well you think it went and how you could improve it.

What makes a good question for a geographical enquiry?

There are a few simple guidelines for devising a question for a geographical enquiry (Figures 27.2 and 27.3). A good enquiry question could be one that:

■   you would really like to know the answer to

■   is linked to the geography that you have studied

■   could be investigated in a particular fieldwork location (it is no good trying to investigate rivers if there is no river!)

You could also base a geographical enquiry on a hypothesis, rather than a question. A hypothesis is a theory, or idea, that you want to test. So, for example, instead of asking the question, 'How does a river change along its course?', you could test the hypothesis, 'A river gets deeper, wider and faster along its course.'

▼ Figure 27.2 Questions you could investigate along a river

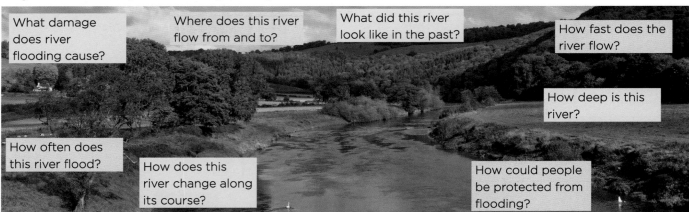

What damage does river flooding cause?

Where does this river flow from and to?

What did this river look like in the past?

How fast does the river flow?

How deep is this river?

How often does this river flood?

How does this river change along its course?

How could people be protected from flooding?

How is this city changing?

What impact do changes in the city have on people?

Where do the people in this city come from?

How many people live in this city?

What inequalities are found in this city?

What challenges are found in different parts of this city?

How sustainable is this city?

How does the city meet the needs of different groups of people?

▲ Figure 27.3 Questions you could investigate in a town or city

How are fieldwork risks assessed?

One of the most important things to think about when planning fieldwork is to make sure it will be *safe*. It is essential to do a **risk assessment**. Your teacher will have done this before you start planning the fieldwork, but there is no harm in you thinking about it too. Some of the dangers you need to consider when planning fieldwork include:

- deep water
- fast-flowing water
- slippery surfaces
- uneven surfaces
- steep drops
- traffic
- using public transport
- falling objects
- landslips
- air pollution
- water pollution
- extreme temperatures
- bad weather
- meeting strangers.

➜ Activities

1 Study Figure 27.2.

   a) Choose the best question for a geographical enquiry about rivers. Justify your choice.

   b) Now, turn your question into a hypothesis.

2 Study Figure 27.3.

   a) Choose the best question for a geographical enquiry about cities. Justify your choice.

   b) Now, turn your question into a hypothesis.

3 Choose two other photos in this book – one of a coastal environment and one of an economic activity, e.g. tourism – that would be suitable locations for fieldwork. Think of a good enquiry question in each location.

4 What are the advantages and disadvantages of doing fieldwork in physical or human environments? For example, one advantage of a human environment might be that it is closer to your school. Draw a large table like this and complete the boxes. Write at least two ideas in each box.

	Advantage	Disadvantage
Physical environment		
Human environment		

5 What are the main risks you would have to consider before doing fieldwork:

   a) along a river?   c) in an urban area?

   b) at the coast?

   In each case, mention three risks and suggest what you could do to reduce them.

An enquiry in a physical environment

Why do a geographical enquiry along a river?

One of the most popular physical environments for geographical enquiry is along a river (Figure 27.4). There are many good reasons for choosing rivers for fieldwork.

- Most areas have a river, so you do not have to travel too far.
- You have studied rivers in geography. You can see for real many of the features and processes you learnt about in the classroom (see Chapter 11).
- There are plenty of opportunities for data collection using a variety of fieldwork methods.
- Rivers are a fun location for fieldwork – but, remember to be careful!

It is difficult to do fieldwork on a river without getting wet! It is important to have done the risk assessment first. Even a shallow river can be dangerous if the water is flowing fast. In most river channels, the water depth varies, so you could be standing up to your knees in water in one place, but be out of your depth in another.

What types of data can be used in a geographical enquiry?

The data you are likely to use in a geographical enquiry along a river will be **primary data**. This is data, or information, that you collect yourself. Primary data includes:

KEY LEARNING

- Why do a geographical enquiry along a river
- What types of data can be used in a geographical enquiry
- What methods can be used to collect data in a physical environment

▲ Figure 27.4 Students doing fieldwork on a river

tape measure

ranging pole

clinometer

chain (or rope)

flow meter

Flooding Questionaire

fishing net

callipers

questionnaire

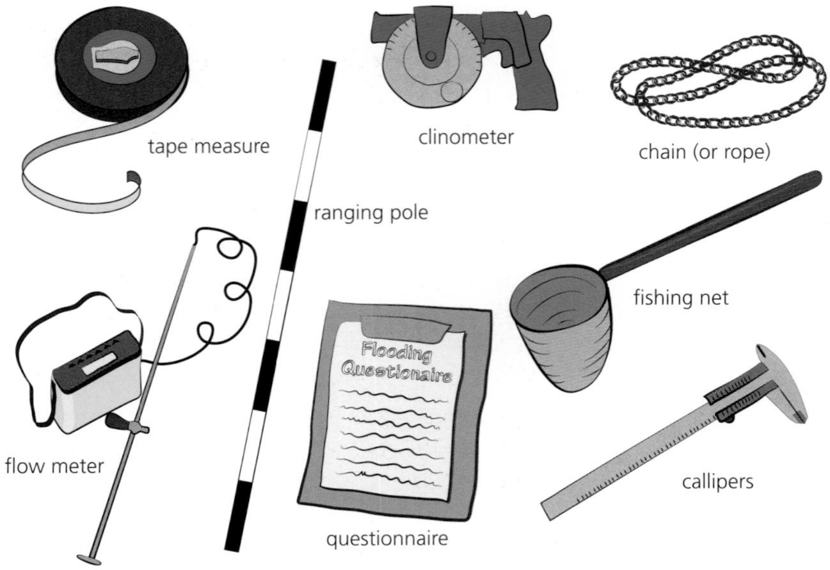

▲ Figure 27.5 Fieldwork equipment for use on a river

- measurements that you make using a variety of equipment (Figure 27.5)
- images, such as photos you take or sketches you draw
- maps or diagrams you complete while you are outside
- responses to questions you ask people through questionnaires or interviews.

But, as part of a geographical enquiry, you can also use **secondary data**. This is data that someone else has collected. It could be data that other students collected on a previous field trip or it could be data published in a book or online. Secondary data can be useful when you want to look at changes over time in a physical environment, such as a river.

Much of the data you collect along a river is likely to be **quantitative data** – numerical data that you get by counting or measuring. You may also collect qualitative data – descriptive data that can't be measured, but can be collected through observation or subjective judgement, and presented through images or text. It includes answers to open questions, field sketches, photos and video diaries.

What methods can be used to collect data in a physical environment?

Channel survey – measure the width, depth and wetted perimeter of the river channel. This could be done at different points along the channel.

River load survey – measure the shapes and sizes of the river's load at different points along, or across, the channel.

Valley slope survey – measure the gradient of the river valley, either along the course of the river or on the valley sides.

Flow survey – measure the velocity of water flowing in the river at different points along, or across, the channel.

Flood risk assessment – identify the factors along the river that affect flood risk and ask people about their experience of flooding.

Water quality survey – measure the number of living organisms in the water at different points along the river.

▲ Figure 27.6 Some river fieldwork methods

→ Activities

1 Look at Figure 27.4. Write a risk assessment for this fieldwork location.

   a) List the main hazards you can see.
   b) Suggest ways to reduce each hazard. Include any special equipment that could reduce the risk.

2 Study the equipment in Figure 27.5.

   a) Match each item with the data it could be used to record:
   - channel width
   - channel depth
   - wetted perimeter (the channel surface in contact with the river)
   - gradient of the river valley
   - velocity of water in the river
   - water quality
   - pebble size
   - people's views on the risk of flooding.
   b) In each case, explain how the equipment could be used.
   c) State whether each method is quantitative or qualitative.

3 You have decided to investigate the question, 'How does a river change along its course?'

   a) Select two fieldwork methods from Figure 27.6 you could use to investigate this question.
   b) Justify your choice of these methods.

4 a) Choose one more question you could investigate along a river. Look back at Figure 27.2 on page 380 for ideas.
   b) Select two more fieldwork methods you could use to collect data and justify your choice.

5 Plan a geographical enquiry in a different physical environment on the coast.

   a) Choose a suitable location for your enquiry somewhere on the UK coast. Look at Chapter 10 for ideas.
   b) Think of a suitable enquiry question or hypothesis for your investigation.
   c) Think of two fieldwork methods you could use to investigate the question. (Hint: many of the fieldwork methods used on rivers could also be used on coasts, but there is no reason why you could not think of your own methods.)

Bringing the enquiry together

How can data be processed and presented?

✪ KEY LEARNING

➤ How to process and present data

➤ How to describe, analyse and explain data

➤ How to reach conclusions and evaluate the enquiry

Once back in the classroom you have to select the best way to process and present the data you have collected. This might involve sharing data with the rest of your class. For example, if your class collected data from different sites on a river (Figure 27.7), you may have to share data and put it into a table like this (Figure 27.8).

This still counts as primary data because, even though you didn't collect all the data yourself, you were working as part of a team. The data in the table can then be used to draw cross-sections of the river at each site (Figure 27.9).

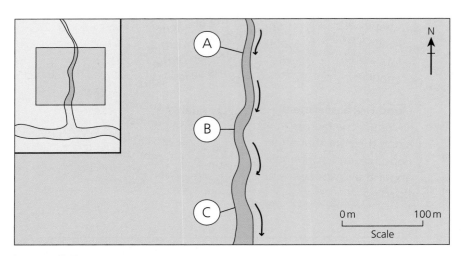

▲ Figure 27.7 Three data collection sites on a river

▼ Figure 27.8 Data collected from three sites on a river on Figure 27.7

Site	Width (metres)	Depth (metres), measured at points across the river								Flow rate (m/sec)	Load size (cm)	River gradient (°)
		0	0.5	1.0	1.5	2.0	2.5	3.0	3.5			
A	1.5	0	0.2	0.2	0					0.25	2.5	5
B	2.5	0	0.3	0.4	0.5	0.3	0			0.40	2.8	4
C	3.5	0	0.3	0.6	0.7	0.6	0.4	0.2	0	0.50	2.4	2

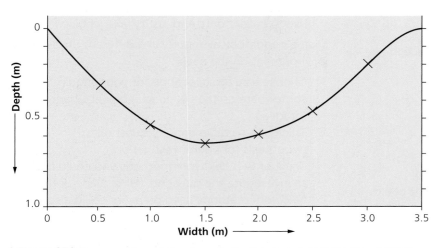

▲ Figure 27.9 Cross-section of the river at site C (using data from Figure 26.8)

▼ Figure 27.10 Ways to present data

Type of data presentation	Examples
Maps	Locational map of study area
	Flow map
	Land use map
Graphs	Bar chart
	Line graph
	Pie chart
	Cross-section
	Scatter graph
Visuals	Field sketch
	Photo
Calculations (for rivers)	Mean depth
	Cross-sectional area
	Discharge

The data can be processed and presented by making further calculations. For example, it is possible to calculate the mean depth of the river from all the depth measurements.

Mean (average) depth = sum of depth measurements / Number of measurements

= 0 + 0.3 + 0.6 + 0.7 + 0.6 + 0.4 + 0.2 + 0 / 8 = 2.8 / 8 = 0.35 m

From the width and mean depth of the river, it is possible to calculate the cross-sectional area.

Cross-sectional area = Mean depth x width = 3.5 x 0.35 = 1.225 m²

From the cross-sectional area and flow, it is possible to calculate the discharge of the river.

Discharge = Cross-sectional area x flow = 1.225 x 0.5 = 0.6125 m³/sec

---

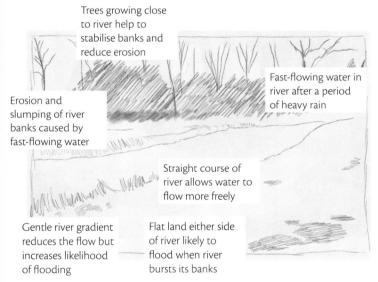

Trees growing close to river help to stabilise banks and reduce erosion

Fast-flowing water in river after a period of heavy rain

Erosion and slumping of river banks caused by fast-flowing water

Straight course of river allows water to flow more freely

Gentle river gradient reduces the flow but increases likelihood of flooding

Flat land either side of river likely to flood when river bursts its banks

▲ Figure 27.11 An annotated sketch to describe and analyse a river

How can data be described, analysed and explained?

- identify patterns and trends
- make links between different sets of data
- identify anomalies
- explain reasons for patterns

You have to describe, analyse and explain the data in your own words. Sometimes, this can be done by annotating a sketch or map you have drawn, or a photo you have taken (Figure 27.11).

How can conclusions be drawn?...

Your conclusion should aim to come back to answer the original enquiry question – in this case, 'How does a river change along its course?' It could be helpful to point to evidence from the data your collected to support your conclusion.

...and the enquiry be evaluated?

In your evaluation, you can talk about the strengths and weaknesses of your enquiry – what went well and what could be improved.

- What problems did you have with data collection?
- What were the limitations of the data you did collect?
- What other data could you have collected?
- How reliable are the conclusions you came to?

Sometimes an enquiry does not go according to plan. It does not matter. The important thing is that you are able to evaluate it. The lessons you learn will help when it comes to planning your next enquiry.

→ Activities

1 Study the data in Figure 27.8 and the cross-section in Figure 27.9. Draw a similar cross-section of the river at site B, using the data from Figure 27.8.

2 Make the following calculations for the river at sites A and B:
   a) mean depth
   b) cross-sectional area
   c) discharge.

3 Study Figure 27.11. Copy the labels from the sketch and underline all the words in each label in two colours to highlight:
   a) description
   b) analysis and explanation

4 You cannot really evaluate a geographical enquiry until you have done it. But you can anticipate some of the problems you might face.
   a) Identify at least three problems you might face when doing a geographical enquiry on a river. Think about the methods you will use and the data you hope to collect.
   b) Suggest solutions for these problems that would help to improve the enquiry.

An enquiry in a human environment

Why do a geographical enquiry in an urban area?

Most of us live in towns or cities (see Chapter 13). Urban areas are also a popular location for geographical enquiries in a human environment. There are several good reasons for choosing an urban area for fieldwork:

- Most schools are located in towns or cities, so they have urban areas on the doorstep
- Most of us are familiar with an urban area, so we have a good insight into what happens there. That helps when it comes to analysing and explaining.
- There are many sources of data, including people, especially if you use a questionnaire (see Figure 27.14)
- Urban areas are often lively, exciting places to be.

How is data collected in a human environment?

One way to collect data in an urban area is to use a **questionnaire** (Figure 27.13). This is a set of questions to ask people about their lives and/or opinions. The same questionnaire can be used to interview lots of people, recording their responses as a tally. Later, in the classroom, the data can be used to generate a range of graphs or maps.

There are a few simple guidelines to follow when devising or using a questionnaire:

- Keep it as short as possible to encourage more people to respond.
- Use **closed questions**, with a limited choice of answers, to obtain quantitative data.
- Use **open questions**, where people can express opinions, to obtain qualitative data.
- Work in pairs when you carry out the questionnaire – never alone and, preferably, not in large groups (you might scare people away!).
- Try to speak to a range of people representing different age groups, ethnic groups, male and female.
- Don't be afraid of rejection, but always be polite!

▲ Figure 27.12 Students doing fieldwork in an urban area

How can primary and secondary data be used?

There are some aspects of urban areas for which it is difficult to obtain primary data. For example, it is difficult to get an accurate picture of the population of an area. You could try counting people in the street, but this may include people who do not live in the area. It will not include other groups, like children at school, people at work or housebound people. In this case, it is better to use secondary data.

Secondary data is also useful when looking at any type of change in urban areas. For example, old photos or maps could be compared with primary data to show changes in land use, or changes in environmental quality brought about by regeneration.

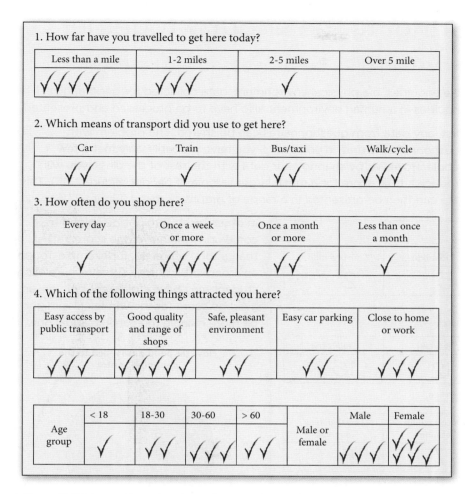

1. How far have you travelled to get here today?

Less than a mile	1-2 miles	2-5 miles	Over 5 mile
✓✓✓✓	✓✓✓	✓	

2. Which means of transport did you use to get here?

Car	Train	Bus/taxi	Walk/cycle
✓✓	✓	✓✓	✓✓✓

3. How often do you shop here?

Every day	Once a week or more	Once a month or more	Less than once a month
✓	✓✓✓✓	✓✓	✓

4. Which of the following things attracted you here?

Easy access by public transport	Good quality and range of shops	Safe, pleasant environment	Easy car parking	Close to home or work
✓✓✓	✓✓✓✓✓	✓✓	✓✓	✓✓✓

	< 18	18-30	30-60	> 60		Male	Female
Age group	✓	✓✓	✓✓✓	✓✓	Male or female	✓✓✓	✓✓✓✓

⋏ Figure 27.13 A questionnaire about shopping in an urban area

Environmental quality survey – give positive or negative scores to different environmental features on a score sheet.

Sustainability survey – similar method to an environmental quality survey, with a focus on features that make the area more or less sustainable

Land use mapping – complete an outline map of the urban area to show how each piece of land is being used.

Traffic/pedestrian survey – count the number of vehicles and/or pedestrians going past within a set time.

Shopping survey – classify and map the different types of shop in a shopping centre.

Urban role play – look at the area through the eyes of different types of people as you walk around, to see how it meets those people's needs.

⋏ Figure 27.14 Some urban fieldwork methods

→ Activities

1 Look at Figure 27.13. Think of at least five aspects of urban areas for which a questionnaire would be a good way to collect data (e.g. shopping). Make a list.

2 Take the question, 'How does a shopping centre meet the needs of different groups of people?'
   a) Select two fieldwork methods you could use
   b) Justify your choice of these methods.
   c) Suggest any secondary data that could be used to help your investigation.

3 Read Figure 27.13. Study how the questionnaire about shopping has been devised. Now, devise a similar questionnaire for one of the other aspects of urban areas you listed in Activity 1. Decide whether you will use closed or open questions and what space you will have for the responses.

4 a) Choose one more question you could investigate in an urban area.

   Look back at Figure 27.3 on page 381 for ideas.
   b) Select two more fieldwork methods you could use to collect data and justify your choices.
   c) Suggest any secondary data that could be used to help your investigation.

5 Plan an enquiry in a different human environment (e.g. a tourist resort in a glaciated upland area).

   a) Choose a suitable location for your enquiry in the UK. Look at Chapter 9 for ideas.
   b) Think of a suitable enquiry question or hypothesis for your investigation.
   c) Think of two fieldwork methods and any secondary data you could use to investigate the question. (Hint: many of the fieldwork methods used in urban areas could also be used in tourist resorts, but there is no reason why you could not think of your own methods.)

Bringing the enquiry together

How can data be processed and presented?

Like enquiries in a physical environment, data collected for geographical enquiries in a human environment also have to be processed and presented.

First, any data from questionnaires can be shared. The problem with questionnaires is that, in your pairs, you have probably only met a few people. However, by sharing your data with the rest of the class, you will produce a total score for a much larger sample of people (Figure 27.15). This data can then be presented in a range of graphs or maps (Figure 27.16).

In other cases, it might be helpful to come up with an average score, rather than a total score. So, for example, for an environmental quality survey where each student may have a different opinion, an average score for the whole class might be more reliable than an individual one.

1. How far have you travelled to get here today?

Less than a mile	1-2 miles	2-5 miles	Over 5 mile
20	13	5	2

2. Which means of transport did you use to get here?

Car	Train	Bus/taxi	Walk/cycle
7	5	12	16

3. How often do you shop here?

Every day	Once a week or more	Once a month or more	Less than once a month
3	25	7	5

4. Which of the following things attracted you here?

Easy access by public transport	Good quality and range of shops	Safe, pleasant environment	Easy car parking	Close to home or work
17	35	13	5	15

	< 18	18-30	30-60	> 60		Male	Female
Age group	4	9	17	10	Male or female	15	25

▲ Figure 27.15 The total scores from a shopping questionnaire

How far people travelled to the shopping centre

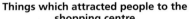

■ Less than a mile   ☐ 2-5 miles
■ 1-2 miles   ☐ Over 5 miles

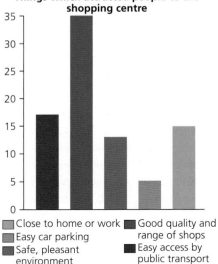

Things which attracted people to the shopping centre

☐ Close to home or work   ■ Good quality and range of shops
■ Easy car parking
■ Safe, pleasant environment   ■ Easy access by public transport

➤ Figure 27.16 Data presented from the results of a shopping questionnaire

How can data be described, analysed and explained?

Photos, combined with a map, are a good way to analyse data. Photos can be located on a map, either manually or using a GIS tool. Always remember to keep a record of where you took your photos, so they can be located on the map later (Figure 27.17).

How can conclusions be drawn?...

Your conclusion should aim to come back to answer the original enquiry question – in this case, 'How does a shopping centre meet the needs of different groups of people?' You should be able to point to evidence from the data you collected to support your conclusion (Figure 27.18).

...and the enquiry evaluated?

Once again, do not forget the all-important evaluation (page 385).

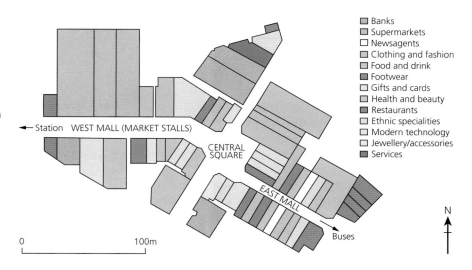

Banks
Supermarkets
Newsagents
Clothing and fashion
Food and drink
Footwear
Gifts and cards
Health and beauty
Restaurants
Ethnic specialities
Modern technology
Jewellery/accessories
Services

← Station   WEST MALL (MARKET STALLS)

CENTRAL SQUARE

EAST MALL

Buses

N

0   100m

▲ Figure 27.17 Map of a shopping centre

> Our enquiry showed that a shopping centre met the needs of different groups of people. Most of the people using the shopping centre travelled less than two miles, showing that is mainly meeting the needs of local people. We found that:
>
> - young people were attracted by fashion clothing and modern technology shops
> - families and older people use the supermarket and food shops for regular food shopping
> - ethnic minorities are able to find speciality shops catering for their particular needs
> - the shopping centre is all on one level with no steps, making it easier for disabled people.

▲ Figure 27.18 Extract from a conclusion to an urban investigation

→ Activities

1   Study the data in Figure 27.15 and the two graphs in Figure 27.16.

   a) Can you think of any different ways of presenting the same data? Draw your own two graphs.

   b) Which way do you think is the best way of presenting the data – the ones shown here or the ones you have drawn? Give reasons.

2   Draw two more graphs to show:

   a) What means of transport people use to get to the shopping centre.

   b) How often people use the shopping centre.

   In each case, say why you chose this type of graph.

3   Study Figure 27.17.

   a) Describe the data shown on the map. For example, describe the numbers of each type of shop and any patterns you can see.

   b) Analyse the data shown on the map. For example, suggest why some types of shop are more common and why they are distributed in this way.

4   Read the extracts from the conclusion in Figure 27.18. What evidence could you give to support each conclusion? Use Figures 27.15 and 27.17 to help you.

5   a) Identify at least three problems for doing a geographical enquiry in an urban area. Think about methods to use and the data to collect.

   b) Suggest solutions for these problems that would help to improve the enquiry.

   c) To what extent do you think the results and conclusions from the enquiry would be reliable? Give reasons.

Glossary

Abiotic – relating to non-living things.

Abrasion (or corrasion) – the wearing away of cliffs by sediment flung by breaking waves.

Abrasion – rocks carried along by the river wear down the river bed and banks.

Active layer – the top layer of soil in an area where permafrost is present. Each summer the icy soil briefly melts before refreezing.

Active volcano – a volcano that is currently erupting or showing signs of eruption, such as earthquakes or new gas emissions.

Adaptation – actions taken to adjust to natural events such as climate change, to reduce potential damage, limit the impacts, take advantage of opportunities or cope with the consequences.

Aeroponics – growing plants in an air or mist environment without the use of soil.

Agribusiness – application of business skills to agriculture.

Air pollution – the presence of chemicals and particles in the air that can be harmful to people or the environment.

Appropriate technology (or intermediate technology) – technology that is suited to the needs, skills, knowledge and wealth of local people in the environment in which they live. It usually combines simple ideas with cheap and readily available materials, especially for use in poorer countries, and is environmentally friendly.

Arch – a wave-eroded passage through a small headland. This begins as a cave formed in the headland, which is gradually widened and deepened until it cuts through.

Arctic Circle – a major circle of latitude that runs 66° north of the Equator.

Arête – a sharp, knife-like ridge formed between two corries cutting back by processes of erosion and freeze-thaw.

Arid – a climate where there is not enough precipitation to support vegetation growth. An arid climate receives less than 250 mm of rain annually. The definition encompasses both hot deserts and some polar regions ('cold deserts').

Assisted area – an area of the UK selected by the government to receive financial support to encourage new business.

Asthenosphere – the upper layer of the Earth's mantle.

Attrition – erosion caused when rocks and boulders transported by waves bump into each other and break up into smaller pieces.

Bar – where a spit grows across a bay, a bay bar can eventually enclose the bay to create a lagoon. Bars can also form offshore due to the action of breaking waves.

Bays – a rocky coastal promontory made of rock that is resistant to erosion; headlands lie between bays of less resistant rock where the land has been eroded back by the sea.

Beach – the zone of deposited material that extends from the low water line to the limit of storm waves. The beach or shore can be divided in the foreshore and the backshore.

Beach nourishment – the addition of new material to a beach artificially, through the dumping of large amounts of sand or shingle.

Beach reprofiling – changing the profile or shape of the beach. It usually refers to the direct transfer of material from the lower to the upper beach or, occasionally, the transfer of sand down the dune face from crest to toe.

Biodiesel – alternative energy that is produced from plant material so is carbon neutral.

Biodiversity – the variety of life in the world or a particular habitat.

Biomass – renewable organic materials, such as wood, agricultural crops or wastes, especially when used as a source of fuel or energy. Biomass can be burned directly or processed into biofuels such as ethanol and methane.

Biome – a large plant and animal community covering a large area of the Earth's surface.

Biotechnology – the manipulation (through genetic engineering) of living organisms to produce useful commercial products (such as pest-resistant crops and new bacterial strains).

Biotic – relating to living things.

Birth rate – the number of births in a year per 1,000 of the total population.

BRICs – four of the world's fastest-growing economies: Brazil, Russia, India and China. See also MINTs.

Brownfield site – land that has been used, abandoned and now awaits some new use. Commonly found across urban areas, particularly in the inner city.

Bulldozing – ice pushes material of all shapes and sizes as it moves slowly forward.

Business park – purpose-built area of offices and warehouses, often at the edge of a city and close to a main road.

Carbon capture and storage (CCS) – the process of capturing or trapping carbon dioxide produced by burning fossil fuels, transporting the carbon dioxide and then storing the carbon dioxide emissions in such a way that it is unable to affect the atmosphere (such as deep underground).

Carbon footprint – a measurement of all the greenhouse gases we individually produce, through burning fossil fuels for electricity or transport expressed as tonnes (or kg) of carbondioxide equivalent.

Carrying capacity – the maximum number of people an area of land can support before environmental damage occurs.

Cave – a large hole in the cliff caused by waves forcing their way into cracks in the cliff face.

Central business district – central part of the city where most shops and businesses are located.

(Channel) straightening – removing meanders from a river to make the river straighter. Straightening the river (also called channelising) allows it to carry more water quickly downstream, so it doesn't build up and is less likely to flood.

Chemical weathering – the decomposition (or rotting) of rock caused by a chemical change within that rock; sea water can cause chemical weathering of cliffs.

Cliff – a steep high rock face formed by weathering and erosion along the coastline.

Climate change – a long-term change in the Earth's climate, especially a change due to an increase in the average atmospheric temperature.

Closed question – a question with a limited choice of answers.

Coalfield – an area of coal-bearing rock where coal is mined.

Combined heat and power – a more efficient way of supplying energy by using the same source of energy to generate electricity and provide heat.

Commercial farming – farming to sell produce for a profit to retailers or food-processing companies.

Commonwealth – the Commonwealth is a voluntary association of 53 independent and equal sovereign states, which were mostly territories of the former British Empire. It is home to 2.2 billion citizens. Member states have no legal obligation to one another. Instead, they are united by language, history, culture, and their shared values of democracy, human rights and the rule of law.

Commute – travel to and from work on a daily basis, often from outside a city to the centre.

Composite volcano – a steep-sided, dome-shaped volcano that erupts a variety of materials such as sticky acidic lava and ash. Occurs at destructive plate margins.

Conservation – managing the environment in order to preserve, protect or restore it.

Conservative plate margin – tectonic plate margin where two tectonic plates slide past each other.

Constructive plate boundary – tectonic plate margin where rising magma adds new material to plates that are diverging or moving apart.

Consumer – creature that eats herbivores and/or plant matter.

Continental crust – the thicker part of the Earth's crust.

Convection – the transfer of heat energy in liquids and gases.

Convection cell – when differences in air temperature lead to the formation of areas of high and low air pressure; they become linked together by flows of warmer and cooler air.

Convection currents – the circular movement of magma within the Earth's mantle.

Core – the centre of the Earth, mainly made of iron and nickel, and with a solid inner core and liquid outer core.

Coriolis effect – the deflection, or bending, of the wind due to the rotational spin of the Earth.

Corrie (also called cirque) – armchair-shaped hollow in the mountainside formed by glacial erosion, rotational slip and freeze-thaw weathering. This is where the valley glacier begins. When the ice melts, it can leave a small circular lake called a tarn.

Counter-urbanisation – movement out of cities back into the countryside.

Cross profile – the side to side cross-section of a river channel and/or valley.

Crust – the outer layer of the Earth's structure.

Cyclone – see Tropical storm.

Dam and reservoir – a barrier (made of earth, concrete or stone) built across a valley to interrupt river flow and create a man-made lake (reservoir) which stores water and controls the discharge of the river.

Death rate – the number of deaths in a year per 1,000 of the total population.

Debt crisis – a situation whereby a country cannot pay its debts, often leading to calls to other countries for assistance. Many low-income countries are facing severe debt problems.

Debt reduction – countries are relieved of some of their debt in return for protecting their rainforests.

Debt relief – when high-income countries write-off some low-income countries' debt, or lower interest rates, so the low-income country has to pay back less.

Decomposer – an organism such as a bacterium or fungus, that breaks down dead tissue, which is then recycled to the environment.

Deforestation – the chopping down and removal of trees to clear an area of forest.

De-industrialisation – the decline of a country's traditional manufacturing industry due to exhaustion of raw materials, loss of markets and competition from NEEs.

Democracy – a system of government by the whole population through elected representatives.

Demographic transition model – a generalised model linking population changes with development changes over time.

Deposition – occurs when material being transported by the sea is dropped due to the sea losing energy.

Dereliction – abandoned buildings and wasteland.

Desalinisation – the process of removing salts and minerals from sea water.

Desertification – the process by which land becomes drier and degraded, as a result of climate change or human activities, or both.

Destructive plate margin – tectonic plate margin where two plates are converging or coming together and the oceanic plate is subducted. It can be associated with violent earthquakes and explosive volcanoes.

Development – the progress of a country in terms of economic growth, the use of technology and human welfare.

Development gap – the difference in standards of living and well-being between the world's richest and poorest countries (between high- and low-income countries).

Discharge – the quantity of water that passes a given point on a stream or river-bank within a given period of time.

Diurnal temperature range – the difference between the highest and lowest temperatures in a 24-hour period.

Drought – a long period of low rainfall.

Drumlin – a hill made of glacial till deposited by a moving glacier, usually elongated or oval in shape, with the longer axis parallel to the former direction of ice.

Dune regeneration – action taken to build up dunes and increase vegetation to strengthen the dunes and prevent excessive coastal retreat. This includes the re-planting of marram grass to stabilise the dunes, as well as planting trees and providing boardwalks.

Earthquake – a sudden or violent movement within the Earth's crust followed by a series of shocks.

Ecological footprint – the area of land or sea needed to produce all the resources a city uses and to dispose of its waste.

Economic impact – the effect of an event on the wealth of an area or community.

Economic migrant – someone who migrates with the main purpose of finding work or escaping poverty.

Economic opportunities – chances for people to improve their living standards through employment.

Ecosystem – a community of plants and animals that interact with one another and their physical environment.

Ecosystem services – the benefits human beings gain from leaving the world's ecosystems intact. These benefits can be given a financial value (for example, a forest that soaks up rainwater is worth the same money as flood defences).

Ecotourism – responsible travel to natural areas that conserves the environment, sustains the wellbeing of the local people, and may involve education. It is usually carried out in small groups and has minimal impact on the local ecosystem.

Embankments – raised banks constructed along the river; they effectively make the river deeper so it can hold more water. They are expensive and do not look natural but they do protect the land around them.

Energy conservation – reducing energy consumption through using less energy and becoming more efficient in using existing energy sources.

Energy consumption – the amount of energy that is used during a given period, normally one year. Most often, it is calculated as the average amount of energy consumed per head of population of a country, region or city. Per capita energy consumption is often taken as a measure of development.

Energy exploitation – developing and using energy resources to the greatest possible advantage, usually for profit.

Energy gap – the difference between a country's rising demand for energy and its ability to produce that energy from its own resources.

Energy insecurity – a situation where a country has to rely on others to supply most of its energy. This dependence makes a country politically vulnerable.

Energy mix – the range of energy sources of a region or country, both renewable and non-renewable.

Energy security – uninterrupted availability of energy sources at an affordable price.

Enhanced greenhouse effect – the warming of the Earth's atmosphere due to human activity increasing the layer of greenhouse gases.

Environment Agency (EA) – the organisation responsible for tackling environmental threats like pollution and flooding in the UK. It is directly funded by the government.

Environmental impact – the effect of an event on the landscape and ecology of the surrounding area.

Erosion – the wearing away and removal of material by a moving force, such as a breaking wave.

Erratics – rocks that have been transported and deposited by a glacier some distance from their source region.

Estuary – the tidal mouth of a river where it meets the sea; wide banks of deposited mud are exposed at low tide.

European Union – an international organisation of 28 European countries, including the UK, formed to reduce trade barriers and increase cooperation among its members. Seventeen of these countries also share the same type of money: the euro. A person who is a citizen of a European Union country can live and work in any of the other 27 member countries without needing a work permit or visa.

Evacuate – to leave an area due to dangers posed to lives.

Extreme weather – this is when a weather event is significantly different from the average or usual weather pattern, and is especially severe or unseasonal. This may take place over one day or a period of time. A severe snow blizzard or heatwave are two examples of extreme weather in the UK.

Fairtrade – when producers in LICs are given a better price for the goods they produce. Often this is from farm products like cocoa, coffee or cotton. The better price improves income and reduces exploitation.

Famine – a widespread, serious, shortage of food. In the worst cases it can lead to starvation and even death.

Flood – occurs when river discharge exceeds river channel capacity and water spills out of the channel onto the floodplain and other areas.

Flood plain – the relatively flat area forming the valley floor on either side of a river channel, which is sometimes flooded.

Flood plain zoning – this attempts to organise the flood defences in such a way that land that is near the river and often floods is not built on. This could be used for pastoral farming or playing fields. The areas that rarely get flooded would therefore be used for houses, transport and industry.

Flood relief channels – building new artificial channels which are used when a river is close to maximum discharge. They take the pressure off the main channels when floods are likely, therefore reducing flood risk.

Flood risk – the predicted frequency of floods in an area.

Flood warning – providing reliable advance information about possible flooding. Flood warning systems give people time to remove possessions and evacuate areas.

Fluvial processes – processes relating to erosion, transport and deposition by a river.

Food insecurity – being without reliable access to a sufficient quantity of affordable, nutritious food. More than 800 million people live every day with hunger or food insecurity.

Food miles – the distance covered supplying food to consumers.

Food security – when people at all times have access to sufficient, safe, nutritious food to maintain a healthy and active life.

Foreign direct investment (FDI) – sums of money a transnational corporation spends on building or buying up operations in a foreign country.

Forest degradation – a reduction in the ability of a forest to provide goods and services (see Figure 6.17 on page 76).

Formal economy – this refers to the type of employment where people work to receive a regular wage and are assured certain rights, e.g. paid holidays, sickness leave. Wages are taxed.

Fossil fuel – a natural fuel such as coal or gas, formed in the geological past from the remains of living organisms.

Fragile environment – an environment that is both easily disturbed and difficult to restore if disturbed. Plant communities in fragile areas have evolved in highly specialised ways to deal with challenging conditions. As a result, they cannot tolerate environmental changes.

Freeze-thaw weathering (also called frost-shattering) – it occurs in cold climates when temperatures are often around freezing point and where exposed rocks contain many cracks. Water enters the cracks during the warmer day and freezes during the colder night. As the water turns into ice it expands and exerts pressure on the surrounding rock, causing pieces to break off.

Gabion – steel wire mesh filled with boulders used in coastal defences.

Geothermal energy – energy generated by heat stored deep in the Earth.

Glacial episode – a colder period of time, with ice expansion, lasting approximately 100,000 years.

Glacial trough – a river valley widened and deepened by the erosive action of glaciers; it becomes 'U'-shaped instead of the normal 'V'-shape of a river valley.

Global atmospheric circulation – the worldwide system of winds, which transports heat from tropical to polar latitudes. In each hemisphere, air also circulates through the entire depth of the troposphere which extends up to 15 km.

Global ecosystem – very large ecological areas on the Earth's surface (or biomes), with fauna and flora (animals and plants) adapting to their environment. Examples include tropical rainforest and hot desert.

Globalisation – the process that has created a more connected world, with increases in the movements of goods (trade) and people (migration and tourism) worldwide.

Gorge – a narrow, steep sided valley, often formed as a waterfall retreats upstream.

'Grey' water – wastewater from people's homes that can be recycled and put to good use. Uses include water for laundry and toilet flushing. Treated greywater can also be used to irrigate both food and non-food producing plants. The nutrients in the greywater (such as phosphorus and nitrogen) provide an excellent food source for these plants.

Green belt – green open space or land around cities on which there are strict planning controls to prevent urban development in the countryside, and further building development is not allowed.

Greenfield site – a plot of land, often in a rural or on the edge of an urban area, that has not yet been subject to any building development.

Gross domestic product (GDP) – the total value of the goods and services produced in a country.

Gross national income (GNI) – a measurement of economic activity that is calculated by dividing the gross (total) national income by the size of the population. GNI takes into account not just the value of goods and services, but also the income earned from investments overseas.

Groundwater – water found underground in pores and cracks in the rock.

Groundwater management – regulation and control of water levels, pollution, ownership and use of groundwater.

Growth corridor – an area of the country where the economy is growing, often along a major transport route linking two or more cities.

Groyne – a wooden barrier built out into the sea to stop the longshore drift of sand and shingle, and so cause the beach to grow. It is used to build beaches to protect against cliff erosion and provide an important tourist amenity. However, by trapping sediment it deprives another area, down-drift, of new beach material.

Hanging valley – a tributary valley to the main glacier, too cold and high up for ice to be able to move easily. It therefore was not eroded as much as the lower main valley, and today is often the site for a waterfall crashing several hundred metres to the main valley floor.

Hard engineering – the use of concrete and large artificial structures by civil engineers to defend land against natural erosion processes.

Hazard risk – the probability or chance that a natural hazard may take place.

Headlands – see Bays.

Hi-tech industry – industry that uses the most advanced technology; technology always changes, though, so it may not remain hi-tech for ever.

High-speed rail – a railway on which trains operate at speeds in excess of 250 km/hr on a separate track to other trains.

Holocene epoch – the period of geological time from 12,000 years ago, which continues to the present day.

Hot desert – parts of the world that have high average temperatures and very low precipitation.

Human Development Index (HDI) – a method of measuring development in which GDP per capita, life expectancy and adult literacy are combined to give an overview. This combined measure of development uses economic and social indicators to produce an index figure that allows comparison between countries.

Human resources – people, their labour power and their ingenuity.

Hurricane – see Tropical storm.

Hydraulic action – the force of the river against the banks can cause air to be trapped in cracks and crevices. The pressure weakens the banks and gradually wears it away.

Hydraulic power – the process by which breaking waves compress pockets of air in cracks in a cliff. The pressure may cause the crack to widen, breaking off rock.

Hydro (electric) power – electricity generated by turbines that are driven by moving water.

Hydrograph – a graph which shows the discharge of a river, related to rainfall, over a period of time.

Hydro-meteorological hazard – natural hazard caused by atmospheric processes and any associated flooding.

Hydroponics – a method of growing plants using mineral nutrient solutions, in water, without soil.

Ice core – a cylinder of ice removed by drilling into a glacier or ice sheet, such as in Antarctica, to provide climate data.

Immediate responses – the reactions of people as a disaster happens and in the immediate aftermath.

Impermeable – a material (e.g. rock) that does not allow water to infiltrate or pass through it.

Industrial structure – the relative proportion of the workforce employed in different sectors of the economy (primary, secondary, tertiary and quaternary).

Inequalities – differences between poverty and wealth, as well as in peoples' wellbeing and access to things like jobs, housing and education. Inequalities may occur in housing provision, access to services, access to open land, safety and security.

Infant mortality – the average number of deaths of infants under 1 year of age, per 1000 live births, per year.

Informal economy (or informal employment) – this type of employment comprises work done without the official knowledge of the government and therefore without paying taxes. It is common in many low-income countries.

Information technologies – computer, internet, mobile phone and satellite technologies – especially those that speed up communication and the flow of information.

Infrastructure – the basic equipment and structures (such as roads, utilities, water supply and sewage) that are needed for a country or region to function properly.

Insolation – the amount of solar radiation (sunlight) an area receives over a specified amount of time.

Integrated transport systems – when different transport methods connect together, making journeys smoother and therefore public transport more appealing. Better integration should result in more demand for public transport and should see people switching from private car use to public modes of transport, which should be more sustainable. It may also lead to a fall in congestion due to less road users.

Interglacial episode – a warmer period of time, with less ice, lasting for only around 10,000 years between two glacial periods.

Interlocking spurs – a series of ridges projecting out on alternate sides of a valley and around which a river winds its course.

Intermediate technology – the simple, easily learned and maintained technology used in a range of economic activities serving local needs in LICs.

International aid – money, goods and services given by the government of one country, or a multilateral institution such as the World Bank or International Monetary Fund, to help the quality of life and economy of another country.

Irrigation – Applying water to land in order to supply crops and other plants with the necessary water.

Jet stream – a fast-flowing (200 km/hr) current of air that circles the planet at a height of 10 km.

Lahar – a mudflow formed when ash and water mix.

Landfill site – a site for the disposal of solid waste, often used to reclaim low-lying ground.

Landscape – an extensive area of land regarded as being visually and physically distinct.

Landslide – a rapid mass movement of surface material down a slope.

Land use conflicts – disagreements which arise when different users of the land do not agree on how it should be used.

Lateral erosion – sideways erosion by a river on the outside of a meander channel. It eventually leads to the widening of the valley and contributes to the formation of the flood plain.

Latitude – a line drawn from west to east on a map showing where places lie relative to the Equator and poles.

Leisure – non-work time that people spend on their hobbies and recreation.

Levées – embankment of sediment along the bank of a river. It may be formed naturally by regular flooding or be built up by people to protect the area against flooding.

Life expectancy – the average number of years a person might be expected to live.

Literacy rate – the percentage of people who have basic reading and writing skills.

Lithosphere – the more rigid outer part of the Earth.

Local food sourcing – a method of food production and distribution that is local, rather than national and/or international. Food is grown (or raised) and harvested close to consumers' homes, then distributed over much shorter distances.

Logging – the business of cutting down trees and transporting the logs to sawmills.

Long profile – the gradient of a river, from its source to its mouth.

Longshore drift – the zigzag movement of sediment along a shore caused by waves going up the beach at an oblique angle (wash) and returning at right angles (backwash). This results in the gradual movement of beach materials along the coast.

Long-term responses – later reactions that occur in the weeks, months and years after an event.

Managed retreat – allowing cliff erosion to occur as nature taking its course: erosion in some areas, deposition in others. Benefits include less money spent and the creation of natural environments. It may involve setting back or realigning the shoreline and allowing the sea to flood areas that were previously protected by embankments and seawalls.

Management strategies – techniques of controlling, responding to, or dealing with an event.

Mantle – the area beneath the Earth's crust.

Mass movement – the downhill movement of weathered material under the force of gravity. The speed can vary considerably.

Meander – a pronounced bend in a river.

Mechanical weathering – weathering processes that cause physical disintegration or break up of exposed rock without any change in the chemical composition of the rock, for instance freeze thaw.

Mega-cities – an urban area with a total population in excess of ten million people.

Microfinance loans – very small loans which are given to people in the LICs to help them start a small business.

Migration – when people move from one area to another. In many LICS people move from rural to urban areas (rural–urban migration).

Mineral extraction – the removal of solid mineral resources from the earth. These resources include ores, which contain commercially valuable amounts of metals, such as iron and aluminium; precious stones, such as diamonds; building stones, such as granite; and solid fuels, such as coal and oil shale.

MINTs – four more fast-growing economies (see also BRICs): Mexico, Indonesia, Nigeria and Turkey.

Mitigation – action taken to reduce or eliminate the long-term risk to human life and property from natural hazards, such as building earthquake-proof buildings or making international agreements about carbon reduction targets.

Monitoring – recording physical changes, such as earthquake tremors around a volcano, to help forecast when and where a natural hazard might strike.

Moraine – frost-shattered rock debris and material eroded from the valley floor and sides, transported and deposited by glaciers.

Multiplier effect – the positive spin-off effects that follow on from an initial investment (e.g. a new factory) in a region. Other firms gain business by supplying parts. The wages of factory workers help local shops to grow their businesses too.

National park – an area of outstanding natural beauty, preserved for people to enjoy.

Natural decrease – population decline due to the birth rate being lower than the death rate.

Natural hazard – a natural event (for example an earthquake, volcanic eruption, tropical storm, flood) that threatens people or has the potential to cause damage, destruction and death.

Natural increase – the birth rate minus the death rate of a population.

Natural resources – things found in the natural environment, like minerals or plants, that humans make use of to improve their standard of living.

The new green revolution – a combination of modern technology, traditional knowledge and an emphasis on farming, social and agro-ecological systems as well as yields, especially in poorer countries. At the same time, it emphasizes alternative approaches and improved farm management and information systems in order to minimise environmental damage from external inputs and benefit poor farmers and marginal areas bypassed by the original green revolution.

Newly emerging economies (NEE) – countries that have begun to experience high rates of economic development, usually with rapid industrialisation. They differ from low-income countries in that they no longer rely primarily on agriculture, have made gains in infrastructure and industrial growth, and are experiencing increasing incomes and high levels of investment; for example, the so-called BRICs countries.

North–South divide (UK) – economic and cultural differences between Southern England (the South-East, Greater London, the South-West and parts of the East) and Northern England (the North-East, West and Yorkshire and the Humber). There are clear differences in health conditions, house prices, earnings and political influence.

Nuclear power – the energy released by a nuclear reaction, especially by fission or fusion. Nuclear energy uses fuel made from mined and processed uranium to make steam and generate electricity.

Nutrient cycling – a set of processes whereby organisms extract minerals necessary for growth from soil or water, before passing them on through the food chain, and ultimately back to the soil and water.

Oceanic crust – the thinner part of the Earth's crust.

Ocean ridge – an underwater mountain range caused by two tectonic plates moving away from each other at constructive plate margins.

Open question – a question to which people can respond with their opinions.

Orbit – the path an object takes around a point in space, such as planet Earth around the Sun.

Organic produce – food which is produced using environmentally and animal friendly farming methods on organic farms. Artificial fertilisers are banned and farmers develop fertile soil by rotating crops and using compost, manure and clover. It must be free of synthetic additives like pesticides and dyes.

Outwash – material, chiefly sand or gravel, deposited by meltwater streams in front of, and underneath, a glacier. The material is sorted and rounded by water action.

Over-abstraction – when water is used more quickly that it is replaced.

Over-cultivation – exhausting the soil by over-cropping the land.

Overgrazing – grazing too many livestock for too long on the land, so it is unable to recover its vegetation.

Ox-bow lake – an arc-shaped lake which has been cut off from a meandering river.

Permaculture – a system of agricultural and social design principles based upon or directly using patterns and features observed in natural ecosystems.

Permafrost – permanently frozen ground, found in polar and tundra regions.

Permeable – allowing water to soak through.

Photosynthesis – the process by which green plants create food for themselves by converting sunlight, water and carbon dioxide into sugars.

Pioneer species – simple, tough plants that can survive in places where most others cannot due to a lack of soil or extreme climate.

Planning – actions taken to enable communities to respond to, and recover from, natural disasters, through measures such as emergency evacuation plans, information management, communications and warning systems.

Plastic flow – changes caused by stressing a material beyond its elastic limit. The change remains when the stress is removed.

Plate boundary – the boundary or margin between two tectonic plates.

Plate margin – the margin or boundary between two tectonic plates.

Pleistocene epoch – the period of geological time from about 2.6 million years ago to 12,000 years ago.

Plucking – a type of erosion where melt water in the glacier freezes onto rocks, and as the ice moves forward it plucks or pulls out large pieces along the rock joints.

Polar – the regions of Earth surrounding the North and South Poles. These regions are dominated by Earth's polar ice caps, the northern resting on the Arctic Ocean and the southern on the continent of Antarctica.

Pollution – the presence of chemicals, noise, dirt or other substances that have harmful or poisonous effects on an environment.

Post-industrial economy – the economy of many economically developed countries where most employment is now in service industries.

Precipitation – moisture falling from the atmosphere as rain, hail, sleet or snow.

Prediction – attempts to forecast when and where a natural hazard will strike, based on current knowledge. This can be done to some extent for volcanic eruptions (and tropical storms), but less reliably for earthquakes.

Primary data – data, or information, you collect yourself.

Primary effects – the initial impact of a natural event on people and property, caused directly by it; for instance, buildings collapsing following an earthquake.

Primary products – unprocessed raw materials and food produced by mining, farming, forestry and fishing.

Producer – an organism that is able to absorb energy from the sun through photosynthesis.

Protection – actions taken before a hazard strikes to reduce its impact, such as educating people or improving building design.

Proxy data – natural recordings used to estimate what conditions were like.

Pyramidal peak – where several corries cut back to meet at a central point, the mountain takes the form of a steep pyramid.

Qualitative data – descriptive data you get by observing or asking people's opinions; for example, an environmental quality survey.

Quality of life – the wide range of human needs that should be met alongside income growth.

Quantitative data – numerical data you get by counting or measuring.

Quaternary industry – work in the 'knowledge economy' that involves providing information and the development of new ideas.

Quaternary period – the period of geological time from about 2.6 million years ago to the present. It is characterised by the appearance and development of humans, and includes the Pleistocene and Holocene epochs.

Questionnaire – a set of questions to ask people about their lives and/or opinions.

Radiometer – device that measures infrared radiation.

Rain shadow – an area found on the far side of a mountain range that receives little rainfall thanks to descending air having already shed its moisture over the mountains.

Reclaimed water – sewage water that has been treated to remove solids and impurities so it can be recycled.

Reforestation – the replanting and restocking of existing forests and woodlands that have been depleted, usually through deforestation.

Remittances – the money that international migrants send home to their families and communities.

Renewable energy sources – a resource which is not diminished when it is used; it recurs and cannot be exhausted (for example wind and tidal energy).

Resource management – the control and monitoring of resources so that they do not become depleted or exhausted.

Retrofit – adding something that was not originally there when an object was manufactured or built.

Ribbon lake – a long, narrow lake found in glaciated valleys formed in locations where the glacier had more erosive power, e.g. in areas of softer rock, where the valley gradient temporarily steepened or a tributary glacier joined the main valley.

Richter scale – a logarithmic scale used for measuring the magnitude (strength) of an earthquake.

Ridge push – when gravity causes the ridge to push on the lithosphere and move tectonic plates.

Rift valley – a steep-sided valley caused by two tectonic plates moving away from each other on land at constructive plate margins.

Risk assessment – the identification of the potential risks of doing an activity, and strategies to minimise the risks.

Rock armour – large boulders dumped on the beach as part of the coastal defences.

Rotational slip – this occurs when the ice moves in a circular motion. This process can help to erode hollows in the landscape, and deepen hollows into bowl shapes.

Rural–urban fringe – a zone of transition between the built-up area and the countryside, where there is often competition for land use. It is a zone of mixed land uses, from out-of-town shopping centres and golf courses to farmland and motorways.

Rural–urban migration – movement of people from the countryside to cities.

Saltation – particles bouncing down the river bed.

Sand dune – coastal sand hill above the high tide mark, shaped by wind action and covered with grasses and shrubs.

Sanitation – measures designed to protect public health, including the provision of clean water, and the disposal of sewage and waste.

Science and business parks – business parks are purpose-built areas of offices and warehouses, often at the edge of a city and on a main road. Science parks are often located near university sites, and high-tech industries are established. Scientific research and commercial development may be carried out in co-operation with the university.

Sea wall – a concrete wall which aims to prevent erosion of the coast by providing a barrier which reflects wave energy.

Secondary data – data, or information, someone else has collected.

Secondary effects – the after-effects that occur as indirect impacts of a natural event, sometimes on a longer timescale; for instance, fires due to ruptured gas mains resulting from the ground shaking.

Selective logging – the cutting out of trees which are mature or inferior, to encourage the growth of the remaining trees in a forest or wood.

Semi-arid – a semi-arid climate receives between 250 mm and 500 mm of rain annually.

Service industries (tertiary industries) – the economic activities that provide various services – commercial (shops and banks), professional (solicitors and dentists), social (schools and hospitals), entertainment (restaurants and cinemas), and personal (hairdressers and fitness trainers).

Shield volcano – a wide, low volcano that erupts basic runny lava. Occurs at constructive plate margins.

Slab pull – when the weight of a dense tectonic plate is subducted into the mantle.

Slash and burn – an agricultural technique in which existing vegetation or forest is cut down and burned off to clear land before new seeds are sown.

Sliding – occurs after periods of heavy rain when loose surface material becomes saturated and the extra weight causes the material to become unstable and move rapidly downhill, sometimes in an almost fluid state.

Slumping – rapid mass movement which involves a whole segment of the cliff moving down-slope along a saturated shear-plane or line of weakness.

Social deprivation – the degree to which an individual or an area is deprived of services, decent housing, adequate income and local employment.

Social impact – the effect of an event on the lives of people or a community.

Social opportunities – chances for people to improve their quality of life through services like education and healthcare.

Soft engineering – managing erosion by working with natural processes to help restore beaches and coastal ecosystems.

Soil erosion – removal of topsoil faster than it can be replaced, due to natural (water and wind action), animal and human activity. Topsoil is the top layer of soil and is the most fertile because it contains the most organic, nutrient-rich materials.

Solar energy – the Sun's energy exploited by solar panels, collectors or cells to heat water or air or to generate electricity.

Solifluction – the flow of saturated surface soil down a slope under the influence of gravity.

Solution – soluble particles are dissolved into the river.

Sparsely populated – low population density.

Spit – a depositional landform formed when a finger of sediment extends from the shore out to sea, often at a river mouth. It usually has a curved end because of opposing winds and currents.

Squatter settlement – an area of poor-quality housing, lacking in amenities such as water supply, sewerage and electricity, which often develops spontaneously and illegally in a city in a low-income country.

Stack – an isolated pillar of rock left when the top of an arch has collapsed. Over time further erosion reduces the stack to a smaller, lower stump.

Storm surge – when a storm creates strong waves and a rise in sea level, leading to coastal flooding.

Straightening – see Channel straightening

Subduction zone – the area in which an oceanic plate is pushed under a continental plate at a destructive plate margin.

Subsistence farming – a type of agriculture producing food and materials for the benefit only of the farmer and his family.

Sunspots – dark patches on the Sun's surface, caused by magnetic activity inside the Sun. More patches mean the solar output is greater.

Suspension – fine solid material held in the water while the water is moving.

Sustainability – actions and forms of progress that meet the needs of the present without reducing the ability of future generations to meet their needs.

Sustainable development – development that meets the needs of the present without limiting the ability of future generations to meet their own needs.

Sustainable energy supply – energy that can potentially be used well into the future without harming future generations. Sustainable energy is the combination of energy savings, energy efficiency measures and technologies, as well as the use of renewable energy sources.

Sustainable food supply – food that is produced in ways that avoid damaging natural resources, provide social benefits such as good quality food and safe and healthy products, and contribute to local economies.

Sustainable urban living – a sustainable city is one in which there is minimal damage to the environment, the economic base is sound with resources allocated fairly and jobs secure, and there is a strong sense of community, with local people involved in decisions made. Sustainable urban living includes several aims including the use of renewable resources, energy efficiency, use of public transport, accessible resources and services.

Sustainable water supply – meeting the present-day need for safe, reliable and affordable water, which minimises adverse effects on the environment, while enabling future generations to meet their requirements.

Tectonic hazard – a natural hazard caused by movement of tectonic plates (including volcanoes and earthquakes).

Tectonic plate – a rigid segment of the Earth's crust which can 'float' across the heavier, semi-molten rock below. Continental plates are less dense, but thicker than oceanic plates.

Temperature range – the difference between the highest and lowest temperatures over a period of time.

Thermal expansion – when the sea expands and becomes larger as a result of increased temperature.

Thermal growing season – the period of time when temperatures are above 6°C and plants can grow.

Thermokarst – an uneven landscape of mounds and hollows, some of which may be water filled.

Till – an unsorted mixture of sand, clay and boulders carried by a glacier and deposited as ground moraine over a large area.

Traction – the rolling of boulders and pebbles along the river bed.

Trade – the buying and selling of goods and services between countries.

Traffic congestion – occurs when there is too great a volume of traffic for roads to cope with, so traffic jams form and traffic slows to a crawl.

Transnational corporation (TNC) – a company that has operations (factories, offices, research and development, shops) in more than one country. Many TNCs are large and have well-known brands.

Transpiration – the process by which plants lose water vapour through their leaves. Strong winds increase transpiration.

Transportation – the movement of eroded material.

Tropical storm (hurricane, cyclone, typhoon) – an area of low pressure with winds moving in a spiral around the calm central point called the 'eye' of the storm. Winds are powerful and rainfall is heavy.

Truncated spur – a former river valley spur that has been sliced off by a valley glacier, forming cliff-like edges.

Tundra – the flat, treeless Arctic regions of Europe, Asia and North America, where the ground is permanently frozen. Lichen, moss, grasses and dwarf shrubs can grow here.

Typhoon – see Tropical storm.

Undernutrition – this occurs when people do not eat enough nutrients to cover their needs for energy and growth, or to maintain a healthy immune system.

Urban farming – the growing of fruits, herbs, and vegetables and raising animals in towns and cities, a process that is accompanied by many other activities such as processing and distributing food, collecting and reusing food waste.

Urban greening – the process of increasing and preserving open space such as public parks and gardens in urban areas.

Urbanisation – the process by which an increasing percentage of a country's population comes to live in towns and cities. Rapid urbanisation is a feature of many LICs and NEEs.

Urban regeneration – the revival of old parts of the built-up area by either installing modern facilities in old buildings (known as renewal) or opting for redevelopment (i.e. demolishing existing buildings and starting afresh).

Urban sprawl – the unplanned growth of urban areas into the surrounding countryside.

Vertical erosion – downward erosion of a river bed.

Volcano – an opening in the Earth's crust from which lava, ash and gases erupt.

Waste recycling – the process of extracting and reusing useful substances found in waste.

Waterborne diseases – diseases caused by microorganisms that are transmitted in contaminated water. Infection commonly results during bathing, washing, drinking, in the preparation of food, or the consumption of infected food, for example, cholera, typhoid, botulism.

Water conflict – disputes between different regions or countries about the distribution and use of freshwater. Conflicts arise from the gap between growing demands and diminishing supplies.

Water conservation – the preservation, control and development of water resources, both surface and groundwater, and prevention of pollution.

Water deficit – this exists where water demand is greater than supply.

Waterfall – sudden descent of a river or stream over a vertical or very steep slope in its bed. It often forms where the river meets a band of softer rock after flowing over an area of more resistant material.

Water footprint – a water footprint is the amount of water you use in and around your home, school or office throughout the day. It includes the water you use directly (e.g. from a tap to drink or to shower). It also includes the water it took to produce the food you eat, the products you buy, the energy you consume and even the water you save when you recycle.

Water insecurity – when water availability is not enough to ensure the population of an area enjoys good health, livelihood and earnings. This can be caused by water insufficiency or poor water quality.

Water quality – quality can be measured in terms of the chemical, physical, and biological content of water. The most common standards used to assess water quality relate to health of ecosystems, safety of human contact and drinking water.

Water security – the reliable availability of an acceptable quantity and quality of water for health, livelihoods and production.

Water stress – water stress occurs when the demand for water exceeds the available amount during a certain period or when poor quality restricts its use.

Water surplus – this exists where water supply is greater than demand.

Water table – the level below which rock is saturated with water.

Water transfer – water transfer schemes attempt to make up for water shortages by constructing elaborate systems of canals, pipes, and dredging over long distances to transport water from one river basin to another.

Wave cut platform – a rocky, level shelf at or around sea level representing the base of old, retreated cliffs.

Waves – ripples in the sea caused by the transfer of energy from the wind blowing over the surface of the sea. The largest waves are formed when winds are very strong, blow for lengthy periods and cross large expanses of water.

Weathering – when rock is broken down in one place.

Wilderness area – a natural environment that has not been significantly modified by human activity. Wilderness areas are the most intact, undisturbed areas left on Earth –places that humans do not control and have not developed.

Wind energy – electrical energy obtained from harnessing the wind with windmills or wind turbines.

Index

Photo credits